T0122466

Intelligent Systems Reference Library

Volume 124

Series editors

Janusz Kacprzyk, Polish Academy of Sciences, Warsaw, Poland
e-mail: kacprzyk@ibspan.waw.pl

Lakhmi C. Jain, University of Canberra, Canberra, Australia;
Bournemouth University, UK;
KES International, UK
e-mail: jainlc2002@yahoo.co.uk; jainlakhmi@gmail.com
URL: http://www.kesinternational.org/organisation.php

James F. Peters

Foundations of Computer Vision

Computational Geometry, Visual Image Structures and Object Shape Detection

 Springer

James F. Peters
Electrical and Computer Engineering
University of Manitoba
Winnipeg, MB
Canada

ISSN 1868-4394   ISSN 1868-4408   (electronic)
Intelligent Systems Reference Library
ISBN 978-3-319-84912-6   ISBN 978-3-319-52483-2   (eBook)
DOI 10.1007/978-3-319-52483-2

Printed on acid-free paper

This Springer imprint is published by Springer Nature
The registered company is Springer International Publishing AG
The registered company address is: Gewerbestrasse 11, 6330 Cham, Switzerland

This book is dedicated to
Anna Di Concilio, Arturo Tozzi and sweet $\mathscr{P}$
for the many shapes and pullbacks they
have given to me

Preface

This book introduces the foundations of computer vision. The principal aim of computer vision (also, called machine vision) is to reconstruct and interpret natural scenes based on the content of images captured by various cameras (see, e.g., R. Szeliski [191]). Computer vision systems include such things as survey satellites, robotic navigation systems, smart scanners, and remote sensing systems. In this study of computer vision, the focus is on *extracting useful information from images* (see, e.g., S. Prince [162]). Computer vision systems typically emulate human visual perception. The hardware of choice in computer vision systems is some form of digital camera, programmed to approximate visual perception. Hence, there are close ties between computer vision, digital image processing, optics, photometry and photonics (see, e.g., E. Stijns and H. Thienpont [188]).

From a computer vision perspective, **photonics** is the science of light in the capture of visual scenes. **Image processing** is the study of digital image formation (e.g., conversion of analogue optical sensor signals to digital signals), manipulation (e.g., image filtering, denoising, cropping), feature extraction (e.g., pixel intensity, gradient orientation, gradient magnitude, edge strength), description (e.g., image edges and texture) and visualization (e.g., pixel intensity histograms). See, e.g., the mathematical frameworks for image processing by B. Jähne [87] and S.G. Hoggar [82], extending to a number of practitioner views of image processing provided, for example, by M. Sonka and V. Hlavac and R. Boyle [186], W. Burger and M.J. Burge [21], R.C. Gonzalez and R.E. Woods [58], R.C. Gonzalez and R.E. Woods and S.L. Eddins [59], V. Hlavac [81], and C. Solomon and T. Breckon [184]. This useful information provides the bedrock for the focal points of computer visionists, namely, image object shapes and patterns that can be detected, analyzed and classified (see, e.g., [142]). In effect, **computer vision** is the study of digital image structures and patterns, which is a layer of image analysis above that of image processing and photonics. Computer vision includes image processing and photonics in its bag of tricks in its pursuit of image geometry and image region patterns.

In addition, it is helpful to cultivate an intelligent systems view of digital images with an eye to discovering hidden patterns such as repetitions of convex enclosures of image regions and embedded image structures such as clusters of points in image regions of interest. The discovery of such structures is made possible by quantizers. A **quantizer** restricts a set of values (usually continuous) to a discrete value. In its simplest form in computer vision, a quantizer observes a particular target pixel intensity and selects the nearest approximating values in the neighbourhood of the target. The output of a quantizer is called a codebook by A. Gersho and R.M. Gray [55, §5.1, p. 133] (see, also, S. Ramakrishnan, K. Rose and A. Gersho [164]).

In the context of image mesh overlays, the Gersho–Gray quantizer is replaced by geometry-based quantizers. A **geometry-based quantizer** restricts an image region to its shape contour and observes in an image a particular target object shape contour, which is compared with other shape contours that have approximately the same shape as the target. In the foundations of computer vision, geometry-based quantizers observe and compare image regions with approximately the same regions such as mesh maximal nucleus clusters (MNCs) compared with other nucleus clusters. A **maximal nucleus cluster** (MNCs) is a collection of image mesh polygons surrounding a mesh polygon called the nucleus (see, e.g., J.F. Peters and E. İnan on Edelsbrunner nerves in Voronoï tessellations of images [150]). An **image mesh nucleus** is a mesh polygon that is the centre of a collection of adjacent polygons. In effect, every mesh polygon is a nucleus of a cluster of polygons. However, only one or more mesh nuclei are maximal.

A **maximal image mesh nucleus** is a mesh nucleus with the highest number of adjacent polygons. MNCs are important in computer vision, since what we will call a MNC contour approximates the shape of an underlying image object. A **Voronoï tessellation** of an image is a tiling of the image with polygons. A Voronoï tessellation of an image is also called a Voronoï mesh. A sample tiling of a musician image in Fig. 0.1.1 is shown in Fig. 0.1.2. A sample nucleus of the musician image tiling is shown in Fig. 0.2.1. The red ● dots inside each of the tiling polygons are examples of Voronoï region (polygon) generating points. For more about this, see Sect. 1.22.1. This musician mesh nucleus is the centre of a maximal nucleus cluster shown in Fig. 0.2.2. This is the only MNC in the musician image mesh in Fig. 0.1.2. This MNC is also an example of a Voronoï mesh nerve. The study of image MNCs takes us to the threshold of image geometry and image object shape detection. For more about this, see Sect. 1.22.2.

Each **image tiling polygon** is a convex hull of the interior and vertex pixels. A **convex hull** of a set of image points is the smallest convex set of the set of points. A set of image points A is a **convex set**, provided all of the points on every straight line segment between any two points in the set A is contained in the set. In other words, knowledge discovery is at the heart of computer vision. Both knowledge and understanding of digital images can be used in the design of computer vision systems. In vision system designs, there is a need to understand the composition and structure of digital images as well as the methods used to analyze captured images.

The focus of this volume is on the study of raster images. The sequel to this volume will focus on vector images, which are composed of points (vectors), lines and curves. The basic content of every raster image consists of pixels

0.1.1: Muscian   0.1.2: Muscian tiling

Fig. 0.1 Voronoï tessellation of musician image

0.2.1: Musician mesh nucleus   0.2.2: Musician maximal nucleus cluster

Fig. 0.2 Maximal nucleus cluster on musician image

(e.g., distinguished pixels called sites or mesh generating points), edges (e.g., common, parallel, intersecting, convex, concave, straight, curved, connected, unconnected), angles (e.g., vector angle, angle between vectors, pixel angle), image geometry (e.g., Voronoï regions [141], Delaunay triangulations [140]), colour, shape, and texture. Many problems in computer vision and scene analysis are solved by finding the most probable values of certain hidden or unobserved image variables and structures (see, e.g., P. Kohli and P.H.S. Torr [96]). Such structures and variables include the topological neighbourhood of a pixel, convex hulls of sets of pixels, nearness (and apartness) of image structures and pixel gradient distributions as well as feature vectors that describe elements of captured scenes.

Other computer vision problems include image matching, feature selection, optimal classifier design, image region measurement, interest point identification, contour grouping, segmentation, registration, matching, recognition, image clustering, pattern clustering in F. Escolono, P. Suau, B. Bonev [45] and in N. Paragios, Y. Chen, O. Faugeras [138], landmark and point shape matching, image warping, shape gradients [138], false colouring, pixel labelling, edge detection, geometric structure detection, topological neighbourhood detection, object recognition, and image pattern recognition.

In computer vision, the focus is on the detection of the basic geometric structures and object shapes commonly found in digital images. This leads into a study of the basics of image processing and image analysis as well as vector space and computational geometry views of images. The basics of image processing include colour spaces, filtering, edge detection, spatial description and image texture. Digital images are examples of Euclidean spaces (both 2D and 3D). Hence, vector space views of digital images are a natural outcome of their basic character. A **digital image structure** is basically a geometric or a visual topological structure. Examples of image structures are image regions, line segments, generating points (e.g. Lowe keypoints), set of pixels, neighbourhood of a pixel, half spaces, convex sets of pixels and convex hulls of sets of image pixels. For example, such structures can be viewed in terms of image regions nearest selected points or collections of image regions with a specified range of diameters. An **image region** is a set of image points (pixels) in the interior of a digital image. The **diameter** of any image region is the maximum distance between a pair of points in the region). Such structures can also be found in line segments connected between selected points to form triangular regions in 2D and 3D images.

Such structures are also commonly found in 2D and 3D images in the intersection of closed half spaces to form either convex hulls of a set of points or what G.M. Ziegler calls **polytopes** [221]. An **image half space** is the set of all points either above or below a line. In all three cases, we obtain a regional view of digital images. For more about polytopes, see Appendix B.15.

Every image region has a shape. Some region shapes are more interesting than others. The interesting image region shapes are those containing objects of interest. These regional views of images leads to various forms of image segmentations that have practical value when it comes to recognizing objects in images. In addition, detection of image region shapes of interest views lead to the discovery of image patterns that transcend the study of texels in image processing. A **texel** is an image region represented by an array of pixels. For more about shapes, see Appendix B.18 on shape and shape boundaries.

Image analysis focuses on various digital image measurements (e.g., pixel size, pixel adjacency, pixel feature values, pixel neighbourhoods, pixel gradient, closeness of image neighbourhoods). Three standard region-based approaches in image analysis are isodata thresholding (binarizing images), watershed segmentation (computed using a distance map from foreground pixels to background regions), and non-maximum suppression (finding local maxima by suppressing all pixels that are less likely than their surrounding pixels) [212].

In image analysis, object and background pixels are associated with different adjacencies (neighbourhoods) by T. Aberra [3]. There are three basic types of neighbourhoods, namely, Rosenfeld adjacency neighbourhoods [171, 102], Hausdorff neighbourhoods [74, 75] and descriptive neighbourhoods in J.F. Peters [142] and in C.J. Henry [77, 76]. Using different geometries, an adjacency neighbourhood of a pixel is defined by the pixels adjacent to a given pixel. An image **Rosenfeld adjacency neighbourhood** of a pixel p is a set of pixels that are adjacent to p. Adjacency neighbourhoods are commonly used in edge detection in digital images.

A **Hausdorff neighbourhood** of a point p is defined by finding all pixels whose distance from p is less that a positive number r (called the neighbourhood radius). A **descriptive neighbourhood** of a pixel p (denoted by $N(img\ (x, y), r)$ is the set of pixels with feature vectors that match or are similar to the feature vector that describes $img(x, y)$ (the neighbourhood 'centre' of a digital image img) and which are within a prescribed radius r.

Unlike an adjacency neighbourhood, a descriptive neighbourhood can have holes in it, i.e., pixels with feature vectors that do not match the neighbourhood centre and are not part of the neighbourhood. Other types of descriptive neighbourhoods are introduced in [142, Sect. 1.16, pp. 29–34].

The chapters in this book grew out of my notes for an undergraduate class in Computer Vision taught over the past several years. Many topics in this book grew out my discussions and exchanges with a number of researchers and others, especially, S. Ramanna (those many shapes, especially in crystals), Anna Di Concilio (those proximities, region-free geometry, and seascape shapes like those in Fig. 0.3), Clara Guadagni (those flower nerve structures), Arturo Tozzi (those Borsuk-Ulam Theorem insights and Gibson shapes, Avenarius shapes), Romy Tozzi (remember 8, ∞), Zdzisław Pawlak (those shapes in paintings of the Polish countryside), Lech Polkowski (those mereological, topological and rough set structures), Piotr Artiemjew (those dragonfly wings), Giangiacomo Gerla (those tips (points)–vertices–of UNISA courtyard triangles and spatial regions), Gerald Beer (those moments in Som Naimpally's life), Guiseppe Di Maio (those insights about proximities), Somashekhar (Som) A. Naimpally (those topological structures), Chris Henry (those colour spaces, colour shape sets), Macek Borkowski (those 3D views of space), Homa Fashandi, Dan Lockery, Irakli Dochviri, Ebubekir İnan (those nearness relations and near groups), Mehmet Ali Öztürk (those beautiful algebraic structures), Mustafa Uçkun, Nick Friesen (those shapes of dwellings), Özlem Umdu, Doungrat Chitcharoen, Çenker Sandoz (those Delaunay triangulations), Surabi Tiwari (those many categories), Kyle Fedoruk (application of computer vision: Subaru EyeSight®), Amir H. Meghdadi, Shabnam Shahfar, Andrew Skowron (those proximities at Banacha), Alexander Yurkin, Marcin Wolksi (those sheaves), Piotr Wasilewski, Leon Schilmoeler, Jerzy W. Grzymala-Busse (those insights about rough sets and LATEX hints), Zbigniew Suraj (those many Petri nets), Jarosław Stepaniuk, Witold Pedrycz, Robert Thomas (those shapes of tilings), Marković G. oko (polyforms), Miroslaw Pawlak, Pradeepa Yahampath, Gabriel Thomas, Anthony (Tony) Szturm, Sankar K. Pal, Dean McNeill, Guiseppe (Joe) Lo Vetri, Witold Kinsner, Ken Ferens, David Schmidt (set theory), William Hankley (time-based specification), Jack Lange (those chalkboard topological doodlings), Irving Sussman (gold nuggets in theorems and proofs) and Brian Peters (those fleeting glimpses of geometric shapes on the walls).

Fig. 0.3 Seascape shapes along the coastline of Vietri, Italy

A number of our department technologists have been very helpful, especially, Mount-First Ng, Ken Biegun, Guy Jonatschick and Sinisa Janjic.

And many of my students have given important suggestions concerning topics covered in this book, especially, Drew Barclay, Braden Cross, Binglin Li, Randima Hettiarachchi, Enoch A-iyeh, Chidoteremndu (Chido) Chinonyelum Uchime, D. Villar, K. Marcynuk, Muhammad Zubair Ahmad, and Armina Ebrahimi.

Chapter problems have been classified. Those problems that begin with 🚲 are the kind you can run with, and probably will not take much time to solve. Problems that begin with ☕ are the kind you can probably solve in about the time it takes to drink a cup of tea or coffee. The remaining problems will need varying lengths of time to solve.

Winnipeg, Canada   James F. Peters

Contents

Chapter 1
Basics Leading to Machine Vision

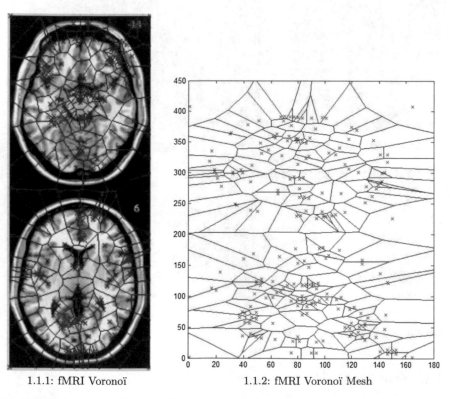

1.1.1: fMRI Voronoï   1.1.2: fMRI Voronoï Mesh

Fig. 1.1 Voronoï geometric views of image structures

© Springer International Publishing AG 2017   1
J.F. Peters, *Foundations of Computer Vision*, Intelligent Systems
Reference Library 124, DOI 10.1007/978-3-319-52483-2_1

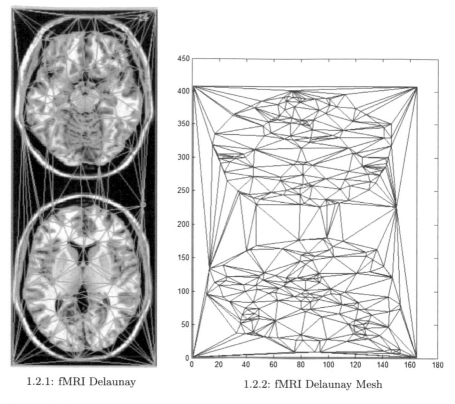

1.2.1: fMRI Delaunay   1.2.2: fMRI Delaunay Mesh

Fig. 1.2 Delaunay geometric views of image structures

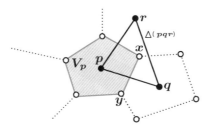

Fig. 1.3 $p, q \in S$, $\Delta(pqr)$ = Delaunay triangle

1.1   What Is Computer Vision?

The principal aim of computer vision is to reconstruct and interpret natural scenes based on the content of images captured by digital cameras [190]. A **natural scene** is that part of visual field that is captured either by human visual perception or by optical sensor arrays.

A optical sensor array-based natural scene is either as a single digital image captured by camera or as video frame image captured by a video camera such as a webcam.

The basic content of every image scene consists of pixels (e.g., adjacent, non-adjacent), edges (e.g., common, parallel, intersecting, convex, concave, straight, curved, connected, unconnected), angles (e.g., vector, between vectors, pixels), image geometry (e.g., Voronoï regions [141], Delaunay triangulations [140]), colour, shape, and texture.

1.2 Divide and Conquer Approach

The reconstruction and interpretation of natural scenes is made easier by tiling (tessellating) a scene image with known geometric shapes such as triangles (Delaunay triangulation approach) and polygons (Voronoï diagram approach). This is a divide-and-conquer approach. Examples of this approach in computer vision are found in

Fig. 1.4 Triangulated video frame from [214]

Shape detection: Video frame shape detection using Delaunay triangulation is given in C.P. Yung, G.P.-T. Choi, K. Chen and L.M. Lui [214] (see, e.g., Fig. 1.4).
Silhouettes: Use silhouettes to find epipolar lines to calibrate a network of cameras, an approach used by G. Ben-Artzi, T. Halperin, M. Werman and S. Peleg in [14]. The basic goal here is to achieve binocular vision and determine the scene position of a 3D object point via triangulation on a pair of 2D images (what each single camera sees). Points called epipoles are used to extract 3D objects from a pair of 2D images. An **epipole** is the point of intersection of the line joining optical centers with an image plane. The line between optical centers is called a **baseline**. An **epipolar plane** is the plane defined by a 3D point m and the optical

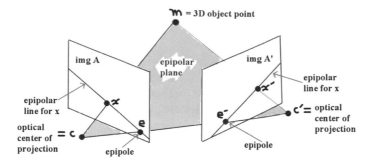

Fig. 1.5   Epipoles and epipolar lines

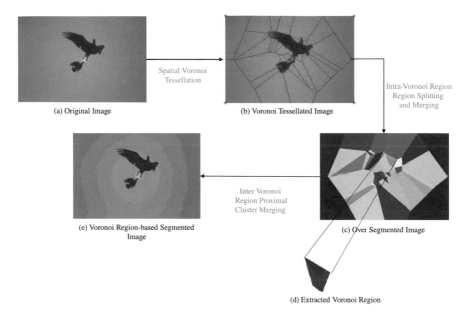

Fig. 1.6   Stages of the Voronoï segmentation method from [79]

centers C and C'. See, for example, the pair of epipoles and epipolar lines in
Fig. 1.5.

Video stippling: Stippling renders an image using point sets, elementary shapes
and colours. The core technique in video stippling is Voronoï tessellation of video
frames. This is the approach by T. Houit and F. Nielsen in [85]. This article contain
a good introduction to Voronoï diagrams superimposed on video frame images (see
[85, Sect. 2, pp. 2–3]). Voronoï diagrams are useful in segmenting images. This
leads to what are known as Dirichlet tessellated images, leading a new form of k-
means clusters of image regions (see Fig. 1.6 for steps in the Voronoï segmentation
method). This form of image segmentation uses cluster centroid proximity to
find image clusters. This is approach used by R. Hettiarachchi and J.F. Peters in

[79]. Voronoï manifolds are introduced by J.F. Peters and C. Guadagni in [146]. A **manifold** is a topological space that is locally Euclidean, i.e., around every point in the manifold there is an open neighbourhood. A nonempty set X with a topology τ on it, is a **topological space**. A collection of open sets τ on a nonempty open set X is a *topology* on X, provided it has certain properties (see Appendix B.19 for the definitions of open set and topology). An **open set** is a nonempty set of points A in space X contains all points sufficiently close to A but does not include its boundary points.

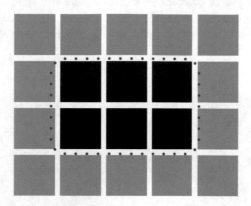

Fig. 1.7 Open set $A = \{\blacksquare, \blacksquare, \blacksquare, \blacksquare, \blacksquare, \blacksquare\}$

Example 1.1 **Sample open sets**.

apple pulp   Apple without its skin.
egg interior   Egg yoke without its shell.
wall-less room   Room without its walls.
Open Subimage   A subimage that does include its boundary points. A sample 2×3 subimage A in a tiny 4×5 digital image Img is shown in Fig. 1.7. The set A is open, since it contains only black squares $\blacksquare$ and does not contain the gray pixels along its boundary represented by $\blacksquare$.
    ∎

A manifold M is a **Voronoï manifold**, provided M is a Voronoï diagram. Any digital image or video frame image with a topology defined on it, is a Voronoï manifold. This is important in computer vision, since images that are Voronoï manifolds have geometric structures that are an aid in the study of the character of image shapes and objects.

Combined Geodesic Delaunay and Vororoni Tessellation: Delaunay triangulation and Voronoï diagrams are combined in the study of geodesic lines and graphs is introduced by Y.-J. Lin, C.-Xu Xu, D. Fan and Y. He in [112]. A graph G is a

geodetic graph, provided, for any two vertices p, q on G, there is at most one shortest path between p and q. A **geodetic line** is a straight line, since the shortest path between the endpoints of a straight line is the line itself. For more about this, see J. Topp [195]. For examples, see Appendix B.7.

Convex Hulls: A **convex hull of a set of points** A (denoted by convh$A$) is the smallest convex set containing A. A nonempty set A in an n-dimensional Euclidean space is a **convex set** (denoted by conv A), provided every straight line segment between any two points in the set is also contained in the set. Voronoï tessellation of a digital image results in image region clusters that are helpful in shape detection and the analysis of complex systems such as the cosmic web. This approach is used by J. Hidding, R. van de Weygaert, G. Vegter, B.J.T. Jones and M. Teillaud in [80]. For a pair of 3D convex hulls, see Fig. 1.8. For more about convex hulls, see Appendix B.3.   ∎

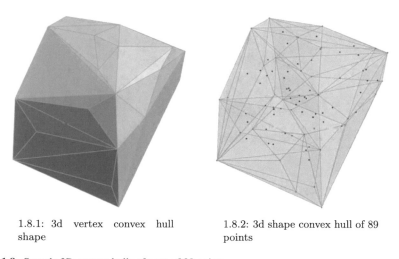

1.8.1: 3d   vertex   convex   hull shape

1.8.2: 3d shape convex hull of 89 points

Fig. 1.8 Sample 3D convex hulls of a set of 89 points

These methods use image areas instead of pixels to extract image shape and object information. In other words, we use computational geometry in the interpretation and analysis of scene images.

1.3   Voronoï Diagrams Superimposed on Images

Let S be any set of selected pixels in a digital image and let $p \in S$. The pixels in S are called **sites** (or **generating points**) to distinguish them from other pixels in an image. Recall that Euclidean distance between a pair of points x, y in the Euclidean plane is denoted by $\|x - y\|$ and defined by

$$\|x - y\| = \sqrt{\sum_{i=1}^{n} x_i^2 - y_i^2}$$

A *Voronoï region* of $p \in S$ (denoted V_p) is defined by

$$V_p = \{x \in E : \|x - p\| \le \|x - q\| \text{ for all } q \in S\}.$$

Every site in S belongs to only one Voronoï region. A digital image covered with Voronoï regions is called a **tessellated image**. Notice that each Voronoï region is a **convex polygon**. This means that all of the points on a straight edge connecting any pair of points in a Voronoï region belongs to the region. And a complete set of Voronoï regions covering an image is called a **Voronoï diagram** or **Voronoï mesh**.

Example 1.2 **Sample Voronoï and Delaunay Image Meshes**.
Sample Voronoï regions on an fMRI image are shown in Fig. 1.1.1 with the extracted Voronoï mesh shown in Fig. 1.1.2. In this case, each Voronoï region is a convex polygon in the mesh that surrounds an image corner. A Delaunay triangle is formed by connecting the site points of neighbouring Voronoï regions. A sample Delaunay triangulation mesh is shown in Fig. 1.2.1. This gives us another view of image geometry formed by the interior points surrounding a mesh generator of each Delaunay triangle. The extracted Delaunay triangulation is shown in Fig. 1.2.2. ∎

Many problems in computer vision and scene analysis are solved by finding the most probable values of certain hidden or unobserved image variables and structures [96]. Such structures and variables include Voronoï regions, Delaunay triangles, neighbourhoods of pixels, nearness (and apartness) of image structures and pixel gradient distributions as well as values of encoded desired properties of scenes.

Other computer vision problems   include image matching, feature selection, optimal classifier design, image region measurement, interest point, contour grouping, segmentation, registration, matching, recognition, image clustering, pattern clustering [45, 138], landmark and point shape matching, image warping, shape gradients [138], false colouring, pixel labelling, edge detection, geometric structure detection, topological neighbourhood detection, object recognition, and image pattern recognition. Typical applications of computer vision are in digital video stabilization [49, Sect. 9, starting on p. 261] and in robot navigation [93, Sect. 5, starting on p. 109].

The term **camera** comes from Latin *camera obscura* (dark chamber). Many different forms of cameras provide a playground for computer vision, e.g., affine camera,pinhole camera, ordinary digital cameras, infrared cameras (also thermographic camera), gamma (tomography) camera devices (in 3D imaging). An affine camera is a linear mathematical model that approximates the perspective projection derived from an ideal pinhole camera [218]. A pinhole camera is a perspective projection device, which is a box with light-sensitive film on its interior back plane and which admits light through a pinhole.

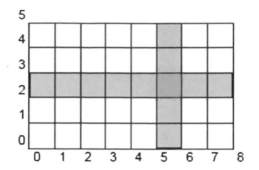

Fig. 1.9 Pixel centered at (5.5, 2.5) in a very small image grid

In this work, the focus is on the detection of the basic content and structures in digital images. An interest in **image content** leads into a study of the basics of image processing and image analysis as well as vector space and computational geometry views of images. The basics of image processing include colour spaces, filtering, edge detection, spatial description and image texture. The study of **image structures** leads to a computational geometry view of digital images. The basic idea is to detect and analyze image geometry from different perspectives.

Digital images are examples of subsets of Euclidean spaces (both 2D and 3D). Hence, vector space views of digital images are a natural outcome of their basic character. Digital image structures are basically geometric structures. Such structures can be viewed in terms of image regions nearest selected points (see, e.g., the tiny region nearest the highlighted pixel centered at (5.5, 2.5) in Fig. 1.9). Such structures can also viewed with respect to line segments connection between selected points to form triangular regions. Both a regional view and a triangulation view of image structures leads to various forms of image segmentations that have practical values when it comes to recognizing objects in images and classifying images. In addition, both regional and triangle views lead to the discovery of patterns hidden in digital images.

Basic Approach in Image Computational Geometry

The basic approach is to describe a digital image object with a known geometric structure.

1.4   A Brief Look at Computational Geometry

To analyze and understand image scenes, it is necessary to identify the objects in the scenes. Such objects can be viewed geometrically as collections of connected edges (e.g., skeletonizations or edges belonging to shapes or edges in polygons) or image regions viewed as sets of pixels that are in some sense near each other or set of points near a fixed point (e.g., all points near a site (also, seed or generating point) in a Voronoï region [38]). For this reason, it is highly advantageous to associate geometric structures in an image with mesh-generating points (sites) derived from the fabric of an image. Image edges, corners, centroids, critical points, intensities, and keypoints (image pixels viewed as feature vectors) or their combinations provide ideal sources of mesh generators as well as sources of information about image geometry.

Computational geometry is the brain child of A. Rosenfeld, who suggested approaching image analysis in terms of distance functions in measuring the separation between pixels [168] and image structures such as sets of pixels [169, 170]. Rosenfeld's work eventually led to the introduction of topological algorithms useful in image processing [99] and the introduction of a full-scale digital geometry in picture analysis [94].

Foundations of Scene Analysis

The foundations for digital image scene analysis are built on the pioneer-ing work by A. Rosenfeld work on digital topology [98, 168–172] (later called digital geometry [94]) and others [39, 99, 102, 104, 105]. The work on digital topology runs parallel with the introduction of compu-tational geometry by M.I. Shamos [175] and F.P. Preparata [158, 159], building on the work on spatial tessellations by G. Voronoï [201, 203] and others [27, 53, 64, 103, 124, 196].

Computational geometry (CG) is an algorithmic approach   in the study of geometric structures. In CG, algorithms (step-by-step methods) are introduced to construct and analyze the lines and surfaces of objects, especially real world objects. The focus in CG is on how points, lines, polygons, smooth curves (in 2D) and polyhedra and smooth surfaces (in 3D) are either constructed or detected and analyzed by a computer. For a more general view of CP from a line geometry perspective, see, for example, H. Pottmann and J. Wallner [157].

In the context of digital images, computational geometry focuses on the con-struction and analysis of various types of mesh overlays on images. On the ground floor, the two main types of meshes result from Delaunay triangulations and Voronoï tessellations on sets of image pixels. A **Delaunay triangulation** is a covering of a digital image with triangles with non-intersecting interiors. With Delaunay triangu-lation, the focus is on constructing meshes of triangles derived from selected sets of pixels called sites or generators and which cover either a 2D or 3D digital image. A principal benefit of image triangulations is the detection of image object shapes covered by mesh triangles. Thanks to the known properties of triangles (e.g., uniform shape, sum of the interior angles, perimeter, area, lengths of sides), object shapes can be described in a very accurate fashion. For more about Delaunay triangulation, see J.A. Baerentzen, J. Gravesen, F. Anton and H. Aanaes [8, Sect. 14].

Image object shapes can also be closely approximated by collections of Voronoï polygons (also called Voronoï regions) in a Voronoï tessellation of an image. A **2D Voronoï diagram** represents a tessellation of the plane region by convex polygons. A **3D Voronoï diagram** represents a tessellation of a 3D surface region by convex polygons.

A convex polygon is an example of a convex set. A **convex set** of points A (denoted by conv A) has the property that, for each pair of points p, q in conv A, all of the points on the straight line segment connected between p and q also belong to conv A. For more about convex sets, see Appendix B.3.

Since object shapes tend to irregular, the varying shapes of polygons in a typical Voronoï image covering an image give a more precise view of the shapes of image objects. It is important to notice that Delaunay triangles have empty interiors (only the sides of the triangles are known). By contrast, Voronoï polygons have non-empty interiors. This means that we know both the sides as well as the content of the interior of each Voronoï polygon. A principal benefit of image Voronoï tessellation is the detection of image object shapes covered by mesh polygons. Thanks to the known properties of Voronoï polygons (e.g., shape, interior angles, edge pixel gradient orientation, perimeter, diameter, area, lengths and number of sides), object shapes can be described in a very accurate fashion. For this reason, Voronoï polygons that cover an image neighbourhood containing an object provide a very detailed view of image object shapes and content.

A sample digital image geometry algorithm useful in either triangulating or tessellating a digital image is given in Algorithm 1.

Algorithm 1: Digital Image Geometry via Mesh Covering Image

   Input   : Read digital image img.
   Output: Mesh $\mathcal{M}$ covering an image.
1 $MeshSite \leftarrow MeshGeneratingPointType$;
2 $img \longmapsto MeshSitePointCoordinates$;
3 $S \leftarrow MeshSitePointCoordinates$;
4 /* S contains MeshSitePointType coordinates used as mesh generating points (seeds or sites). */ ;
5 $MeshType \leftarrow MeshChoice$;
6 /* $MeshType$ identifies a chosen form of mesh, e.g., Voronoï, Delaunay, polynomial. */ ;
7 $S \longmapsto MeshType\ \mathcal{M}$;
8 $MeshType\ \mathcal{M} \longmapsto img$;
9 /* Use $\mathcal{M}$ to gain information about image geometry. */ ;

Algorithm 1 leads to a mesh covering a digital image. Image meshes can vary considerably, depending on the type of image and the type mesh generating points that are chosen. Image geometry tends to be revealed, whenever the choice of generating points accurately reflects the image visual content and the structure of the objects in an image scene. For example, corners would be the logical choice for image scenes containing buildings or objects with sharply varying contours such as hands or facial profiles.

1.10.1: Corner-Based Voronoï mesh

1.10.2: Corner-Based Delaunay mesh

Fig. 1.10 Hunting grounds for scene information: corner-based Delaunay and Voronoï meshes

Example 1.3 **Meshes Covering a Salerno Poste Auto Scene**.
A corner-based Voronoï mesh covering an image scene containing a Poste auto parked outside the train station in Salerno, Italy is shown in Fig. 1.10.1. This Voronoï mesh is also called a *Dirichlet tessellation*. Using the same set of corner generating points, a Delaunay triangulation cover in the Poste auto scene is shown in Fig. 1.10.2. For

a Delaunay triangulation view of fMRIs (functional Magnetic Resonance Images), see Fig. 1.2. For more about Delaunay triangulation, see Sect. 6.1.

For a Voronoï tessellation of the same fMRIs, see Fig. 1.1. One important thing to look for in Voronoï tessellation of an image is the presence of clusters of mesh polygons, each with a central polygon that has a maximum number of adjacent polygons. The central polygon of a mesh cluster is called the **cluster nucleus**. In that case, the cluster is called a **maximal nucleus cluster** (MNC). Image mesh MNCs approximate the shape of the underlying image object covered by the MNC. For more about MNCs, see Sect. 7.5. ∎

1.5 Framework for Digital Images

A **digital image** is a discrete representation of visual field objects that have spatial (layout) and intensity (colour or grey tone) information.

From an appearance point of view, a **greyscale digital image**[1] is represented by a 2D light intensity function $I(x, y)$, where x and y are spatial coordinates and the value of I at (x, y) is proportional to the intensity of light that impacted on an optical sensor and recorded in the corresponding picture element (pixel) at that point.

If we have a multicolour image, then a pixel at (x, y) is 1×3 array and each array element indicates a red, green or blue brightness of the pixel in a colour band (or colour channel). A greyscale digital image I is represented by a single 2D array of numbers and a colour image is represented by a collection of 2D arrays, one for each colour band or channel. This is how, for example, Matlab represents colour images. A **binary image** consists entirely of black pixels (pixel intensity $= 0$) and white pixels (pixel intensity $= 1$). For simplicity, we use the term **binary image** to refer to a black and white image. By contrast, a **greyscale image** is an image that consists entirely of pixels with varying shades of black, grey tones and white (grey tones).

Binary images and greyscale images are 2-dimensional intensity images. By contrast, an RGB (red green blue) **colour image** (is a 3-dimensional or **multidimensional image**) image, since each colour pixel is represented by 3 colour channels, one channel for each colour. RGB images live in a what is known as an RGB colour space. There are many other forms of colour spaces. The most common alternative to an RGB space is the HSV (Hue, Saturation, Value) space implemented by Matlab or the HSB (Hue, Saturation, Brightness) space implemented by Mathematica.

---

[1] A **greyscale image** is an image containing pixels that are visible as black or white or grey tones (intermediate between black and white).

1.11.1: Sample image gems2.jpg   1.11.2: Sample image cup.jpg

Fig. 1.11 Sample images to compare using cpselect

Fig. 1.12 Pixels as tiny squares on the edges of the maple leaf

An **image point** (briefly, **point**) in a digital image is called a **picture point** or **pixel** or **point sample**.

A **pixel** is a physical point in a raster image. In this book, the terms *picture point, point* and *pixel* are used interchangeably. Each pixel has information that represents the response of an optical sensor to a particle of light (photon) reflected by a part of an object within a field of view (also field of vision). A **visual field** or field of view is the extent of the observable world that is seen (part of scene in front of a camera) at any given moment. In terms of a digitized optical sensor value, a **point sample** is a single number in a greyscale image or a set of 3 numbers in a colour image. It is common to use the little square model, which represents a pixel as a geometric square.

Example 1.4 **Jaggies**

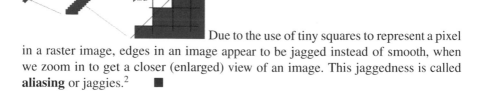

Due to the use of tiny squares to represent a pixel in a raster image, edges in an image appear to be jagged instead of smooth, when we zoom in to get a closer (enlarged) view of an image. This jaggedness is called **aliasing** or jaggies.[2] ∎

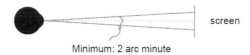

Fig. 1.13 Relation between human eye and TV screen

Example 1.5 **Experiment with cpselect from Matlab**.
Try the following Matlab® experiment:

$$\gg cpselect(\,'gems2.jpg'\,,'\,cup.jpg')$$

using a pair of colour images like the ones shown in Fig. 1.11. To see this, move the cpselect window over the maple leaf in the sample image in Fig. 1.11.1 and set the zoom at 600%. Then notice the tiny squares along the edges of the zoomed-in maple leaf in Fig. 1.12. Try a similar experiment with a second image such as the one in Fig. 1.11.2 (or the same image) in the right-hand cpselect window. There are advantages in choosing the same image for the right-hand cpselect window, since this makes it possible to compare what happens while zooming in by different amounts in relation to a zoomed-in image in the left-hand cpselect window.

Note The human eye can identify 120 pixels per degree of visual arc, i.e., if 2 dots are closer than $\frac{1}{120}$ degree, then our eyes cannot tell the difference. At a distance of 2 m (normal distance to a TV), our eyes cannot differentiate 2 dots 0.4 mm apart (see, for example, Fig. 1.13). ∎

   In other words, for example, a pixel p centered at location (i, j) in a digital image is identified with an area of the plane bounded by a square with sides of length 0.5 mm, i.e.,

---

[2]Doug Baldwin, http://cs.geneseo.edu/~baldwin/reference/images.html.

$$p = \{(x, y) : i - 0.5 \le x \le i + 0.5, \, j - 0.5 \le y \le j + 0.5\}.$$

See, e.g., the sample pixel p centered at $(5.2, 2.5)$ represented as a square in Fig. 1.9, where

$$p = \{(x, y) : i - 0.5 \le 5.5 \le i + 0.5, \, j - 0.5 \le 2.5 \le j + 0.5\}.$$

A.R. Smith points out that this is misleading [179]. Instead, in a 2D model of an image, a pixel is a point sample that exists only at a point in the plane. For a colour image, pixels contains three point samples, one for each colour channel. Normally, a pixel is the smallest unit of analysis of images. Sub-pixel analysis is also possible. For more about pixels, see Appendix B.15.

In photography, a **visual field** is that part of the physical world that is visible through a camera at a particular position and orientation in space. A visual field is identified with a view cone or angle of view. In Matlab, a greyscale image pixel $I(x, y)$ denotes the light intensity (without colour) at the x row and y column of the image. Values of x and y start at the origin in the upper lefthand corner of an image (see, e.g., the greyscale image of a cameraman in Fig. 1.14).

A sample display of a coordinate system with a greyscale colorbar for an image is shown in Fig. 1.14 using the code in Listing 1.1. The **imagesc** function is used to scale the intensities in a greyscale image. The **colormap(gray)** and **colorbar** functions are used to produce a colorbar to the west of a displayed image.

Fig. 1.14 Scaled image

```
A=imread('cameraman.tif'); % Read in image
figure, imagesc(A);   %scale intensities & display  to  use  colormap
colormap(gray); colorbar;  %
imfinfo('cameraman.tif')
```

Listing 1.1 Matlab code in `eg_01.m` to produce Fig. 1.14.

In Fig. 1.14, the top lefthand corner has coordinates (0, 0), the origin of the array representation of the image. To see the information for the cameraman image, use the **imfinfo** function (see Listing 1.2).

$$\gg \text{imfinfo('cameraman.tif')}$$

to obtain

```
imfinfo('cameraman.tif')

ans =

                  Filename: 'cameraman.tif'
               FileModDate: '20-Dec-2010 09:43:30'
                  FileSize: 65240
                    Format: 'tif'
             FormatVersion: []
                     Width: 256
                    Height: 256
                  BitDepth: 8
                 ColorType: 'grayscale'
           FormatSignature: [77 77 42 0]
                 ByteOrder: 'big-endian'
            NewSubFileType: 0
             BitsPerSample: 8
               Compression: 'PackBits'
    PhotometricInterpretation: 'BlackIsZero'
               StripOffsets: [8x1 double]
            SamplesPerPixel: 1
               RowsPerStrip: 32
            StripByteCounts: [8x1 double]
                XResolution: 72
                YResolution: 72
             ResolutionUnit: 'None'
                   Colormap: []
         PlanarConfiguration: 'Chunky'
                  TileWidth: []
                 TileLength: []
                TileOffsets: []
             TileByteCounts: []
                Orientation: 1
                  FillOrder: 1
           GrayResponseUnit: 0.0100
             MaxSampleValue: 255
             MinSampleValue: 0
               Thresholding: 1
                     Offset: 64872
           ImageDescription: [1x112 char]
```

Listing 1.2 Image information using `imfinfo` in Listing 1.1.

The image in Fig. 1.14 is an example of a greyscale[3] image. A greyscale image A is represented by an array such as the following one with corresponding pixel intensities.

$$
\mathbf{A} =
\begin{bmatrix}
A(1,1) & A(2,1) & \ldots & A(450,1) \\
A(1,2) & A(2,2) & \ldots & A(450,2) \\
\vdots & \vdots & \ddots & \vdots \\
A(1,350) & A(2,350) & \ldots & A(450,350)
\end{bmatrix}
=
\begin{bmatrix}
50 & 52 & \ldots & 50 \\
50 & 152 & \ldots & 250 \\
\vdots & \vdots & \ddots & \vdots \\
100 & 120 & \ldots & 8
\end{bmatrix}.
$$

In image A, notice that pixel $A(450, 350)$ has greyscale intensity 8 (almost black). And the pixel at $A(450, 2)$ has intensity 250 (almost white).

1.6 Digital Visual Space

A **digital visual space** is a nonempty set that consists of points in a digital image. A **space** is a nonempty set with some sort of structure.

Historical Note 1 Visual Space.
J.H. Poincaré introduced sets of similar sensations to represent the results of G.T. Fechner's sensation sensitivity experiments [50] and a framework for the study of resemblance in representative spaces as models of what he termed physical continua [154–156]. Visual spaces are prominent among the types of spaces that Poincaré wrote about.

The elements of a physical continuum (pc) are sets of sensations. The notion of a pc and various representative spaces (tactile, visual, motor spaces) were introduced by Poincaré in an 1894 article on the mathematical continuum [156], an 1895 article on space and geometry [155]. ■

From the Historical Note, the important thing to observe is that a digital image can be viewed as a visual space with some form of structure. Notice that the idea of a digital visual space extends to collections (sets) of digital images, where the structure of each such collection is defined, for example, by the nearness or apartness of image structures such as neighbourhoods of points in the images in the collection. In effect, a collection of digital images with some form of structure constitutes a **visual space**.

---

[3]Or *grayscale* image, using the Mathworks (Matlab) spelling. These notes are written using Canadian spelling.

Digital Image Analysis Secret

One of the important secrets in computer vision and digital image analysis is the discovery of image structures that reveal image patterns.

Remark 1.6 **Tomasi View of a Raster Image.**
View an ordinary camera image *im* as a mapping f from a 2D image location to either a 1-dimensional Euclidean space (for binary or greyscale images) or a 3-dimensional Euclidean space (for colour images), i.e.,

$$f : \mathbb{R}^m \longrightarrow \mathbb{R}^n, \ m = 2, n = 1 \text{ (binary, greyscale)}, n = 3 \text{ (colour)}.$$

For example, if p is the location a pixel in a 2D rgb colour image, then $f(p)$ is a vector with 3 components, namely, the red, green and blue intensity values for the pixel p. This is the Tomasi model of a raster image [194]. ∎

1.7 Creating Your Own Images

Any 2D array of natural numbers in the range $[0, n]$, $n \in \mathbb{N}$ (natural numbers) can be viewed as a greyscale digital image. Each natural number specifies a pixel intensity. The upper limit n on the range of intensities is usually 255.

Here is an example. The greyscale image in Fig. 1.15 (an image that approximates the Mona Lisa painting) is constructed from an array of positive integers, where each integer represents a pixel grayscale intensity. Internally, Matlab represents a single intensity as tiny subimage (each pixel in the subimage has the same intensity).

```
% sample digital image

132  128  126  123  137  129  130  145  158  170  172  161  153  158  162  172  159  152;
139  136  127  125  129  134  143  147  150  146  157  157  158  166  171  163  154  144;
144  135  125  119  124  135  121  62   29   16   20   47   89   151  162  158  152  137;
146  132  125  125  132  89   17   19   11   8    6    9    17   38   134  164  155  143;
142  130  124  130  119  15   46   82   54   25   6    6    11   17   33   155  173  156;
134  132  138  148  47   92   208  227  181  111  33   9    6    14   16   70   180  178;
151  139  158  117  22   162  242  248  225  153  62   19   8    8    11   13   159  152;
153  135  157  46   39   174  207  210  205  136  89   52   17   7    6    6    70   108;
167  168  128  17   63   169  196  211  168  137  121  88   21   9    7    5    34   57;
166  170  93   16   34   63   77   140  28   48   31   25   17   10   9    8    22   36;
136  111  83   15   48   69   57   124  55   86   52   112  34   11   9    6    15   30;
49   39   46   11   83   174  150  128  103  199  194  108  23   12   12   10   14   34;
26   24   18   14   53   175  153  134  98   172  146  59   13   14   13   12   12   46;
21   16   11   14   21   110  126  47   62   142  85   33   10   13   13   11   11   15;
17   14   10   11   11   69   102  42   39   74   71   28   9    13   12   12   11   18;
18   19   11   12   8    43   126  69   49   77   46   17   7    14   12   11   12   19;
24   30   17   11   12   6    73   165  79   37   15   12   10   12   13   10   10   16;
24   40   18   9    9    2    2    23   16   10   9    10   10   11   9    8    6    10;
43   40   25   6    10   2    0    6    20   8    10   16   18   10   4    3    5    7;
39   34   23   5    7    3    2    6    77   39   25   31   36   11   2    2    5    2;
```

```
17  16  9  4  6  5  6  36  85  82  68  75  72  27  5  7  8  0;
4  8  5  6  8  15  65  127  135  108  120  131  101  47  6  11  7  4;
2  9  6  6  7  74  144  170  175  149  162  153  110  48  11  12  3  5;
11  9  3  7  21  127  176  190  169  166  182  158  118  44  10  11  2  5;
8  0  5  23  63  162  185  191  186  181  188  156  117  38  11  12  25  33;
3  5  6  64  147  182  173  190  221  212  205  181  110  33  19  42  57  50;
5  3  7  45  160  190  149  200  253  255  239  210  115  46  30  25  9  5;
9  4  10  16  24  63  93  187  223  237  209  124  36  17  4  3  2  1;
7  8  13  8  9  12  17  19  26  41  42  24  11  5  0  1  7  4;
%
```

Listing 1.3 Matlab code in `lisa.m` to produce Fig. 1.15.

can be viewed as a set of grey level intensities (see Fig. 1.15).

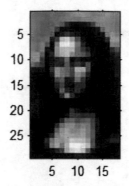

Fig. 1.15 Greyscale image

Here is the code used to produce the image in Fig. 1.15.

```
% Creating an image

fid=fopen('lisa2.txt');
A = textscan(fid,'%d','delimiter','\b\t;');
B=reshape(A{1},[18,29]);
B=double(B);
B=B';
figure; imshow(B,[]);
```

Listing 1.4 Matlab code in `readTxt.m` to produce Fig. 1.15.

Thought Problem 1 ☕ Path Leading to a UXW

We know that unbounded digital images are possible. This takes along the path of UXWs (UneXplored Worlds). Let $\mathbb{N}$ denote the set of natural numbers from 0 to ∞ (infinity). In some UXW, there are digital cameras that produce images that are $\mathbb{N} \times \mathbb{N}$ arrays, where $m \times n$ image has $m \in (0, \infty]$ and $n \in (0, \infty]$. Incidentally, city buses in a UXW have an unlimited number of seats. Here is the thought problem. Design a digital camera that produces $m \times n$ in a UXW. ∎

Generating $n \times n$ Images Using rand(n)

The Matlab **rand(n)** function (n a positive integer) produces pseudo-random numbers in the range [0,1] in a $n \times n$ array. In Matlab,

$$rand(n). * m$$

produces random numbers in the range from [0,m] in an $n \times n$ array. The values produced by **rand(n)** are of type double. By varying n in **rand(n)**, we can vary the size of the arrays that are realized in Matlab as $n \times n$ images.

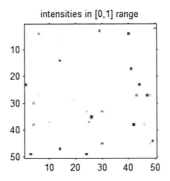

Fig. 1.16 Greyscale image with sparse background

1.8   Randomly Generated Images

This section illustrates the use of random numbers to generate a digital image. This is done with the **rand** function in Matlab. The image in Fig. 1.16 represents an array of randomly generated intensities in the range from 0 to 100, using the code in Listing 1.5. Here is a sample of 8 numbers of type double produced by the code:

81.4724   70.6046   75.1267   7.5854   10.6653   41.7267   54.7009   90.5792

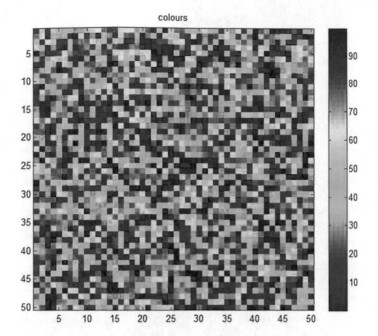

Fig. 1.17 rgb image

```
% Generate  random  array  of  numbers  in  range  0  to  max

I = rand(50).*100; % max = 100
%I = rand(100).*1000; % max = 1000
%I = rand(150).*1000;
%I = rand(256).*1000;
figure, imshow(I); title('bw');
figure, image(I,'CDataMapping','scaled');
axis image; title('colours');
colorbar
```

Listing 1.5 Matlab code in eg_02.m to produce Figs. 1.16 and 1.17.

In Matlab, the **image** function displays a matrix *I* as an image. Using this function, each element of matrix *I* specifies the colour of a rectangular patch (see Fig. 1.17 to see how the image function displays a matrix of random number values as an image containing color patches). Listing 1.5 contains the following line of code that produces the greyscale image in Fig. 1.16. Because the second image produced by the code in Listing 1.5 has been scaled, the colourbar to the right of the image in Fig. 1.17 shows intensities (colour amounts) in the range from 0 to 100, instead of the range 0 to 1 (as in Fig. 1.16).

```
figure, imshow(I); title('intensities in [0,1] range');
```

Listing 1.6 Matlab code in eg_02.m to produce Fig. 1.16.

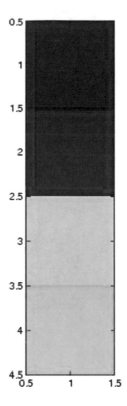

Fig. 1.18 Small rgb image

The **CDataMapping** parameter tells Matlab to select a matrix element colour using a colormap entry. For a **scaled** CDataMapping, values of a matrix I are treated as colormap indices.

Problem 1.7

(**image**.1) 🚲 Write Matlab code to produce the image in Fig. 1.18.
(**image**.2) 🚲 Write Matlab code to display the Mona Lisa shown in Fig. 1.15 as a colour image instead of a greyscale image. Your colour image should be similar to colour image in Fig. 1.17. ■

For Problem 1.7, you may find the following trick useful. To transform the color array in Fig. 1.19.1 to a narrower display like the one in Fig. 1.19.1, try
≫ **axis** image.

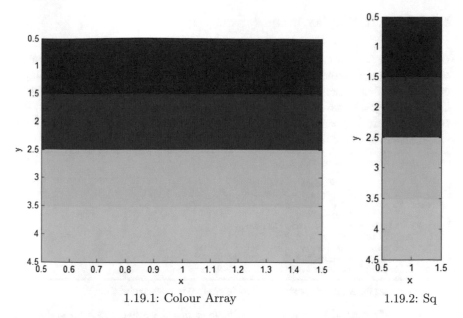

1.19.1: Colour Array   1.19.2: Sq

Fig. 1.19 Set aspect ratio to square pixels

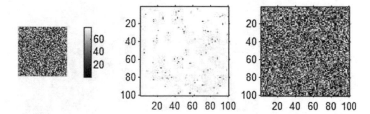

Fig. 1.20 Image displays

Generating $n \times n$ Images Using randi(n,m)

The Matlab function **randi** produces pseudo-random numbers
that are positive integers. To obtain an image with, for example,
exactly 80 intensities in a 100×100 array, try

$$I = \text{randi}(80,100) - 1;$$

where $I = 100 \times 100$ is an array with values in [0,79].

1.9 Ways to Display Images

By controlling or not controlling the scaling of an image, many different displays
of an image are possible. Here is an example. The images in Fig. 1.20 are displayed
using the Matlab code in Listing 1.7. Notice the use of the **subplot(r,c,i)** to display
a row of images. In general, subplot(r,c,i) displays r rows of images (or plots) in c
columns, where $1 \leq i \leq r * c$. Listing 1.7 uses the **rand** function. This function is
written **rand(n)** to produce n random numbers in [0, 1].

```
% Display image contrast

                    % what's happening?
I = rand(100).*80; %generate random image array
% with 80 intensities in range 0...100
subplot(1,3,1); imshow(I);
imagesc(I); % scale colormap to data
axis image; axis off; %range
colormap(gray); colorbar; % produce colorbar
subplot(1,3,2); imshow(I); % do not specify range
subplot(1,3,3); imshow(I,[0 80]); % specify range
```

Listing 1.7 Use the Matlab code in eg_03.m to produce the images in Fig. 1.20.

By way of a second sample of image displays, consider displaying *three different
images in three different ways*. The method used to produce Fig. 1.21 is given in
Listing 1.8.

Fig. 1.21 Three image displays (cell, spine, and onions)

```
% Display multiple images

                    % what's happening?
I = imread('cell.tif'); % choose .tif file
J = imread('spine.tif'); % choose 2nd .tif file
K = imread('onion.png'); % choose .png file
%
subplot(1,3,1); imagesc(I); axis image; % scale image
axis image; axis off; % display first image
colormap(gray); colorbar; % produce colorbar
subplot(1,3,2); imagesc(J); axis image; % 2nd image
axis off; colormap(jet); % set colormap to jet (false colour)
subplot(1,3,3); imshow(K); % display colour image
```

Listing 1.8 Use the Matlab code in eg_04.m to produce the images in Fig. 1.21.

1.10   Digital Image Formats

There are a number of important, commonly available digital image formats, briefly summarised as follows.

(**Format**.1) .bmp (bit mapped picture) basic image format, limited generally, loss-less compression (lossy variants exist). .bmp originated in the development of Microsoft Windows.

(**Format**.2) .gif (graphics interchange format) Limited to 256 colours (8 bit), loss-less compression. **Lossless data compression** is the name of a class of compression algorithms that make it possible to reconstruct the exact original data from the compressed data. By contrast, **lossy data compression** only permits an approximation of the original data from the compressed data.

(**Format**.3) .jpg, .jpeg (joint photographic experts group) Most commonly used file format, today (e.g., in most cameras), lossless compression (lossy variants exist).

(**Format**.4) .png (portable network graphics) .png is a bit mapped image format that employs lossless data compression.

(**Format**.5) .svg (scalable vector graphics) Instead of a raster image format (describes the characteristics of each pixel), a vector image format gives a geometric description that can be rendered smoothly at any display size. An .svg image provides a versatile, scriptable and all-purpose vector format for the web and other publishing applications. To gain some experience working with vector images, download and experiment with the public domain tool named **Inkscape**.

(**Format**.6) .tif, .tiff (tagged image file format) Highly flexible, detailed, adaptable format, compressed and uncompressed variants exist.

> .png was designed to replace .gif, new lossless compression format. The acronym png can be read p*ng* n*ot* g*if*. This format was approved by the internet engineering steering group in 1996 and was published as an ISO/IEC standard in 2004. This format supports both greyscale and full colour (rgb) images. This format was designed for internet transfer and not for professional quality print graphics. See http://linuxgazette.net/ issue13/png.html for a history of the .png format.

The need to transmit images over networks and the need to recognize bodies of numerical data as corresponding to digital images has led to the introduction of a number of different image formats. Among these formats, .jpg and .tif are the most popular. In general, .jpg and .tif are better suited for photographic images. The .gif and .png formats are better suited for images with limited colour, detail, e.g., logos, handwriting, line drawings, text.

Fig. 1.22 Greyscale image with 16 bit intensity values

1.11 Image Data Types

The choice of an image format can be determined, for the most part, not just by image contents but also by the image data type required for storage. Here are a number of distinct image types.

(**Type**.1)   **Binary** (logical) images. A **binary image** is represented by a 2D array, where each array element assigns to each pixel a number from {0, 1}. Black corresponds to a 0 (off or background pixel). White corresponds to 1 (on or foreground pixel). A fax image is an example of a binary image. For an rgb image I, **im2bw('I.rgb')** is used by Matlab to convert I to a binary image. For the details, type the following line the Matlab workspace:

$$\gg \text{ help im2bw}$$

(**Type**.2)   **Intensity** (greyscale) images are 2D arrays, where each array element assigns one numerical value from $\mathbb{N}^{0+}$ (natural numbers plus zero, usually natural numbers in the 8 bit range from 0 to 255 (or scaling in the range from 0.0 to 1.0). For a greyscale image with 16 bit integer intensities from 0 to 65535, try the Matlab code in Listing 1.9 using the **im2uint16** function and the **rgb2gray** function to convert a colour image to a greyscale image. Using **imtool**, you can inspect the resulting image intensities (see, e.g., Fig. 1.22, with an intensity equal to 6842 for the pixel at (2959, 1111)).

```
% 16 bit greyscale image

g = imread('workshop.jpg'); % a 4.7 MB colour image
g = im2uint16(g);
g = rgb2gray(g);
imtool(g)
```

Listing 1.9 Use the Matlab code in eg_im2uint16.m to produce the images in Fig. 1.22.

Each intensity assigned to a pixel represents the light intensity at a particular point, i.e., an intensity represents the amount of light reflected from a point in a visual field and captured by a camera sensor. Notice that many Matlab functions representing image features such as edge detection and pixel gradient methods require greylevel intensities as input (this means, for example, that an rgb image must be converted to a greyscale.

For an rgb image I, **rgb2gray('I.rgb')** is used by Matlab to convert I to a greyscale image before pixel gradients can be computed. From a computer vision point of view for robot navigation systems, see http://homepages.inf.ed.ac.uk/rbf/CVonline/LOCAL_COPIES/DIAS2/ for details. This is important in detecting image shapes and in implementing corner detection methods studied by R.M. Haralick and L.G. Shapiro in 1993 [70].

(**Type**.3) *True colour*[4] images (e.g., *.rgb* images) assign 3 numerical values to each pixel, where each assigned value corresponds to a colour amount for a particular colour channel (either red or green or blue image channel).

(**Type**.4) *False colour* image depicts an object in colours that differ from those a photograph (a true colour image) would show. The term **false colour** refers to a group of colour rendering methods used to display images in colours recorded in the visual of non-visual parts of the electromagnetic spectrum. **Pseudo colour**, **density slicing** and **choropleths** are variants of false colour used for information visualization for objects in a single greyscale channel such as relief maps and tissue types in magnetic resonance imaging. In a topology of digital images, it is helpful to render the picture points in the neighbourhood of a point with a single colour, making it possible to distinguish one neighbourhood of a point from other neighbourhoods of points in a picture (see, e.g., false colour-rendering of the picture points in an unbounded descriptive neighbourhood in Fig. 1.23).

---

[4]This is the same type **true color** (U.S. spelling) images commonly found in the literature and also supported by Matlab and devices such as printers.

(**Type**.5)   *Floating point* images are very different from the other image types. Instead of integer pixel values, the pixels in these image store floating point numbers representing intensity.

Fig. 1.23 Unbounded descriptive Nbd of $I(x, y)$ with $\varepsilon = 10$

Thought Problem 2 ☕
The code in Listing 1.9 leads us to a path of a UXW, namely, the world of greyscale images with intensities in the range $[0, \infty]$. *To inspect images in this unexplored world, we just need to invent a* **im2uint∞** *matlab function that gives us an unbounded range of intensities for greyscale images.*   ■

Problem 1.8 Pick several different points in a colour image and display the colour channel values for each of the selected points. How would you go about printing (displaying) all of the values for the red colour channel for this image using only one command in Matlab? If you get it right, you will be displaying only the red colors in the peppers.png image.   ■

Problem 1.9 To see an example of a floating point image, try out the code in Listing 1.10.

```
% Generating a floating point image

C = rand(100,2);
figure, image(C,'CDataMapping','scaled')
axis image
imtool(C)
```

Listing 1.10 Use the Matlab code to produce a floating point image.

Modify the code to produce an image with width = 3, then 4, then 10. To find out how **rand** works in Matlab, enter

$$\gg \textbf{help rand}$$

Also enter
$$\gg \textbf{help image}$$

to find out what the **'CDataMapping** and **'scaled'** parameters are doing to produce the image when you run the code in Listing 1.10. For more information about Matlab image formats, type[5]

$$\gg help\,imwrite \text{ or } \gg help\,imformats \,.\quad \blacksquare$$

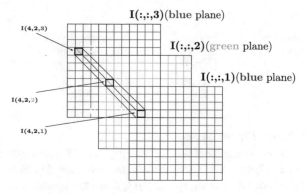

Fig. 1.24 Sample pixels in 2D planes for color image I

---

[5]For example, **imwrite** is useful, if there is a needed to create a new image file containing a processed image such as the image **g** in Listing 1.9. To see this, try $\gg$ imwrite(g,'greyimage.jpg');.

1.12   Colour Images

This section focuses on the use of a colour lookup table by Mathworks (in Matlab) in correlating pixel intensities with the amounts of pixel colour in the rgb colour space. An overview of the different colour spaces is given by R.C. Gonzalez and R.E. Woods [58, Chap. 6, p. 394ff]. For a Matlab-oriented view of the colour spaces, see R.C. Gonzalez, R.E. Woods and S.L. Eddins [59, Chap. 5, p. 272ff].

1.12.1   Colour Spaces

This section briefly introduces an approach to producing sample colours in the RGB, HSB and CIE LUV colour spaces.

Remark 1.10 **Colour Spaces and Photonics**.
A thorough study of various colour spaces, starting with the RBG colour space, is presented by M. Sonka, V. Hlavac, and R. Boyle [185, Sect. 2.4.3]. The physics of colour is briefly considered in [185, Sect. 2.4.1] and thoroughly presented from a photonics perspective in E. Stijns and H. Thienpont [188, Sect. 2].   ∎

Fig. 1.25 Samples from three colour spaces via mathematica

1.25.1:   1.25.2:   1.25.3:
RGB   HSB   LUV

Example 1.11 **Three Colour Spaces**.
The HSB (hue, saturation, brightness) is a variation of HSV (hue, saturation, value) colour space. In 1931, the CIE (Commission Internationale de l'Eclairage) defined three standard primaries X, Y, Z to replace R, G, B and introduced the LUV (*L* stands for luminance) colour space (Fig. 1.25).   ∎

1.12.2   Colour Channels

Mathematically, the colour channels for a colour image are represented by three distinct 2D arrays with dimension $m \times n$ for an image with m rows and n columns with one array for each colour, red (colour channel 1), green (colour channel 2), blue (colour channel 3). A **pixel colour** is modelled as 1×3 array. In Fig. 1.24, for example,

$$I(2, 4, :) = (I(2, 4, 1), I(2, 4, 2), I(2, 4, 3)) = \left(\blacksquare , \blacksquare , \blacksquare \right),$$

where, for instance, the numerical value of $I(4, 2, 1)$ is represented by $\blacksquare$. For a colour image I in Matlab, the colour channel values for a pixel with coordinates (x, y) are displayed using

$\gg I(x, y, 1)\% =$ red colour channel value for image I,

$\gg I(x, y, 2)\% =$ green colour channel value for image I,

$\gg I(x, y, 3)\% =$ blue colour channel value for image I.

Three 2D colour planes for an image I are shown in Fig. 1.24. A sample pixel at location $(x, y) = (2, 4)$ (column 2, row 4) is also shown in each colour plane, i.e.,

$\gg I(2, 4, 1)\% =$ **red** colour channel pixel value,

$\gg I(2, 4, 2)\% =$ green colour channel pixel value,

$\gg I(2, 4, 3)\% =$ **blue** colour channel pixel value.

The rgb colour cube in Fig. 1.26 is generated with the Matlab code in Listing 1.11.

Fig. 1.26 rgb colour cube

```
function rgbcube(x,y,z)
vertices = [0 0 0;0 0 1;0 1 0;0 1 1;1 0 0;1 0 1;1 1 0;1 1 1];
faces = [1 5 6 2;1 3 7 5;1 2 4 3;2 4 8 6;3 7 8 4;5 6 8 7];
colors = vertices;
patch('vertices',vertices,'faces',faces,...
    'FaceVertexCData',colors,'FaceColor','interp',...
    'EdgeAlpha',0)
if nargin == 0
  x = 10;y = 10;z = 4;
elseif nargin ~= 3
  error('wrong no. of inputs')
end
axis off
view([x,y,z])
axis square
%
% >> rgbcube    %sample use of this function
%
```

Listing 1.11 Use the Matlab code in `rgbcube.m` to produce the image in Fig. 1.26.

The Matlab code Listing 1.11 uses the **patch** function. Briefly, **patch(X,Y,C)** adds a 'patch' or filled 2D polygon defined by the vectors X and Y to the current axes. If X and Y are matrices of the same size, one polygon per column is added. The parameter C specifies the colour of the added face. To obtain a detailed explanation of the patch function, type

$$\gg doc\ patch$$

Fig. 1.27 Green-cyan-white-yellow colour plane

Problem 1.12 Use the rgbcube function to the following colour planes, using the specific values for (x, y, z) as an argument for the function:

$$\textbf{Colour plane} = (x, y, z)$$

$$\textbf{blue-magenta-white-cyan} = (?, ?, ?), \; -10 \le x, y, z \le 10,$$

$$\textbf{red-yellow-white-magenta} = (?, ?, ?),$$

$$\textbf{green-cyan-white-yellow} = (0, 10, 0),$$

$$\textbf{black-red-magenta-blue} = (?, ?, ?),$$

$$\textbf{black-blue-cyan-green} = (?, ?, ?),$$

$$\textbf{black-red-yellow-green} = (0, 0, -10).$$

For example, the green-cyan-white-yellow colour plane in Fig. 1.27 is produced using

$$\gg \; figure, rgbcube(0, 10, 0)$$

To solve this problem, give the missing values for each (?, ?, ?) and display and name the corresponding colour plane. ∎

Table 1.1 shows the six main colours in the visible spectrum, along with their typical wavelength and frequency[6] ranges. The wavelengths are given in nanometres (10^{-9} m) and the frequencies are given in terahertz (10^{12} Hz).

Table 1.1 Wavelength interval & frequency interval

Colour	Wavelength (10^{-9} m (nm))	Frequency (10^{12} Hz (THz))
Red	$\sim 700\,nm - \sim 635\,nm$	$\sim 430\,THz - \sim 480\,THz$
Orange	$\sim 635\,nm - \sim 590\,nm$	$\sim 480\,THz - \sim 510\,THz$
Yellow	$\sim 590\,nm - \sim 560\,nm$	$\sim 510\,THz - \sim 540\,THz$
Green	$\sim 560\,nm - \sim 490\,nm$	$\sim 540\,THz - \sim 610\,THz$
Blue	$\sim 490\,nm - \sim 450\,nm$	$\sim 610\,THz - \sim 670\,THz$
Violet	$\sim 450\,nm - \sim 400\,nm$	$\sim 670\,THz - \sim 750\,THz$

[6]Frequencies:
High frequency RF signals (**3–30 GHz**) and Extreme High Frequency RF signals (**30–300 GHz**) interacting with an electron-hole plasma in a semiconductor [178, Sect. 1.2.1, p. 10] (see, also, [213]), Picosecond photoconducting dipole antenna illuminated with femtosecond optical pulses, radiate electrical pulses with frequency spectra from **dc-to-THz** [178, Sect. 1.4.1, p. 16].

1.13   Colour Lookup Table

The Matlab function **colormap** is used to specify one of the built-in colour tables used to define colour. Mathematically, a colormap is an $m \times 3$ matrix of real numbers between 0.0 and 1.0 (or between 0 and 255 on an integer scale). The kth row defines the kth colour in an image. To obtain a detailed explanation of the colormap function, type

$$\gg doc\ colormap$$

To see sample colormaps, type

$\gg I = imread('peppers.jpg');$

$\gg I = rgb2gray(I);$

$\gg figure, image(I)$

$\gg colormap(bone)\%$ greyscale colour map

$\gg colormap(pink)\%$ pastel shades of pink colormap

$\gg colormap(copper)\%$ colours from black to bright copper

A colour translation table or a colour lookup table (LUT) associates a pixel intensity value (0 to 255) to a colour value. This colour value is represented as a triple (i, j, k); this is much like the representation of a colour using one of the colour models. Once a desired colour table or LUT has been set up, any image displayed on a colour monitor automatically translates each pixel value by the LUT into a colour that is then displayed at that pixel point. In Matlab, a LUT is called a colourmap.

A sample colour lookup table is given in Table 1.2. This table is based on the RGB colour cube and goes gradually from black (0, 0, 0) to yellow (255, 255, 0) to red (255, 0, 0) to white (255, 255, 255). Thus, the pixel value (the subscript to the colour table) for black is 0, for yellow is 84 and 85, for red is 169 and 170, and for white is 255. In Matlab, the subscripts to the colour table would run from 1 to 256.

Notice that not all the possible 256^3 RGB hues are represented, since there are only 256 entries in the table. For example, both blue (0, 0, 255) and green (0, 255, 0) are missing. This is due to the fact that a pixel may have only one of 256 8-bit values. This means that the user may choose which hues to put into his colour table.

A colour translation table based on the hue saturation value (hsv) or hue saturation lightness (hsl) colour models would have entries with real numbers instead of integers. The basic representation for hsv uses cylindrical coordinate representations of points in an rgb colour model. The first column would contain the number of degrees (0–360°) needed to specify the hue. The second and third columns would contain values between 0.0 and 1.0.

Table 1.2   Colour table

Intensity	Red ■	Green ■	Blue ■
0 (0.0)	0	0	0
1	3	3	0
2	6	9	0
3	9	9	0
4	12	12	0
⋮	⋮	⋮	⋮
81	245	245	0
82	248	248	0
83	251	251	0
84	255	255	0
85	255	251	0
86	255	248	0
87	255	245	0
⋮	⋮	⋮	⋮
167	255	6	0
168	255	3	0
168	255	0	0
170	255	0	0
171	255	3	3
172	255	6	6
⋮	⋮	⋮	⋮
251	255	248	245
252	255	248	248
253	255	251	251
254	255	255	255
255 (1.0)	255	255	255

Of course, a standard or default colour table is usually provided automatically. Do a

≫ **help color**

within Matlab to see what colour tables or colourmaps it has predefined. For more details, see https://csel.cs.colourado.edu/~csci4576/SciVis/SciVisColor.html.

Remark 1.13 To see the colour channel values for a pixel, try the following experiment.

```
% Experiment with a pixel in a colour image
%
                          % What's happening?
g = imread('rainbow-shoe2.jpg');  % read colour image
figure,imagesc(g), colorbar; % display rainbow image
g(196,320)                 % display red channel value
g(196,320,:)               % display 3 colour channel values
```

Listing 1.12 Use the Matlab code in band.m to produce the image in Fig. 1.28.

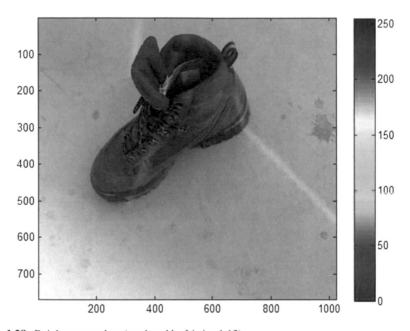

Fig. 1.28 Rainbow on a shoe (produced by Listing 1.12)

Problem 1.14 Do the following:

(**HSV**.1)   ☕ Compare and contrast the rgb and hsv colour models. **Hint**: Check Wikipedia introductions and Matlab documentation for these two colour models.

(**HSV**.2)   ☁ Using the Matlab **rgb2hsv** function, write a program to display the individual hue, saturation and value colour channels of two colour images (peppers.png and your own choice of a colour image). **Hint**: To get started, try out the following in Matlab:

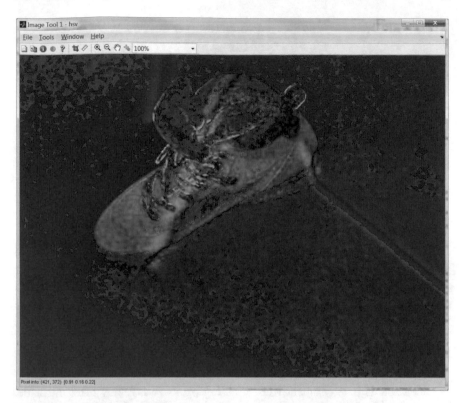

Fig. 1.29 Sample **imtool** hsv image display

$$\gg g = imread('peppers.png');$$
$$\gg hsv = rgb2hsv(g);$$
$$\gg imtool(hsv)$$

When the imtool window is displayed with the hsv version of the pep-pers.png image, move the cursor over the image to see the real values that correspond to each hsv colour. Notice, for example, the hsv color channel values for the pixel at **hsv(355, 10)** in Fig. 1.29. Also, custom interpretations of the hsv color space are possible, e.g., hsv colours represented by integers corresponding to degrees in a circle.

(**HSV**.3)   Using the Matlab **imtool** function to display the colour channel values for the hsv image. Give some sample colour channel values for three of the same pixels in the original rgb image and the new hsv image. ■

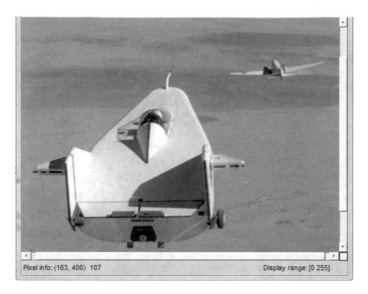

Pixel info: (163, 400) 107   Display range: [0 255]

Fig. 1.30 Sample **imtool** pixel value display

1.14 Image Geometry, a First Look

The idea now is to experiment with accessing pixels and display pixel intensities in an image. The following tricks are used to view individual pixels, modify pixel values and display images.

trick.1   Use % for comments.
trick.2   Matlab = % assignment operator.

Example 1.15 Draw a line segment between the specified coordinates.

$$\gg r1 = 450, c1 = 20, r2 = 30, c2 = 350; \ \% \text{ pixel coordinates}$$

trick.3   Matlab imread(image) % moves image into workspace.
trick.4   Matlab imtool(image) % interactive image viewer.

Example 1.16

$$\gg imtool(imread('liftingbody.png')) \ \% \text{ See Fig. 1.30}$$

trick.5   I(x,y) % displays pixel value at position (x,y).
trick.6   I(x,y,:) % displays colour channel values in rgb image.
trick.7   I(x,y) = 255% assigns maximum intensity to pixel at (x,y).
trick.8   Matlab imshow( ) % displays image.
trick.9   I(25,50,:) % displays pixel colour channel values.

trick.10   I(25,50,:) = 255% assigns white to each colour channel.
trick.11   Matlab line % draws a line segment in the current image.

Fig. 1.31 Sample **line** segment display

Example 1.17 Draw a line segment between the specified coordinates.

$$\gg line([450, 20], [30, 350]); \% \text{ See Fig. 1.31}$$

trick.12   Matlab improfile % computes the intensity values along a line or multiline path in an image.

Example 1.18 Draw a line segment between the specified coordinates.

$$\gg improfile(im, [r1, c1], [r2, c2]); \% \text{ See Fig. 1.32}$$

Listing 1.13 puts together the basic approach to accessing and plotting the pixel values along a line segment. Notice that properties parameters can be added to the line function to change the colour and width of a displayed line segment. For example, try

$$\gg improfile(im, [r1, c1], [r2, c2], 'Color', 'r', 'LineWidth', 3); \% \text{ red line}$$

Notice that the set of pixels in a line segment is an example of a simple convex set. In general, a set of points is a *convex set*, provided the straight line segment connecting

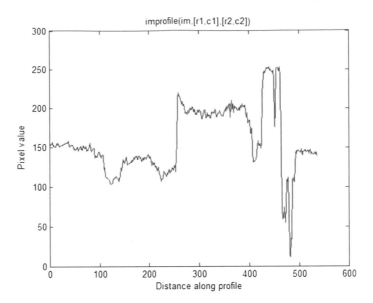

Fig. 1.32 Sample **line** segment intensities

each pair of points in the set is contained in the set. A line segment is an example of one-sided convex polygon. The combination of the **line** and **improfile** functions gives a glimpse of what is known as the texture of an image. Small elementary patterns repeated periodically in an image constitute what is known as *image texture*. The study of image line segments leads to the skeletonization of digital images, which is a stepping to object recognition and the delineation of image regions (e.g., areas of interest in topographic maps), which is an important part of computer vision (see, e.g., [176, Sect. 5.2]). The study of image line segments also ushers in the combination of digital images and computational geometry [41].

```
% pixel intensity profile along a line segment
clc, clear all, close all
im = imread('liftingbody.png');          % built-in greyscale image
image(im), axis on, colormap(gray(256)); % display image
r1 = 450; c1 = 20; r2 = 30; c2 = 350;    % select pixel coords.
line([r1,c1],[r2,c2]);                    % draw line segment
figure,
improfile(im,[r1,c1],[r2,c2]),            % plot pixel intensities
ylabel('Pixel value'),
title('improfile(im,[r1,c1],[r2,c2])')
```

Listing 1.13 Use the Matlab code in findIt.m to produce the images in Fig. 1.32.

A geometric view of digital images approximates what the eye sees and
what a camera captures from a visual scene.

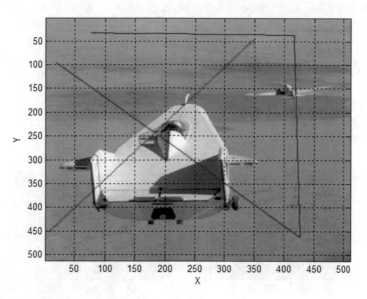

Fig. 1.33 Multiple line segments on an image

Problem 1.19 Working with Line Segments in Images.
Give Matlab code to do the following:

(a) Construct a pair of line segments in a greyscale image.
     Hint. See Fig. 1.33 and Listing 1.14.
(b) Give a plot of the pixel intensities for each line segment in part (a). **Hint**. Insert
     hold off (instead of **hold on**) in Listing 1.14. In a 3D view of pixel intensities,
     be sure to insert an appropriate **xlabel** on the vertical axis and a title for the 3D
     view.
(c) Based on the definition of a convex set, give two examples of convex sets that
     are not straight line segments. ∎

```
% pixel intensity profile along a line segment
clc, clear all, close all
im = imread('liftingbody.png');           % built-in greyscale image
figure
image(im), axis on, colormap(gray(256)); % display image
hold on
seg1 = [19 427 416 77];                   % define line segment 1
seg2 = [96 462 37 33];                    % define line segment 2
r1 = 8; c1 = 350; r2 = 450; c2 = 45;      % select pixel coords.
line([r1,c1],[r2,c2],'Color','r');        % draw line segment
```

```
improfile(im,seg1,seg2), grid on          % multiple line segments
```

Listing 1.14 Use the Matlab code in findLines.m to produce the image in Fig. 1.33.

Problem 1.20 More about Line Segments in Images.
毳 Solve Problem 1.19 with one line of Matlab code.   ■

1.15   Accessing and Modifying Image Pixel Values

It is possible to access, modify and display modified pixel values in an image. The
images in Fig. 1.34 (original image) and Fig. 1.35 (modified image) are produced
using the code in Listing 1.15.

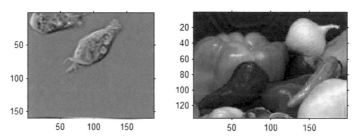

Fig. 1.34 Image display before changing pixel values

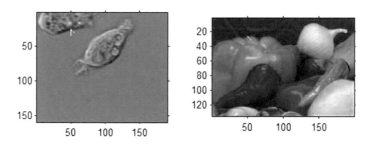

Fig. 1.35 Image display after changing pixel values

```
% Display and modify image rgb pixel values

                           % what's happening?
I = imread('cell.tif'); % choose .tif file
imtool(I);                  % use interactive viewer
%
K = imread('onion.png'); % choose .png file
imtool(K);                  % use interactive viewer
subplot(2,2,1); imshow(I);  % display unmodified greyscale image
subplot(2,2,2); imshow(K);  % display unmodified rgb image
%
I(25,50)                    % print value at (25,50)
I(25,50) = 255;             % set pixel value to white
I(26,50) = 255;             % set pixel value to white
I(27,50) = 255;             % set pixel value to white
I(28,50) = 255;             % set pixel value to white
I(29,50) = 255;             % set pixel value to white
I(30,50) = 255;             % set pixel value to white
I(31,50) = 255;             % set pixel value to white
I(32,50) = 255;             % set pixel value to white
I(33,50) = 255;             % set pixel value to white
I(34,50) = 255;             % set pixel value to white
I(35,50) = 255;             % set pixel value to white
%
I(26,51) = 255;             % set pixel value to white
I(27,52) = 255;             % set pixel value to white
I(28,52) = 255;             % set pixel value to white
I(29,54) = 255;             % set pixel value to white
I(30,55) = 255;             % set pixel value to white
subplot(2,2,3); imshow(I);  % display modified image
imtool(I);                  % use interactive viewer
%
K(25,50,:)                  % print rgb pixel value at (25,50)
K(25,50,1)                  % print red value at (25,50)
K(25,50,2)                  % print green value at (25,50)
K(25,50,3)                  % print blue value at (25,50)
K(25,50,:) = 255;           % set pixel value to rgb white
K(26,50,:) = 255;           % set pixel value to rgb white
K(27,50,:) = 255;           % set pixel value to rgb white
K(28,50,:) = 255;           % set pixel value to rgb white
K(29,50,:) = 255;           % set pixel value to rgb white
K(30,50,:) = 255;           % set pixel value to rgb white
%
K(26,51,:) = 255;           % set pixel value to rgb white
K(27,52,:) = 255;           % set pixel value to rgb white
K(28,52,:) = 255;           % set pixel value to rgb white
K(29,54,:) = 255;           % set pixel value to rgb white
K(30,55,:) = 255;           % set pixel value to rgb white
%K(31,56,:) = 255;            % set pixel value to rgb white
K(25,50,:)
subplot(2,2,4); imshow(K);  % display modified 2nd image
imtool(K);                  % use interactive viewer
```

Listing 1.15 Use the Matlab code in eg_05.m to produce the images in Figs. 1.34 and 1.35.

You will find that running the code in Listing 1.15 will display the following pixel values.

ans(:,:,.1) =   46% unmodified red channel value for pixel I(25, 50)
ans(:,:,.2) =   29% unmodified green channel value for pixel I(25, 50)
ans(:,:,.3) =   50% unmodified blue channel value for pixel I(25, 50)
ans(:,:,.1) =   255% modified red channel value for pixel I(25, 50)
ans(:,:,.2) =   255% modified green channel value for pixel I(25, 50)

ans(:,:,.3) =   255% modified blue channel value for pixel I(25, 50)

In addition, the code in Listing 1.15 displays an image viewer for each (both unmodified and modified). Here is what the image viewer looks like, using **imtool(image)** (Fig. 1.36).

Fig. 1.36 Image viewer for modified rgb onions image (notice hook)

1.16   RGB, Greyscale, and Binary (BW) Images

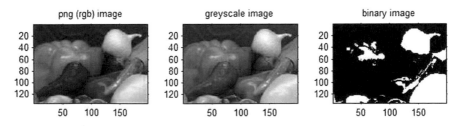

Fig. 1.37 png (rgb) → greyscale and binary images

```
% converting an image to greyscale
                         % What's happening
I = imread('onion.png'); % input png (rgb) image
%
Ig = rgb2gray(I);        % convert to grayscale
Ibw = im2bw(I);          % convert to rgb to binary image
%
subplot(1,3,1); imshow(I); axis image; title('png (rgb) image')
subplot(1,3,2); imshow(Ig); title('greyscale image');
subplot(1,3,3); imshow(Ibw); title('binary image');
```

Listing 1.16 Use the Matlab code in `binary.m` to produce the images in Figs. 1.34 and 1.37.

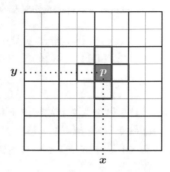

Fig. 1.38 Tiny 3×3 subimage containing a 4-neighbourhood $N_4(p)$

1.17 Rosenfeld 8-Neighbourhood of a Pixel

In the 1970s, A. Rosenfeld [169] introduced what he called an 4-neighbourhood (4-Nbd) of a pixel p, which is a set of 4 pixels adjacent pixels $N_4(p)$. Let p have coordinates (x, y) in an image I. Then $N_4(p)$ is a cross-shaped neighbourhood pixel p defined by

$$N_4(p) = \{p(x, y), p(x - 1, y), p(x + 1, y), p(x, y - 1), p(x, y + 1)\} \text{ (4-Nbd)}.$$

Let B a set of pixels in image I. A sample 4-neighbourhood of a pixel p is shown in 4×4 image in Fig. 1.38. The 4 neighbouring pixels of the pixel p located at (x, y) are outlined in blue. Then $N_4(p) \cup B$ is the union of the pixels in $N_4(p)$ and a second set of pixels B. For example, let B be defined by the corner pixels in Fig. 1.38, i.e.,

$$B = \{p(x - 1, y - 1), p(x + 1, y + 1), p(x + 1, y - 1), p(x - 1, y + 1)\}.$$

That is, B is the set of pixels in the corners of the subimage in I containing the 4-neighbourhood $N_4(p)$ in Fig. 1.38.

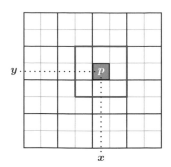

Fig. 1.39 Tiny 3×3 subimage containing an 8-neighbourhood $N_8(p)$

Then a Rosenfeld 8-neighbourhood of the pixel p (denoted $N_8(p)$) is defined as the union of the union of $N_4(p)$ and B (set of corners), i.e.,

$$N_8(p) = N_4(p) \cup B,$$
$$= N_4(p) \cup \{p(x-1, y-1), p(x+1, y+1), p(x+1, y-1), p(x-1, y+1)\},$$
$$= \{p(x, y), p(x-1, y), p(x+1, y), p(x, y-1), p(x, y+1)\} \cup$$
$$\{p(x-1, y-1), p(x+1, y+1), p(x+1, y-1), p(x-1, y+1)\}. \text{ (8-Nbd)}$$

See Fig. 1.39 for a sample 8-neighbourhood. For more about Rosenfeld 4- and 8-neighbourhoods, see R. Klette and A. Rosenfeld [94, Sect. 1.1.4, p. 9].

In an ordinary digital image without zooming, the tiny squares representing individual pixels are usually not visible. To get around this problem and avoid the need to use zooming, the Matlab function **imcrop** is used to select a tiny subimage in an image.

```
% Selecting a tiny subimage using imcrop
clc, clear all, close all
a = imread('peppers.png');            % built-in greyscale image
im = imcrop(a);                       % select tiny subimage
figure                                % display subimage
image(im), axis on, colormap(gray(256));
```

Listing 1.17 Use the Matlab code in `RosenfeldTinyImage.m` to produce the subimage in Fig. 1.41.

Example 1.21 **Extracting a Subimage from an Image**. With, for example, the **peppers.png** image available in Matlab (see Fig. 1.40), use **imcrop** to select a tiny subimage such as (part of the center, lower green pepper in Fig. 1.40). This is done using the Matlab script 1.17. For example, we can obtain the 9×11 subimage in Fig. 1.41. ∎

Fig. 1.40 **peppers.png**

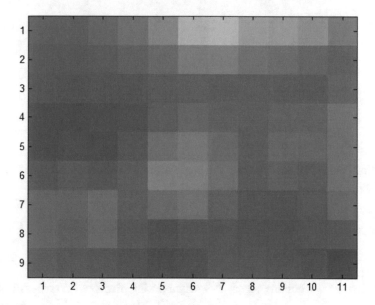

Fig. 1.41 Tiny subimage extracted from **peppers.png**

```
% Rosenfeld 8 neighbours of a pixel
clc, clear all, close all
a = imread('peppers.png');          % built-in greyscale image
im = imcrop(a);                     % select tiny subimage
figure                              % display subimage
image(im), axis on, colormap(gray(256));
row = 4, col = 5;                   % select 8-Nbd center
im(row,col,:) = 255;                % paint center white
im(row-1,col-1:col+1,:) = 155;      % point border grey
```

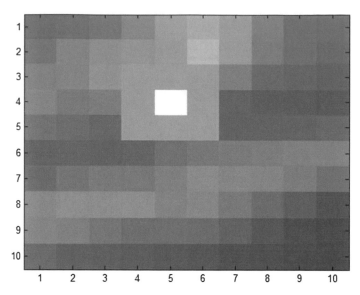

Fig. 1.42 Sample 8-neighbourhood in 10 × 10 subimage in **peppers.png**

```
im(row,  col−1,:)  =  155;
im(row,  col+1,:)  =  155;
im(row+1,col−1:col+1,:)  =  155;
figure                              % display 8−Nbd
image(im),  axis on,  grid on,  colormap(gray(256)); % display image
```

Listing 1.18 Use the Matlab code in `Rosenfeld8Neighbours.m` to display an 8-neighbourhood the subimage in Fig. 1.42.

Putting these ideas together, we can begin finding 8-neighbourhoods of a pixel in any digital image.

Example 1.22 **Displaying an 8-Neighbourhood in a Subimage**.
Several steps are needed to find a visible 8-neighbourhood in an image.

1° Use **imcrop** to extract a tiny subimage in an image.
2° Select the coordinates of the center pixel in an 8-neighbourhood.
3° Assign a false colour to each of 8 neighbouring pixels of the center pixel.

These steps are carried out using the Matlab script Listing 1.18. For example, we can obtain the 8-Neighbourhood displayed with false colours in Fig. 1.42. ∎

Larger Rosenfeld neighbourhoods of pixels are possible. For example, A 24-neighbourhood of a pixel p (denoted by $N_{24}(p)$) in an image contains a center pixel surrounded by 24 neighbouring pixels. In Fig. 1.43, the inner blue box with corners at (2, 2)(2.5,2.5) represents a pixel p. The outer blue box in Fig. 1.43 represents a neighbourhood $N_{24}(p)$ of the pixel p.

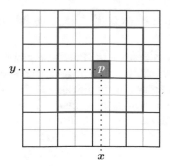

Fig. 1.43 5 × 5 subimage containing a 24-neighbourhood $N_{24}(p)$

Problem 1.23 24-Neighbourhood of a Pixel in an Image.
Give Matlab code to do the following:

(a)   Select a tiny subimage in an image (you choose the image).
(b)   Select the coordinates of the center pixel p in a 24-neighbourhood $N_{24}(p)$ in
       the subimage from part (a).
(c)   Assign a false colour to each of 24 neighbouring pixels surrounding the center
       pixel.
(d)   Display the original image.
(e)   Display the tiny subimage from part (a).
(f)   Display the 24-neighbourhood in the tiny subimage from part (e).   ■

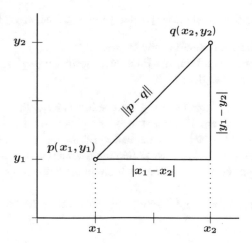

Fig. 1.44 Distances in the Euclidean plane

1.18 Distances: Euclidean and Taxicab Metrics

This section briefly introduces two of the most commonly used means of measuring distance, namely, Euclidean distance metric and Manhattan distance metric. Let $\mathbb{R}^n$ denote the real Euclidean space. In Euclidean space in $\mathbb{R}^n$, a *vector* is also called a *point* (also called a *vector* with n coordinates. The *Euclidean line* (or real line) equals $\mathbb{R}^1$ for $n = 1$, usually written $\mathbb{R}$. A line segment $\overline{x_1 x_2}$ between points x_1, x_2 on the real line has length that is the absolute value $\overline{x_1 - x_2}$ (see, for example, the distance between points on the horizontal axis in Fig. 1.44).

The *Euclidean plane* (or 2-space) $\mathbb{R}^2$ is the space of all points with 2 coordinates. The *Euclidean 3-space* $\mathbb{R}^3$ is the space of all points each with 3 coordinates. In general, the *Euclidean n-space* is the n-dimensional space $\mathbb{R}^n$. The elements of $\mathbb{R}^n$ are points (also called vectors), each with n coordinates.

For example, let points $x, y \in \mathbb{R}^n$ with n coordinates, then $x = (x_1, \ldots, x_n)$, $y = (y_1, \ldots, y_n)$. The norm of $x \in \mathbb{R}^n$ (denoted $\|x\|$) is

$$\|x\| = \sqrt{x_1^2 + x_2^2 + \ldots + x_n^2} \text{ (vector length from the origin).}$$

The distance between vectors x, y is the *norm* of $x - y$ (denoted by $\|x - y\|$). The *Euclidean norm* $\|x - y\|$ in the plane is computed with the *Euclidean metric* defined by

$$\|x - y\| = \sqrt{\sum_{i=1}^{n} \left(x_i^2 - y_i^2\right)} \text{ (Euclidean distance).}$$

Sometimes the Euclidean distance is written $\|x - y\|_2$ (see, e.g., [34, Sect. 5, p. 94]).

Example 1.24 **Euclidean norm in the Plane**.
For the points p, q in Fig. 1.44, the Euclidean norm $\|p - q\|$ is the length of the hypotenuse in a right triangle.   ■

The taxicab metric is computed using the absolute value of the differences between points along the vertical and horizontal axes of a plane grid. Let $|x_1 - x_2|$ equal the absolute value of the distance between x_1 and x_2 (along the horizontal axis of a digital image). The *taxicab metric* d_{taxi}, also called the Manhattan distance between points p at (x_1, y_1) and q at (x_2, y_2), is distance in the plane defined by

$$d_{taxi} = |x_1 - x_2| + |y_1 - y_2| \text{ (Taxicab distance between two vectors in } \mathbb{R}^2).$$

In general, the taxicab distance between two points in $\mathbb{R}^n$ mimics the distance logged by a taxi moving down one street and up another street until the taxi reaches its destination. The taxicab distance between two points $x = (x_1, \ldots, x_n)$, $y = (y_1, \ldots, y_n)$ in n-dimensional Euclidean space $\mathbb{R}^n$ is defined by

$$d_{taxi} = \sum_{i=1}^{n} |x_i - y_i| \quad \text{(Taxicab distance in } \mathbb{R}^n\text{)}.$$

Example 1.25 **Taxicab Distance in the Plane**.
For the points p, q in Fig. 1.44, the taxicab distance is the sum of the lengths of the two sides in a right triangle. ∎

Euclidean distance and the taxicab distance are two of commonest metrics used in measuring distances in digital images. For example, see the sample distances computed in Matlab Listing 1.19.

```
% distance between pixels
clc, clear all, close all
im0 = imread('liftingbody.png'); % built-in greyscale image
image(im0), axis on, colormap(gray(256));  % display image
% select vector components
x1 = 100; y1 = 275; x2 = 325; y2 = 400;
im0(x1,y1),im0(x2,y2),              % display pixel intensities
p = [100 275]; q = [325 400];       % vectors
norm(p), norm(q),                   % 2-norm values
norm(p-q),                          % norm(p-q)=Euclidean dist.
EuclideanDistance = sqrt((x1-x2)^2 +(y1-y2)^2),
ManhattanDistance = abs(x1-x2) + abs(y1-y2)
```

Listing 1.19 Use the Matlab code in distance.m to experiment with distances between pixels.

Example 1.26 **Distance Between Image Pixels**.
Try using Matlab Listing 1.19 to experiment with distances between pixels. ∎

Problem 1.27 ⚙ In computing the Euclidean or Manhattan distance between image pixels, what is the unit of measurement? For example, given pixels p at (x_1, y_1) and q at (x_2, y_2), the distance $\|x - y\|$ is either dimensionless or has a unit measurement. ∎

1.19 False Colours: Pointillist Picture Painting

This section briefly introduces pointillist picture painting approach to modifying the appearance of patterns in a digital image with false colours. Pointillism (from French *pointillisme*) is a form of neo-impressionist art introduced by G. Seurat and P. Signac during the 1880s [174]. *Pointillism* is a form of painting in which small dobs (dots) of pure colour are applied to a canvas and which become blended in a viewer's eye.

$(I(i,j) == I(x,y))$ && $(norm(p-q)<rad))$

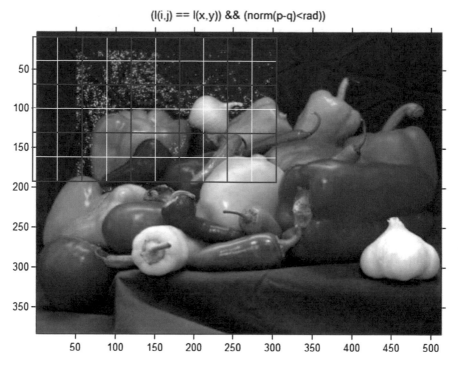

Fig. 1.45 Pixel intensity pattern inside a colour image box

The dobs of pure colour in a pointillist painting are arranged in visual patterns to form a picture. The pointillist approach to painting with dobs of pure colour carries over in the the false-colouring of selected pixels that are part of hidden patterns in digital images.

1.19.1 False-Colour an RGB Image Pattern

With a digital image, the basic approach is to replace selected pixels with a false colour to highlight hidden image patterns. Such an image pattern would normally not be visible without false-colouring. The steps in applying false colours to pixels in a digital image pattern is as follows.

Picture Pattern False Colouring Method for RGB Images.

1° Select a digital image.
2° Choose image pattern to paint with a false colour. To do this, formulate a rule
        that defines some form repetition in a selected image.

Example 1.28 Choose a particular pixel p. **Pattern Rule**. If any other pixel q in the selected image has the same intensity as p, then q belongs to the pattern. ∎

3° Choose a pixel false colour method.
4° Assign values to method parameters such as initial pixel coordinates.
5° Apply the method in Step 2.
6° **False-Colour Step**. If a colour pixel q intensity satisfies the Pattern Rule, then maximize the intensity of q.
7° Display image with false colours.
8° Repeat Step 2 to display a different image pattern. ∎

Example 1.29 **RGB Image Pattern Highlighted with False Colouring**. A pattern (highlighted with a false colour) is shown in the colour image in Fig. 1.45. In this case, the pattern contains all subimage pixels that have the same intensity as a selected pixel. For example, choose the pixel p at (25,50) in Fig. 1.45. In example, $p \longrightarrow$

. What looks like a red polka dot ● in is actually an 8-neighbourhood and each of the neighbourhood pixels is assigned a false colour. The center of this 8-neighbourhood is the only pixel that belongs to the pixel intensity repetition pattern. In this image pattern, each pixel inside a rectangular region of the image is assigned a false colour. Matlab script Listing 1.20 illustrates how this is done. ∎

```
% Some colour image pixels assigned false colours
clc, clear all, close all
I=imread ('peppers.png');
x = 25; y = 50; rad = 250; p = [x y];   % settings
for i = x+1:x+1+rad                     % width of box
   for j = y+1:y+1+rad                  % length of box
      q = [i j];                        % use in norm(p-q)
      if ((I(i,j) == I(x,y)) && (norm(p-q)<rad))
         I(i,j,2) = 255;                % false colour
      end
   end
end
I(x,y,1)=255;I(x-1,y,1)=255;I(x-1,y,1)=255; % 8 Nbd
I(x-1,y+1,1)=255;I(x-1,y-1,1)=255;I(x+1,y+1,1)=255;
I(x+1,y+1,1)=255;I(x,y-1,1)=255;I(x,y+1,1)=255;
figure, imshow(I), axis on,            % show false colors
title('(I(i,j) == I(x,y)) && (norm(p-q)<rad))')
```

Listing 1.20 Use the Matlab code in `falseColourRGB.m` to experiment with false-colouring pixels.

In Listing 1.20, notice that a pixel false colour is obtained by assigning the maximum intensity to one of the colour image channels. In this example, if a pixel intensity $I(i, j)$ matches the intensity of the pixel $I(x, y)$ at the upper lefthand corner of the box and norm of the distance from (x,y) to (i,j) is less than to an upper bound **rad** (e.g., **rad** = 250 pixels), then $I(i, j)$ is painted a false colour. In Listing 1.20, the following assignment is made.

$$I(i,j,2) = 255;$$

Problem 1.30 False-Colouring RGB Image Pixels.
Do the following:

1° 欽 Modify Listing 1.20 so that pixel false colour is changed to red. Display the result.

2° Give a comlete Matlab script that implements the following pattern rule: Choose a particular pixel p. **If any other pixel q in the selected image is less than the intensity of p, then q belongs to the pattern.** Display pixel q with a false colour.

3° Give a comlete Matlab script that implements the following pattern rule: Choose a particular pixel p. **If any other pixel q in the selected image is greater than the intensity of p, then q belongs to the pattern.** Display pixel q with a false colour.

4° Invent your own pixel pattern rule. Give a comlete Matlab script that implements your pattern rule. Display the result. ■

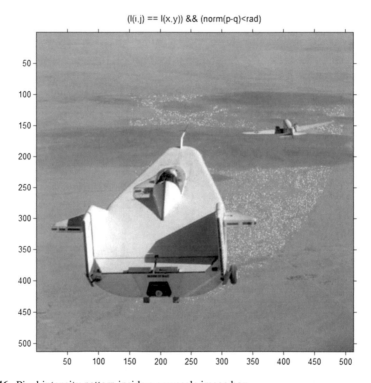

Fig. 1.46 Pixel intensity pattern inside a greyscale image box

1.19.2 False-Colour a Greyscale Image Pattern

False-colouring pixels in a greyscale image works differently, since a greyscale has no colour pixels. In that case, it is necessary to use a greyscale image to build a new image that has colour channels. The new image will be a *pseudo-colour* image with a structure that makes it possible to assign false colours to pixels in an image pattern.

Picture Pattern False Colouring Method for Greyscale Images.

1° Select a digital image.
2° Choose image pattern to paint with a false colour. To do this, formulate a rule that defines some form repetition in a selected image.

Example 1.31   Choose a particular pixel p.
Pattern Rule. If any other pixel q in the selected image has the same intensity as p, then q belongs to the pattern.   ■

3° Choose a pixel false colour method, using the Pattern Rule.
4° **Pseudo-Colour Image Creation Step**. Convert a selected greyscale image to a pseudo-colour image.
5° Assign values to method parameters such as initial pixel coordinates.
6° Apply the Pattern Rule in Step 2.
7° **False-Colour Step**. If a greyscale pixel q intensity satisfies the Pattern Rule, then maximize the intensity of the pseudo-colour channel for q. To see how this is done, see Listing 1.21.
8° Display image with false colours.
9° Repeat Step 2 to display a different image pattern.   ■

Example 1.32   **Greyscale Image Pattern Highlighted with False Colouring**.
A pattern (highlighted with a false colour) is shown in the greyscale image in Fig. 1.46. In this case, the pattern contains all subimage pixels that have the same intensity as a selected pixel. For example, choose the pixel p at (100, 150) in Fig. 1.46. In this

example, $p \longrightarrow$. What looks like a red pokerdot ● in is actually an 8-neighbourhood and each of the neighbourhood pixels is assigned a false colour. This 8-neigbourhood is displayed to make the location of the selected pixel more visible. Also notice that the center of this 8-neighbourhood is the only pixel that belongs to the pixel intensity repetition pattern. In this image pattern, each pixel inside a rectangular region of the image is assigned a false colour. Matlab script 1.21 illustrates how this is done.   ■

Problem 1.33   False-Colouring Greyscale Image Pixels.
˙Do the following:

1° ⚙ Modify Listing 1.21 so that pixel false colour is changed to blue. Display the result.

2° Give a comlete Matlab script that implements the following pattern rule:
Choose a particular pixel p. **If the intensity of any other pixel q in the selected greyscale image is less than the intensity of p, then q belongs to the pattern.**
Display pixel q with a false colour.

3° Give a comlete Matlab script that implements the following pattern rule:
Choose a particular pixel p. **If the intensity of any other pixel q in the selected image is greater than the intensity of p, then q belongs to the pattern.**
Display pixel q with a false colour..

4° Invent your own pixel pattern rule. Give a comlete Matlab script that implements your pattern rule. Display the result.   ∎

```
% Some pixels inside a box region assigned a false colour
clc, clear all, close all
I = imread('liftingbody.png');            % greyscale image
I=double(I);                              % for scaling
I3=zeros(size(I,1),size(I,2),3);          % set up 3 channels
I3(:,:,1)=I;I3(:,:,2)=I;I3(:,:,3)=I;      % channels <- I
I=I3;                                     % I <- channels
x = 100; y = 150; rad = 350; p = [x y];   % settings
for i = x+1:x+1+rad                       % width of box
   for j = y+1:y+1+rad                    % length of box
      q = [i j];                          % q vector
      if ((I(i,j) == I(x,y)) && (norm(p-q)<rad))
         I(i,j,2) = 255;
      end
   end
end
I(x,y,1)=255;I(x-1,y,1)-255,I(x-1,y,1)=255; % 8-neighbourhood
I(x-1,y+1,1)=255;I(x-1,y-1,1)=255;I(x+1,y+1,1)=255;
I(x+1,y+1,1)=255;I(x,y-1,1)=255;I(x,y+1,1)=255;
figure, imshow(I./255), axis on,         % display false colours
title('(I(i,j) == I(x,y)) && (norm(p-q)<rad)')
```

Listing 1.21 Use the Matlab code in `falseColourGrey.m` to experiment with false-colouring pixels.

1.20 Vector Spaces Over Digital Images

A *vector space* is a set of objects or elements that can be added together and multiplied by numbers (the result of either operation is an element of the space) in such a way that the usual calculations hold. For example, the set of all pixels in a 2D digital image form a local vector space that is a subset in $\mathbb{R}^2$ with holes in it. Every pixel has coordinates in the plane that can be treated as vectors in the usual way. Unlike the usual vector space called the Euclidean plane, there are holes in a digital image (between every pair of adjacent pixels, there is no pixel). Similarly, the set of all pixels in a 3D digital image form a local vector space in $\mathbb{R}^3$. Compared with the usual Euclidean dense 3D space, a 3D image looks like a piece of swiss cheese.

1.20.1 Dot Products

Given a pair of vectors x, y, the *dot product* (denoted $x \cdot y$) equals the length of the projection of x onto y. Let θ be the angle between vectors x and y with norms $\|x\|$ and $\|y\|$. Then the dot product is defined by

$$x \cdot y = \|x\| \, \|y\| \cos \theta \text{ (dot product).}$$

This gives us a way to find the angle between a pair of vectors in a digital image, i.e.,

$$\theta = \arccos \left[\frac{x \cdot y}{\|x\| \, \|y\|} \right] \text{ (angle between vectors).}$$

Example 1.34 **Dot products, Angle between vectors**.
Sample dot products and angles between vectors can be found in a digital image using the approach shown in Listing 1.22. ∎

```
% Sample dot product and angle between vectors
clc, clear all, close all
p = [120  150]; q = [100  40];      % pair of 2D vectors
dot(p,q)                            % dot product
X = dot(p,q)./(norm(p)*norm(q));    % ratio to compute angle
acosd(X)                            % angle between p and q
                                    % in degrees
```

Listing 1.22 Use the Matlab code in dotProduct.m to experiment with dot products.

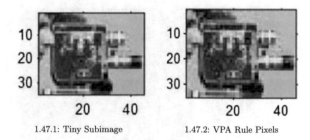

1.47.1: Tiny Subimage   1.47.2: VPA Rule Pixels

Fig. 1.47 Application of vector pair angle rule

Problem 1.35 Let (x, y), (a, b) be a pair of vectors in a digital image of your own choosing and let $\angle((x, y), (a, b))$ be the angle between (x, y) and (a, b). Use false-coloring to display all of the pairs of vectors that satisfy the Vector Pair Angle Rule (VPA Rule). For an example, see Fig. 1.47 for sample false colouring based on the VPA Rule). **Hint:** Solve this problem with a very small subimage.

Vector Pair Angle (VPA) Rule. For each $\angle((r, t), (u, v))$ between pairs of vectors (r, t), (u, v), if $\angle((r, t), (u, v)) = \angle((x, y), (a, b))$, then display the pixels at (r, t), (u, v) with a false colour. ■

1.20.2 Image Gradient

In a 2D image, the *gradient* of vector (location of a pixel intensity) is the slope of the vector. Let f be a 2D image. Also, let $\frac{\partial f}{\partial x}$ be the partial derivative of in the x-direction and let $\frac{\partial f}{\partial y}$ be the partial derivative of in the y-direction. The gradient of f at location (x, y) (denoted by ∇f) is defined as a 2D column vector

$$\nabla f = \begin{bmatrix} \frac{\partial f}{\partial x} \\ \frac{\partial f}{\partial y} \end{bmatrix} \quad \text{(gradient of } f \text{ at (x,y))}.$$

The gradient ∇f points in the direction of the greatest change in f at location (x, y) [58, Sect. 3.6.4, p. 165].

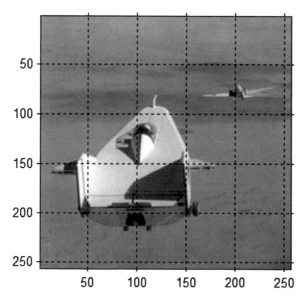

Fig. 1.48 Resized image

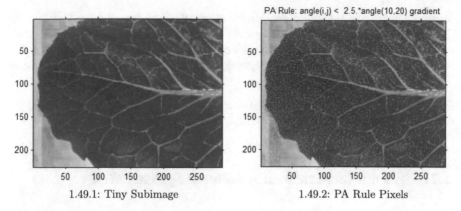

1.49.1: Tiny Subimage   1.49.2: PA Rule Pixels

Fig. 1.49 PA rule application: $\forall r, c \in$ Image, highlight angle(r,c) < 2.5.*angle(10,20)

Example 1.36 **Sample Image Gradients**.
In Matlab, **imresize** is used to shrink an image (see, e.g., Fig. 1.48) and **imgradient**
(computes the gradient angle and gradient magnitudes in the horizontal and vertical
directions) make it possible to inspect the angles of image pixels of interest. See
Listing 1.23 for an example. ∎

```
% image vector x-, y-direction magnitudes and vector angles
clc, clear all, close all
im = imread('liftingbody.png');   % built-in greyscale image
im=imresize(im,0.5);              % shrink image by 50%
imshow(im), axis on, grid on;     % display image
[Gdir,Angle]=imgradient(im);      % vector directions, angles
Angle(150,150)                    % sample angles:
Angle(165,130)
Angle(80,80)
Angle(100,40)
```

Listing 1.23 Use the Matlab code in `vectorDirection.m` to experiment with false-colouring
pixels.

Problem 1.37 ♒ Let $\angle p(x, y)$ be the angle of a pixel $p(x, y)$ with coordinates
(x, y) in a colour image of your choosing. Apply the Pixel Angle (PA) rule: For
$k = 2.5$, false colour all image pixels with coordinates (r, c) so that

$$angle(r, c) < 2.5. * angle(x, y), \text{ where} angle(r, c) = \text{angle of pixel at } (r, c).$$

For an example, see Fig. 1.49 for sample false colouring based on the PA Rule). **Hint:**
Solve this problem by letting i, j range over

$$i = 1 : r \text{ and } j = 1 : c, \text{ where}$$

r, c are number of rows and columns in the selected image, respectively. **Caution**: In Matlab, **imgradient** works with greyscale images, not colour images. Even though the display of false colours is on the selected colour image img, the pixels angles are extracted from $imgGrey$, the greyscale equivalent of the original colour image img.

Pixel Angle (PA) Rule.   Let $k > 0$. For each $\angle q(a, b)$, if $\angle q(a, b) < k * \angle p(x, y)$, then display pixel $q(x, y)$ with a false colour.   ■

1.21   What a Camera Sees: Intelligent Systems View

This section briefly introduces some of the features of camera vision, starting with cameras with some form of low-level intelligence.

1.21.1   Intelligent System Approach in Camera Vision Systems

The intelligent system approach in camera design is part of what is known as intelligent multimedia. *Intelligence* in this context means the ability of a picture-taking device to combine available sensor information to facilitate the capture of an optimal picture. A good discussion on intelligence considered in the context of multimedia is given by M. Ma [37, Sect. 1.1.3, p. 4]. Capturing the underlying geometry of a 3D scene is of the principal problems requiring solution for intelligent camera control (see, e.g., M. Christie, P. Olivier and J.-M. Normand [30]).

Intelligent camera control is central to motion planning in robotic devices (see, e.g., [15]), work by Daimler-Benze on vision-based pixel-classification autonomous driving cite[Fig. 3.2, p. 3]Franke1999 and intelligent vehicle vision systems [109]. A recent survey of hardware accelerated methods for intelligent object recognition in cameras is given by A. Karimaa [90]. Histogram equalization, motion detection, image interpretation and objection recognition are key features that are implemented in intelligent visual surveillance [92].

A good overview of early intelligent camera control is given by S.M. Drucker [36]. *Stability* (steady hold) is a basic feature of cameras that can be considered intelligent. A camera that supports *stability* while capturing an image, compensates for movement (camera shake) while a picture is being taken. This feature is important for both ordinary lens- and macro lens-based picture-taking, since it eliminates (diminishes) image blurring. For example, the Canon®Hybrid IS implements optimal image stabilization.

Another important feature is intelligent cameras is face detection. A camera implements *face detection*, provided the camera selects one or more objects in a scene that resemble faces. For example, Canon cameras implement Intelligent Scene Analysis based on Photographic Space (ISAPS) technology, which predicts the scene in the field of view and selects optimal settings for key functions. Nikon provides a fish-eye lens for its Coolpix®cameras. A *fish-eye lens* is an imaging system with a very short focal length and which gives a hemispherical field of view. Such lenses have good resolution in the central lens region but have poor resolution on the marginal region.

In a human eye, the *fovea centralis* (central zone of the retina) provides high quality vision, whereas the peripheral region of the eye provides less detailed imaging. For this reason, fish-eye lens systems approximate human vision relative to resolution distribution citeOrghidan2005. Subpixel resolution accuracy experiments have been carried out with CCD (charge coupled device) cameras with a Nikon Rayfact lens [191, Sect. 4]. *Sub-pixel* resolution results from estimating values of geometric quantities (points, lines, edges) that are better than pixel-level accuracy (see, e.g., estimating the position of an image corner with sub-level accuracy using pixel gradient directions in the neighbourhood of a corner [110, Sect. 3.4, p. 33]).

1.21.2   Scene Colour Sensing by Cameras

In a colour image, each pixel consists of a combination of primary colours. This section briefly introduces the techniques used to display varying amounts of a single primary colour in a colour image. The idea is to display the amount of a primary colour that corresponds to a single colour channel value for each of the pixels across an entire image.

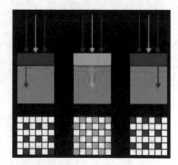

| 1.50.1: Colour filters | 1.50.2: Bayer sensor array |

Fig. 1.50   Colour channel patterns

Bayer filter. Each light-sensitive element of a digital image sensor is fitted with several filters (red, green, or blue filter). For each colour filter in an image sensor, there is a corresponding colour channel in each of the pixels in an image that is captured. In a camera image sensor, there are roughly twice as many green filters as blue and red filters to approximate how an eye perceives colour. The colour arrangement in an image sensor is called a **Bayer pattern colour filter array**, cf. Figure 1.50 (see, e.g., [217]). For example a green channel of a 24-bit pixel (with 8 bits per colour channel) is able to display up to 256 shades of green. Notice that there are other approaches (besides the Bayer filter) to achieve colour separation, e.g., 3CCD (3 sensor array) and Foveon X3 (a special silicon that absorbs different colours).

For example, Fig. 1.51 is produced using the code in Listing 1.24. Each of the colour channels for the rgb image of a workshop is shown in Fig. 1.51. For example, the red colour channel for the workshop is displayed in the second image in the top row of Fig. 1.51. In the second row

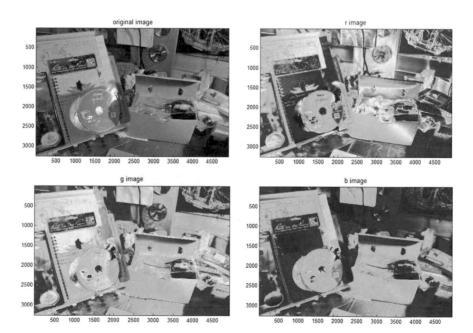

Fig. 1.51   png (rgb) → colour channel images

```
% colour experiments

g = imread('workshop.jpg');
%
gr = g(:,:,1); gg = g(:,:,2); gb = g(:,:,3);
%
subplot(2,2,1);image(g);axis image;
title('original image');
subplot(2,2,2);image(gr,'CDataMapping','scaled');axis image;
title('r image');
subplot(2,2,3);image(gg,'CDataMapping','scaled');
title('g image');
subplot(2,2,4);image(gb,'CDataMapping','scaled');
title('b image');
%
% >> figure, colour  %sample use of colour.m
%
```

Listing 1.24 Use the Matlab code in `colour.m` to produce the images in Fig. 1.34 and Fig. 1.51.

Problem 1.38 Give Matlab code to display an hsv image (converted from rgb to hsv) and the red, green, and blue colour changes for the hsv image (see Fig. 1.53). Do this for a pair of colour (rgb) images by converting the rgb images to hsv images. ■

Problem 1.39 ⚅ Give Matlab code to display the cameraman image as shown in Fig. 1.52 so that concentric circles are drawn on the image. The center of both circles should be positioned at (120,75) with inner circle radius equal to about 30 pixels and outer circle radius equal to about 50 pixels.

Hint: Since you are only dealing with a greyscale image for the cameraman, you do have to worry about color channel values. Use the false-colour approach and change the pixel intensity to maximum intensity for each pixel along the circumference of

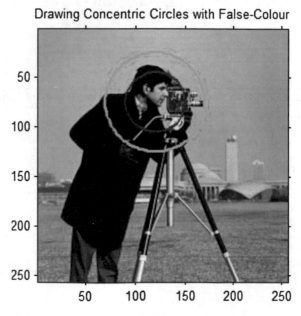

Fig. 1.52 Greyscale image with superimposed circle

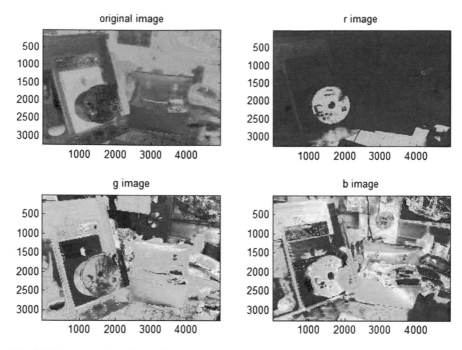

Fig. 1.53   hsv → colour channel images

the circle to be drawn in the cameraman image. Let (xc, yc) be the center of a circle
with radius r. And let $x = 0 : 0.01 : 1$, $y = 0 : 0.01 : 1$. Then false colour each of
the points at

$$(xc + r\cos(2\pi x), yc + r\sin(2\pi y))   \quad \blacksquare$$

Algorithm 2: Construct Corner-Based Voronoï Mesh on a Digital Image

---

   Input   : Read digital image img.
   Output: Image with Superimposed Corner-Based Voronoï Mesh.
1 $img \longmapsto greyscaleImg$;
2 $greyscaleImg \longmapsto cornerCoordinates(greyscaleImg)$;
3 /* $greyscaleImg$ maps to coordinates of image corners */ ;
4 $S \longleftarrow cornerCoordinates(greyscaleImg)$;
5 /* S contains corner coordinates used as mesh generating points (sites). */ ;
6 $Display\ img$;
7 /* Hold on displayed img */ ;
8 $S \longmapsto VoronoiMesh\ \mathcal{M}$;
9 $VoronoiMesh\ \mathcal{M} \longmapsto img$;
10 /* Voronoï regions surrounding each generating point now displayed on image */ ;

Fig. 1.54   Corner-based Voronoï mesh image overlay

Algorithm Symbols.

$\longmapsto$ Maps to.

Example 1.42

$$img \longmapsto greyscaleImg$$

reads **Image** *img* **maps to** *greyscaleImg*(*greyscaleimage*) **(i.e.,** *img* **is converted to a greyscale image)**. ∎

$\longleftarrow\!\mid$ Maps from.

Example 1.43

$$S \longleftarrow\!\mid cornerCoordinates(greyscaleImg)$$

reads **Set** *S* **maps from** *cornerCoordinates*(*greyscaleImg*) **(i.e.,** *S* **gets a copy of the coordinates of corners in the greyscale image)**.
∎

1.22   Image Geometry: Voronoï and Delaunay Meshes on an Image

This section briefly introduces an approach to detecting image geometry using either a Voronoï polygonal mesh or a Delaunay triangulation mesh overlay on a digital image. These meshes can either be view separately or in combination.

1.22.1   Voronoï Mesh on Car Image

This section briefly introduces an approach to creating a Voronoï mesh overlay on a digital image. A Voronoï mesh is also called a Voronoï diagram. The steps to do this are given in Algorithm 2. In every Voronoï mesh, each polygon in the mesh contains all points that are closer to a generating point than to any other generating point.

In general, a **generating point** p (also called a **site**) on a plane or 3D surface is a point that is used used to find all surface points closer to p than to any other generating point on the surface. In Computer Vision, generating points are used or object recognition and pattern recognition purposes to construct either a Voronoï polygonal-shaped region or a Delaunay triangles with vertices that are generating points.

Fig. 1.55 Sample mesh generating points on a cycle image

Example 1.44 Sample mesh generating points are displayed as green stars in

Fig. 1.55. In this case, there are 55 sites scattered across the cycle image. Each indicates a pixel that has gradient orientation angle and gradient magnitude that is different each of the other sites. Such sites are called keypoints. For the details about keypoints, see Sect. 8.8 and Appendix B.10. ∎

In an intelligent systems approach to machine vision, the focus is on the selection of useful generating points found in digital images or in video frames useful in detecting image objects and patterns. An **image generating point** p in a digital image is a pixel used to find all pixels closer to p than to any other generating point in the image. In effect, we always start by identifying the generating points in a digital image. For now, we consider only image corners.

Let $V(p)$ be an **image Voronoï region** of a corner generating point p. When we refer to a Voronoï region, we usually also mention the generating point used to construct the region. The collection of Voronoï regions that cover an image is called a **Dirichlet tessellation**,[7] which is also called a **Voronoï mesh**. By joining pairs of nearest image generating points with straight edges, we obtain a Delaunay triangulation of an image, which is also called a Delaunay tessellation or a Delaunay mesh. A **Delaunay mesh** on an image is collection of triangles that cover the image.

Example 1.45 **Voronoï region and Delaunay Triangle**.
There are two types of mesh polygons important in solving object recognition and pattern recognition problems in either single images or in video frames.

Voronoï region: A sample Voronoï region $V(p)$ of a generating
point p is shown in Fig. 1.3. The points in the interior and along the edges of
$V(p)$ are all those points that are closer to the generating point p than to any
other generating point.

Delaunay triangle: Notice that the generating points p, q, r in
Fig. 1.3 are the vertices of a triangle. This is an example of a Delaunay triangle.

---

[7]This form of tessellation is named after Dirichlet who used Voronoï diagrams in 1850, even though it was René Descartes who first had the idea as early as 1644 in an investigation of quadratic forms. In 1907, it was Voronoï who extended Dirichlet tessellations to higher dimensions. Hence the name Voronoï diagram. For more about this, see http://mathworld.wolfram.com/VoronoiDiagram.html. A complete set of notes on Voronoï diagrams is available at http://www.ics.uci.edu/~eppstein/junkyard/nn.html.

Voronoï regions and Delaunay triangle go hand-in-hand and yield quite different information about a tessellated surface.   ■

Notice again that each Voronoï region (p) is a convex polygon. This means that each straight edge connecting any pair of points in the interior or along the border of a Voronoï region $V(p)$ is contained in that region.

Example 1.46 **Voronoï mesh overlay**.
A sample Voronoï mesh superimposed on an image is shown in Fig. 1.54. Algorithm 2 gives the steps to find a corner-based Voronoï mesh on an image. The Matlab script A.2 that implements Algorithm 2 is given in Appendix A.1. In addition to constructing a Voronoï mesh on an image, this script does a number of other useful things.

1° Displays detailed information about the particular tessellated image.
2° Produces a 3D plot showing the varying intensities is a selected subimage. Notice that this becomes more interesting, when the selected subimage contains the pixels in a particular Voronoï region or in a set of Voronoï regions.
3° Saves a copy of the displayed Voronoï mesh in a .jpg file.   ■

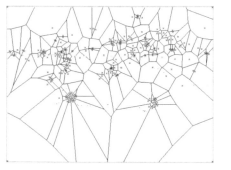

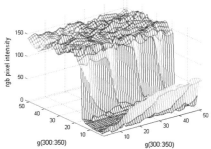

1.56.1: Car Voronoi Mesh

1.56.2: Car Subimage Intensities

Fig. 1.56   Voronoï geometric views of red car image structures

Fig. 1.57   Corner-based car wheels Voronoï mesh

1.22.2 What a Voronoï Image Sub-Mesh by Itself Reveals

A Voronoï mesh extracted from a tessellated digital image tends to reveal image geometry and the presence of image objects.

Image Geometry: The term **image geometry** means geometric shapes such as tiny-, medium- and large-sized polygons that surround image objects.

Example 1.47 **Car Wheels Mesh** An isolated Voronoï image sub-mesh is shown in Fig. 1.57. Notice that the wheels of the tessellated car image in Fig. 1.54 are surrounded by twisting polygons along the borders of the wheels in the mesh in Fig. 1.56.1. This is an example of what is known as a mesh nerve. ∎

Fig. 1.58 Car image geometric nerve

1.23 Nerve Structures

There are four types of nerve structures to consider.

1° A pure, non-image **geometric nerve** is a collection of polygons surrounding a central polygon (nerve nucleus), so that each non-nucleus polygon has an edge in common with the nucleus. See, for example, the nerve in Fig. 1.60.3.

2° An **image geometric nerve** is a collection of polygons derived from a digital image and surrounding a central polygon (nerve nucleus), so that each non-nucleus polygon has an edge in common with the image nerve nucleus. An image geometric nerves approximates the shape of an image object covered by the nerve. A sample image geometric nerve is shown in Fig. 1.58. This sample nerve covers the upper part of the front of a cycle.

3° A **Voronoï mesh nerve** is a collection of polygonal-shaped Voronoï regions derived from generating points on a digital image and surrounding a central polygon (nerve nucleus), so that each non-nucleus Voronoï region polygon has an edge in common with the Voronoï nerve nucleus. Voronoï mesh nerves identify image objects. A sample Voronoï mesh nerve on an image is shown Fig. 1.60.2. The yellow polygon in Fig. 1.60.2 is an example of a Voronoï nerve nucleus. Briefly, a **Voronoï nerve nucleus** is the central polygon in the nerve. Every polygon that has an edge or vertex in common with the nucleus is part of the cluster of polygons in the Voronoï mesh nerve. For example, each of the red polygons has an edge in common with the yellow nucleus in Fig. 1.60.2.

4° A **Delaunay mesh nerve** is a collection of Delaunay triangles with vertices that are generating points on a digital image and surrounding a central triangle (nerve nucleus), so that each non-nucleus Delaunay triangle has either an edge or a vertex in common with the nerve nucleus triangle. A sample Delaunay mesh nerve is shown on a Poste delivery vehicle in Fig. 1.59. The triangle labelled N is an example of a Delaunay nerve center called the **nucleus**. A **Delaunay nucleus** is a central triangle in a Delaunay mesh nerve in which all adjacent triangles along the central triangle border have either a vertex or a side in common with the nucleus. Delaunay mesh nerves are useful in identifying identifying the shapes of image object covered by the nerves. ∎

Fig. 1.59 Poste image Delaunay mesh nerve

Example 1.48 **Voronoï Mesh Nerve that is a Maximal Nucleus Cluster (MNC).**
A sample Voronoï mesh nerve is shown in a mesh with 55 sites (generating points) in Fig. 1.60.2. The nucleus of this nerve is a yellow hexagon covering the upper front

1.60.1: Cycle image

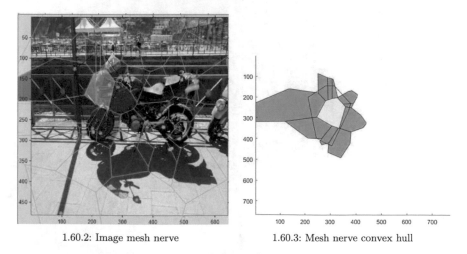

1.60.2: Image mesh nerve   1.60.3: Mesh nerve convex hull

Fig. 1.60 Sample Voronoï mesh nerve

part a motorcycle. This nerve is an example of a maximal nucleus cluster (MNC).
A **maximal nucleus cluster (MNC)** is a mesh nerve that is a cluster of polygons
in which the number of polygons surrounding the nucleus of the nerve is maximal
in this particular Voronoï mesh. For more about MNCs, see Appendix B.12. This

cycle nucleus is surrounded by red polygons. The combination of the yellow polygon nucleus and red adjacent polygons constitutes a mesh nerve. Notice that the sites in the adjacent red polygons can be connected pairwise to form a convex hull of two types of sites, namely, the nucleus site and adjacent polygon sites. Let S be a nonempty set of sites. The smallest convex set containing the set of points in S is the **convex hull** of S. A nonempty set is a **convex set**, provided every straight line segment between any two points in the set is also contained in the set. A sample convex hull is shown in Fig. 1.60.3. In this example, the convex set contains the border points as well as all points inside the borders of blue polygon in Fig. 1.60.3.   ∎

Notice that every polygon in a Voronoï Mesh is the **nucleus of a mesh nerve**, which is a cluster of polygons. Each mesh nerve is a cluster of polygons, containing a nucleus polygon in its center and a collection of polygons that share an edge with the nucleus. Also notice that each polygon in a Voronoï mesh nerve is a Voronoï region of a site used to construct the region polygon. By connecting each neighbouring pair of Voronoï polygons adjacent to the MNC nucleus, we can sometimes obtain a convex hull, which is one of the strongest indications of the shape of an image object covered by the MNC. This is, identifying MNC convex hulls in image nerves is important, since such convex hulls approximate the shape of an object covered by an MNC. For more about MNCs, see Sect. 7.5. For more about convex hulls, see Appendix B.3.

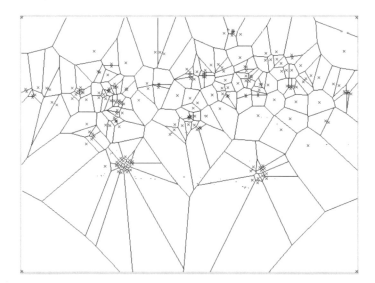

Fig. 1.61 Corner-based generating points on car image

Generating Points: The image geometry revealed by a Voronoï mesh depends on the type of generating point used. So far, we have used only digital image corners as generating points. From the fact that corners are found along the edges in a digital image, we have a convenient means of identifying the geometry of image objects by the shapes of polygon clusters that surround image objects. The corner generating points by themselves (without the Voronoï polygons) tell us a lot about image geometry, since the generating points tend to follow the contours of image objects.

When we use a generating point to construct a particular Voronoï region, the resulting polygon tells about all image pixels that are closer to the particular generating point than to any other pixel that is a generating point in the image. See, for example, the corner-based generating points in Fig. 1.61. ∎

Example 1.49 **Image Corners**.
Up to 1000 corners are displayed on the car image in Fig. 1.62.1. A display of up to 1000 image corners by themselves (extracted from the car image) is shown in Fig. 1.62.2. Matlab script A.1 in Appendix A.1.1 illustrates how to obtain and display image corners in two different ways. ∎

1.62.1: Corners on image   1.62.2: Corners w/o image

Fig. 1.62 Sample image with up to 1000 corners

Problem 1.50 Image Generating Points.
Write a Matlab script to display only the corners extracted from a digital image. Do this for your choice of three colour images of automobiles. In your script, do the following.

1° ⚙ Display original colour image.
2° ⚙ Find the corners in each selected image.
3° ⚙ Display each corner with a red **x**.
4° ⚙ Display–by themselves–up to 1000 image corners extracted from the selected image. In your display, give the image coordinate axes so that the coordinates of the corners can be identified. **Hint**: Use the Matlab option **on** for the axis command.
5° ⚙ Display a count of the total number of image corners found.
6° ☕ Display Voronoï mesh for the corners on **just one of car wheels**. **Hint**: Find the corners in the subimage containing one of the wheels.

7° ☕ Display Voronoï mesh for the corners on **both car wheels**. **Hint**: Find the corners in the subimage containing both the wheels.

8° ☕ Display Voronoï mesh for the corners on **the complete car with the background**. **Hint**: Find the corners in the subimage containing **only the car**. This will be a rectangular-shaped subimage containing the car.   ■

Problem 1.51 Mesh Nerves.

Write a Matlab script to display only the corners extracted from a digital image. Do this for your choice of any three colour images. In your script, do the following.

1° ☏ Repeat the first 4 steps in Problem 1.50.

2° ☕ Use false color to display a mesh nerve in the corner-based Voronoï mesh on the selected image. **Hint**: Find one Voronoï region of a corner in a selected subimage. This selected Voronoï region is the nucleus of a mesh nerve. The use false colouring (try green) to highlight each of the polygons surrounding the selected polygon.

3° ☕ Display the area of the polygon that is nucleus of the nerve.

4° ☕ Display a count of the polygons in the mesh nerve, including the nerve nucleus.

5° ☕ Display the mesh nerve only on the selected image.

6° ☕ Display the mesh nerve by itself (without the selected image).   ■

> Some mesh nerves are more interesting than others. An **interesting nerve** typically has a small polygon as its nucleus and many polygons surrounding the nucleus.   ■

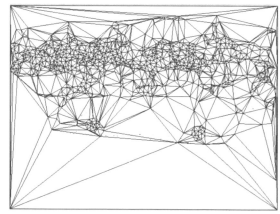

1.63.1: Delaunay Mesh Overlay

1.63.2: Delaunay Mesh

Fig. 1.63 Delaunay geometric views of red car image structures

1.23.1 Delaunay Mesh on Car Image

This section briefly introduces an approach to creating Delaunay mesh overlays on digital images. An advantage in doing this stems from the image geometry revealed by a image mesh overlay. Now, instead of the convex polygons in Voronoï mesh overlay, we obtain the simpler, uniform shapes supplied by the triangles in Delaunay mesh.

Algorithm 3 gives the basic steps in constructing a Delaunay mesh on a digital image. This algorithm is implemented in the Matlab script A.4 given in Appendix A.1.3.

Example 1.52 **Delaunay mesh overlay**.
A sample Delaunay mesh superimposed on an image is shown in Fig. 1.63.1. Algorithm 2 gives the steps to find a corner-based Voronoï mesh on an image. The Matlab script A.4 that implements Algorithm 3 is given in Appendix A.1.3.

Algorithm 3: Construct Corner-Based Delaunay Mesh on a Digital Image

---

Input : Read digital image img.
Output: Image with Superimposed Corner-Based Delaunay Mesh.
1 $img \longmapsto greyscaleImg$;
2 $S \longleftarrow cornerCoordinates(greyscaleImg)$;
3 /* S contains corner coordinates used as mesh generating points (sites). */ ;
4 *Display img*;
5 /* Hold on displayed *img* */ ;
6 $S \longmapsto DelaunayMesh \mathcal{M}$;
7 *DelaunayMesh* $\mathcal{M} \longmapsto img$;
8 /* Delaunay triangulation now displayed on image */ ;

Problem 1.53 Delaunay Mesh on an Image.
For your choice of three colour images, write a Matlab script to do the following.

1° Display original colour image.
2° Find up to 1000 corners in each selected image.
3° ᢒᢖ Display corner-based Delaunay mesh on each selected image.
4° ☕ Display the Delaunay mesh–by itself–extracted from the selected image. In your display, give the image coordinate axes so that the coordinates of the triangle vertices can be identified. **Hint**: Use the Matlab option **tight** for the axis command.
5° Display a count of the total number of image triangles in the Delaunay mesh found.
6° Display the area of the largest triangle in Delaunay mesh.   ■

1.23.2   Combined Voronoï and Delaunay Meshes
on Car Image

This section briefly introduces an approach to combining Voronoï and Delaunay mesh overlays on the same image. Recall that the vertices of each Delaunay triangle are generating points in neighbouring Voronoï regions. As a result, the area of each Delaunay triangle is an indicator of mesh quality and, indirectly, the underlying quality of the underlying image. The more uniform the Delaunay triangle areas, the greater the quality of the subimage covered by the triangles. The area of each Delaunay triangle tells us about the extent of the coverage of a subimage occupied by each trio of Voronoï regions.

Algorithm 4 gives the basic steps in constructing a combined Delaunay on Voronoï mesh on a digital image. This algorithm is implemented in the Matlab script A.6 given in Appendix A.1.4.

Example 1.54 **Delaunay on Voronoï mesh overlay**.
A sample Delaunay-on-Voronoï mesh superimposed on an image is shown in Fig. 1.64.1. In addition, the Matlab script A.7 given in Appendix A.1.4 extracts and displays the Delaunay-on-Voronoï mesh by itself. ■

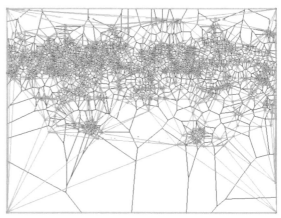

1.64.1: Voronoï + Delaunay

1.64.2: Voronoï + Delaunay Mesh

Fig. 1.64 Delaunay on Voronoï geometric views of red car image structures

---

Algorithm 4: Construct Corner-Based Delaunay on Voronoï Mesh on a Digital Image

---

Input : Read digital image *img*.
Output: Image with Corner-Based Delaunay on Voronoï Mesh Overlay.
1 $img \longmapsto greyscaleImg$;
2 $S \longleftarrow cornerCoordinates(greyscaleImg)$;
3 /* S contains corner coordinates used as mesh generating points (sites). */ ;
4 *Display img*;
5 /* Hold on displayed *img* */ ;
6 $S \longmapsto VoronoiMesh \mathcal{M}$;
7 *VoronoiMesh* $\mathcal{M} \longmapsto img$;
8 $S \longmapsto DelaunayMesh \mathcal{M}$;
9 *DelaunayMesh* $\mathcal{M} \longmapsto img$;
10 /* Delaunay triangulation on Voronoï mesh now displayed on image */ ;

Problem 1.55 Delaunay-on-Voronoï Mesh on an Image.
Write a Matlab script for your choice of three colour images to do the following.

1° Display original colour image.
2° Find up to 1000 corners in each selected image.
3° ⚙ Display corner-based Delaunay-on-Voronoï mesh on each selected image.
4° ⚙ Display the Delaunay-on-Voronoï mesh–by itself–extracted from the selected image. In your display, give the image coordinate axes so that the coordinates of the triangle vertices can be identified. **Hint**: Use the Matlab option **tight** for the axis command.
5° Display a count of the total number of image triangles in the Delaunay mesh found.
6° Display a count of the total number of image polygons in the Voronoï mesh found.
7° Display the area of the largest triangle in Delaunay mesh.
8° Display the number of sides of the largest polygon in Voronoï mesh.   ∎

Problem 1.56 Delaunay-on-Voronoï Mesh Nerve on an Image.
Write a Matlab script for your choice of three colour images to do the following.

1° Display original colour image.
2° Find up to 1000 corners in each selected image.
3° ⚙ Display corner-based Delaunay-on-Voronoï mesh on each selected image.
4° ⚙ Display the Delaunay-on-Voronoï mesh–by itself–extracted from the selected image. In your display, give the image coordinate axes so that the coordinates of the triangle vertices can be identified. **Hint**: Use the Matlab **tight** for the axis command.

5° ☕ Display–by itself–a Delaunay mesh nerve. A **Delaunay mesh nerve** is a collection of triangles that surround a single triangle called the nerve nucleus. The triangles in a Delaunay mesh nerve either have a common edge or a common vertex with the nerve nucleus. **Hint:** Extract and display from a Delaunay mesh those triangles that are contained in a Delaunay mesh nerve.

6° Display–by itself–a Voronoï mesh nerve. Recall that a **Voronoï mesh nerve** is a collection of triangles that surround a single triangle called the nerve nucleus. The polygons in a Voronoï mesh nerve have a common edge with the nerve nucleus. **Hint:** Extract and display from a Voronoï mesh those polygons that are contained in a Voronoï mesh nerve.

7° Display a count of the total number of image triangles in the Delaunay mesh found.

8° Display a count of the total number of image triangles in the Delaunay mesh nerve found.

9° Display a count of the total number of image polygons in the Voronoï mesh found.

10° Display a count of the total number of image triangles in the Voronoï mesh nerve found.

11° Display the area of the largest triangle in Delaunay mesh.

12° Display the area of the triangle that is the Delaunay mesh nerve nucleus.

13° Display the number of sides of the largest polygon in Voronoï mesh.

14° Display the area of the polygon that is the Voronoï mesh nerve nucleus.   ∎

1.24   Video Frame Mesh Overlays

This section briefly introduces video frame mesh overlays. The basic approach is to detect the geometry of objects in digital images by covering each video frame image with mesh polygons surrounding (in the vicinity of) image objects. Image mesh polygons tend to reveal the shapes and identity of objects. For a sample Dirichlet tessellation of a video frame image, see Fig. 1.65.1. A **Dirichlet tessellation** of a plane surface is a tiling of the surface.

Seed points (also called sites or generator) provide a basis for generating Dirichlet tessellations (also called Voronoï diagrams) and Delaunay triangulations of sets of points, providing a basis for the construction of meshes that cover a set with clusters of polygonal shapes. In general, a **tessellation** of a plane surface is a tiling of the surface. A **plane tiling** is a plane-filling arrangement of plane figures that cover the plane without gaps or overlaps [63]. The plane figures are closed sets. Taken by themselves or in combination, pixel intensity, corner, edge, centroid, salient, critical and key points are examples of seed points with many variations.

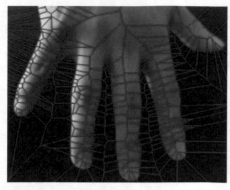

1.65.1: Offline Video Frame 1

1.65.2: Offline Video Frame 1

Fig. 1.65 Sample offline video frames 1 & 2

Problem 1.57 A sample 8×8 grid is shown in Fig. 1.67. The red ● dots indicate interior corners and outer box corners. Using pencil and paper, do the following:

1° Draw a corner-based Voronoi mesh on the grid.
2° Draw a corner-based Delaunay mesh on the grid.   ■

1.24.1 Offline Video Frame Processing

This section briefly introduces an approach to offline video processing. This form of video processing has three basic steps.

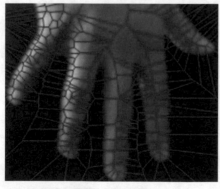

1.66.1: Offline Video Frame 3

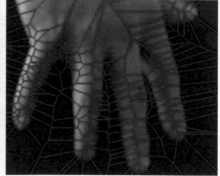

1.66.2: Offline Video Frame 4

Fig. 1.66 Sample offline video frames 3 & 4

Algorithm 5: Construct Corner-Based Voronoï Mesh on offline video frame Images

---

Input : Changing visual scene $View$ and offline $Video$.
Output: Video frames, each with corner-based Voronoï mesh overlay.

1 $View \longmapsto Video$;
2 $Frame \longleftarrow Video$;
3 $Continue \longmapsto False$;
4 **while** $(Continue \neq \emptyset$ and $Frame \in Video)$ **do**
5      $Use\ Algorithm\,2\ to\ overlay\ Voronoï\ mesh\ on\ Frame$;
6      $Video \longleftarrow Video \setminus Frame$;
7   /* $Video \setminus Frame$ reads $Video$ without the current $Frame$. */ ;
8   /* i.e., Remove tessellated $Frame$ from the set of frames in the $Video$ */ ;
9   **if** $Video \neq \emptyset$ **then**
10 | $Continue \longmapsto True$;
11   **else**
12   └ $Continue \longmapsto False$;

Basic Steps in Offline Video Processing

1° Capture complete video of a changing scene.

2° Extract a frame from the video.

3° Extract information from the video frame, e.g., discover image geometry by covering the frame with a Voronoï mesh. Each polygon in the mesh tells us something about the geometric structure of an image such as all image pixels that are close to a particular image corner. A **corner** in an image is a pixel that

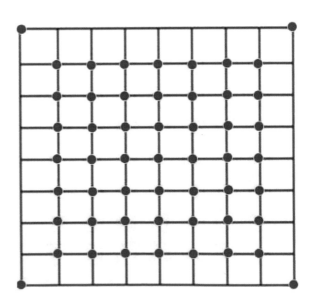

Fig. 1.67 8×8 grid with red ● corners

has a gradient orientation that is sharply different from the gradient orientation of its neighbouring pixels. The *gradient orientation* of a pixel is the angle of the tangent to the edge containing the pixel.

4° Repeat Step 2 until all of the frames in the video have been processed.   ∎

Algorithm 5 illustrates a particular form of video processing. This algorithm produces an offline video in which a corner-based Voronoï mesh is superimposed on each video frame image.

A Matlab script that implements Algorithm 5 is given in Appendix A.1.5. This algorithm uses Algorithm 2 to overlay a Voronoï mesh on each frame in the video. This is done offline, i.e., after a complete video has been captured. In the offline mode, each video frame is treated as an ordinary image.

Example 1.58 The Matlab script A.8 in Appendix A.1.5 creates an .mp4 video file. This script overlays a corner-based Voronoï mesh on each of the frames in each video that is captured by the particular webcam that is used. For example, image corners are used as seed points to construct the image Voronoï mesh in Fig. 1.54 and video frame Voronoï meshes in Figs. 1.65 and 1.66. Notice that as the hand moves, the Voronoï mesh polygons change. The changes in the polygons are a result of the changing positions of the corners found in the image.   ∎

Problem 1.59 Offline Video Production of Corner Delaunay Image Frame Meshes.

⬦ Write a Matlab script to do the following.

1° Create a video of a moving hand.
2° Offline, find and display a corner-based Delaunay mesh on each video frame. Mark the corners with a red × symbol.
3° Create an .avi file showing the production of video frames containing a corner-based Delaunay on each video frame image.   ∎

Problem 1.60 Offline Video Production of Corner Delaunay Head Image Frame Meshes.

⬦ Write a Matlab script to do the following.

1° Create a video of a moving head.
2° Offline, find and display a corner-based Delaunay mesh on each video frame. Mark the corners with a red × symbol.
3° Create an .avi file showing the production of video frames containing a corner-based Delaunay on each video frame image.   ∎

Problem 1.61 Offline Video Production of Corner Voronoï Hand Image Frame Meshes.
⬦ Write a Matlab script to do the following.

1° Create a video of a moving hand.

2° Offline, find and display a corner-based Voronoï mesh on each video frame. Mark the corners with a red × symbol.

3° Create an .avi file showing the production of video frames containing a corner-based Voronoï on each video frame image.   ∎

Problem 1.62 Offline Video Production of Corner Voronoï Head Image Frame Meshes.
🚲 Write a Matlab script to do the following.

1° Create a video of a moving head.

2° Offline, find and display a corner-based Voronoï mesh on each video frame. Mark the corners with a red × symbol.

3° Create an .avi file showing the production of video frames containing a corner-based Voronoï on each video frame image.   ∎

Problem 1.63 Offline Video Frame Corner Generating Points.
Write a Matlab script to do the following.

1° Create a video of a moving hand.

2° Offline, find and display the corners in each video frame. Mark the corners with a red × symbol.

3° Offline, display only the corners found in each video frame.

4° Create an .avi file showing the production of video frames containing video corners on the video frame images.   ∎

1.24.2   Real-Time Video Processing

This section briefly introduces an approach to real-time video processing. This form of video processing has three basic steps.

| **Basic Steps in Real-Time Video Processing** |

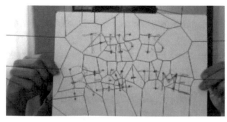   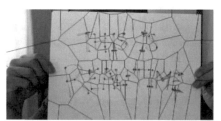

1.68.1: Real-Time Video Frame 1   1.68.2: Real-Time Video Frame 1

Fig. 1.68 Sample real-time video frames 1 & 2

1.69.1: Real-Time Video Frame 3   1.69.2: Real-Time Video Frame 4

Fig. 1.69 Sample real-time video frames 3 & 4

Algorithm 6: Construct Corner-Based Voronoï Mesh on real-time video frame Images

---

Input : Changing visual scene $View$ and real-time $Video$.
Output: Video frames, each with corner-based Voronoï mesh overlay.
1 $View \longmapsto Video$;
2 $Frame \longleftarrow Video$;
3 $LastFrame \longmapsto False$;
4 **while** $(LastFrame \neq True)$ **do**
5   Use Algorithm 2 to overay Voronoï mesh on Frame;
6     $Video \longleftarrow Video \setminus Frame$;
7   /* $Video \setminus Frame$ reads $Video$ without the current $Frame$. */ ;
8   /* i.e., Remove tessellated $Frame$ from the set of frames in the $Video$ */ ;
9   **if** $Video \neq \emptyset$ **then**
10       $Frame \longleftarrow Video$;
11   **else**
12       $LastFrame \longmapsto True$;

1° Start video capture of a changing scene.
2° Select current frame from the video.
3° Extract information from the current frame, e.g., discover video frame geometry by covering the frame with a Voronoï mesh.
4° Repeat Step 2 until the video is completed. ∎

Example 1.64 The Matlab script A.9 in Appendix A.1.6 creates an .mp4 video file for a sample form of real-time video processing. This script overlays a corner-based Voronoï mesh on each of the frames *during video capture* by the particular webcam that is used. For example, image corners are used as seed points to construct the image Voronoï mesh in Fig. 1.68 and video frame Voronoï meshes in Fig. 1.65 and in Fig. 1.69. Notice again that as the hand moves, the Voronoï mesh polygons change in real-time. The changes in the polygons are a result of the changing positions of the corners found in the image. ∎

Problem 1.65 Real-Time Video Production of Corner-Based Delaunay Hand Image Frame Meshes.
🚲 Write a Matlab script to do the following.

1° Create a video of a moving hand.
2° In real-time, find and display a corner-based Delaunay mesh on each video frame. Mark the corners with a red × symbol.
3° Create an .avi file showing the production of video frames containing a corner-based Delaunay on each video frame image.   ■

Problem 1.66 Real-Time Video Production of Corner-Based Delaunay Head Image Frame Meshes.
🚲 Write a Matlab script to do the following.

1° Create a video of a moving head.
2° In real-time, find and display a corner-based Delaunay mesh on each video frame. Mark the corners with a red × symbol.
3° Create an .avi file showing the production of video frames containing a corner-based Delaunay on each video frame image.   ■

Problem 1.67 Real-Time Video Production of Corner-Based Voronoï Hand Image Frame Meshes.
🚲 Write a Matlab script to do the following.

1^n Create a video of a moving hand.
2° In real-time, find and display a corner-based Voronoï mesh on each video frame. Mark the corners with a red × symbol.
3° Create an .avi file showing the production of video frames containing a corner-based Voronoï on each video frame image.   ■

Problem 1.68 Real-Time Video Production of Corner-Based Voronoï Head Image Frame Meshes.
🚲 Write a Matlab script to do the following.

1° Create a video of a moving head.
2° In real-time, find and display a corner-based Voronoï mesh on each video frame. Mark the corners with a red × symbol.
3° Create an .avi file showing the production of video frames containing a corner-based Voronoï on each video frame image.   ■

Problem 1.69 Real-Time Video Frame Corner Generating Points.
Write a Matlab script to do the following.

1° Create a video of a moving hand.
2° In real-time, find and display the corners in each video frame. Mark the corners with a red × symbol.
3° In real-time, display only the corners found in each video frame.
4° Create an .avi file showing the production of video frames containing video corners on the video frame images.   ■

Problem 1.70 Real-Time Video Frame Subimage Corner Generating Points.
Write a Matlab script to do the following.

1° Create a video of a moving hand.

2° In real-time, find and display the corners in a subimage of each video frame.
      Mark the corners of the subimage with a red × symbol.

3° In real-time, display the corners found in each video frame subimage.

4° Create an .avi file showing the production of video frames containing video
      corners on the video frame subimages.   ∎

Chapter 2
Working with Pixels

Fig. 2.1 Lower versus Higher resolution image pixels

2.1 Picture Elements

A **pixel** (*aka* picture element) is an element at position (r, c) (row, column) in a digital image I. A pixel represents the smallest constituent element in a digital image. Typically, each pixel in a raster image is represented by a tiny square called a **raster image tile**. Raster image technology has its origins in the raster scan of cathode ray tube (CRT) displays in which images are rendered line-by-line by magnetically steering a focused electron beam. Usually, computer monitors have bitmapped displays

© Springer International Publishing AG 2017
J.F. Peters, *Foundations of Computer Vision*, Intelligent Systems
Reference Library 124, DOI 10.1007/978-3-319-52483-2_2

in which each screen pixel corresponds to its **bit depth**, i.e., number of pixels used to render pixel colour channels.

By zooming in on (also, **resample**) an image at different magnification levels, these tiny pixel squares become visible.

Example 2.1 **Inspecting Raster Image Pixels**.
Four views of a raster image are shown in Fig. 2.1:

1° Lower-left panel: hand-held camera with pixel inspection window:

. This is a movable window that makes it possible to inspect different parts of the image.

2° Upper-left panel: pixels   (zoomed in at 800%) inside the inspection window.

3° Lower-right panel: hand-held camera with pixel inspection window:

. This is a second movable window that makes it possible to inspect different parts of the image.

4° Upper-left panel: pixels   (zoomed in at 400%) inside the inspection window.

See MScript A.10 in Appendix A.2.1 to experiment with other zoomed-in levels and inspect pixels in other images.   ∎

Example 2.2 **Inspecting Raster Image Pixels**.
Four views of a raster image are shown in Fig. 2.2:

1° Lower-right panel: hand-held camera with pixel inspection window:

. This is a second movable window that makes it possible to inspect different parts of the image.

2° Upper-left panel: pixels   (zoomed in at 100%) inside the inspection window.

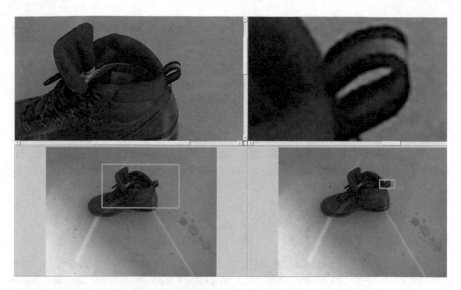

Fig. 2.2 Zoom in at 100 and 800%, exhibiting colour image pixels

3° Lower-left panel: hand-held camera with pixel inspection window: . This is a movable window that makes it possible to inspect different parts of the image.

4° Upper-left panel: pixels (zoomed in at 800%) inside the inspection window.

See MScript A.10 in Appendix A.2.1 to experiment with other zoomed-in levels and inspect pixels in other images. ∎ ∎

Each colour or greyscale or binary image pixel carries with it several numerical values. There are a number of cases to consider.

Binary image pixel values: 1 for a white pixel and 0 for a black pixel.

Greyscale image pixel values: Commonly 0–255, for the pixel greyscale intensity. Each greyscale pixel value quantizes the magnitude of white light for a pixel.

RGB image pixel values: Each colour pixel value quantizes the magnitude of a particular colour channel brightness for a pixel. A **colour channel** is particular colour component of an image and corresponds to a range of visible light wavelengths. Each color pixel contains intensities for three colour channels. For colour pixel with a bit depth equal to 8, we have the following range of intensity (brightness) values for each colour channel.

Red: 0–255, for red pixel intensity (brightness).
Green: 0–255, for green pixel intensity (brightness).
Blue: 0–255, for blue pixel intensity (brightness).

Let $I^k(u, v)$ be the intensity of the k colour channel at camera image cell (u, v), Λ, the set of wavelengths in the visible spectrum, p_0^k, a scaling factor, λ, a particular wavelength, $E_{u,v}(\lambda)$, the amount of incoming light at image cell (u, v), $\tau^k(\lambda)$, the filter transmittance for the k colour channel, and $s(\lambda)$, the spectral responsivity of a camera optical sensor. The final colour pixel value $I^k(u, v)$ is defined by

$$I^k(u, v) = p_0^k \int_\Lambda E_{u,v}(\lambda) \tau^k(\lambda) s(\lambda) d\lambda.$$

In a typical RGB camera, $k \in \{r, g, b\}$. Recently, color pixel values have been used extensively in image segmentation [135, Sect. 2.1, p. 666] and for visual object tracking [32, Sect. 2.1, p. 666].   ■

Separating and Modifying Colour Image Channels

Colour channel values can be separated and pixel values can be modified.

2.2 Separating Colour Image Channels

It is a straightforward task to separate the colour channels in a raster colour image. We illustrate this using notation from Matlab.

Image and Colour Channel Notation

Let img be a $m \times n$ colour image. In that case, the pixels in img can be accessed in the following ways.

> $img(:,:) =$ pixel values in all rows and all columns in img.
> $img(r,:) =$ all pixel values in row r in img, $1 \leq r \leq m$.
> $img(:,c) =$ all pixel values in column c in img, $1 \leq c \leq n$.
> $img(:,:,1) =$ all red channel values in all rows and all columns in img.
> $img(:,:,2) =$ all green channel values in all rows and all columns in img.
> $img(:,:,3) =$ all blue channel values in all rows and all columns in img.

The notation $img(:, :)$, $img(r, :)$, $img(:, c)$ can be used to inspect and change pixel intensities in binary, greyscale or colour images.

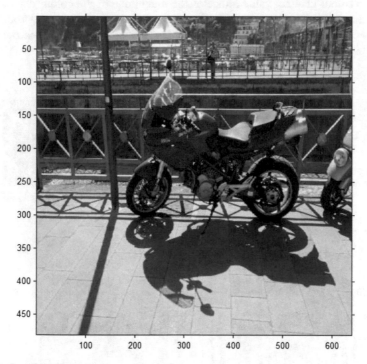

Fig. 2.3 Sample colour image

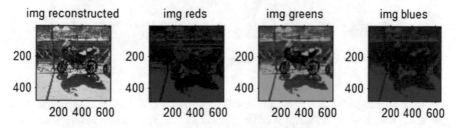

Fig. 2.4 Separated and combined colour image channels

Example 2.3 **Separating Colour Image Channel values**.
A sample colour image is given in Fig. 2.3. Using the MScript A.11 in Appendix A.2.1, the colour channels exhibited by Fig. 2.3 are separated and then recombined in Fig. 2.4. ■

2.3   Colour to Greyscale Conversion

Ideally, a colour channel value indicates the magnitude of the colour channel light recorded by an optical sensor used in pixel formation of a colour image produced by a digital camera. Let I be a colour image. It is possible to convert the colour image I to a greyscale image I_{gr} using the Matlab function $rgb2gray(I)$.

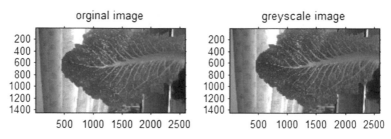

Fig. 2.5   Colour to greyscale conversion

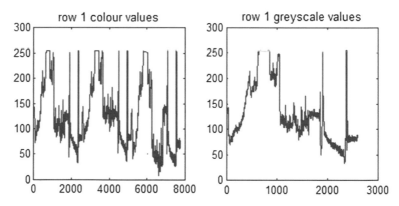

Fig. 2.6   Row of colour and greyscale leaf image intensities

Example 2.4 Figure 2.5 shows the result of converting a leaf colour image to a greyscale image using MScript A.12 in Appendix A.2.3. Contrast between the two forms of images becomes clearer when we zoom in on subimages of the original image and its greyscale counterpart.

Colour Subimage
In this leaf image segment, there is a visible mixture various shades of greens.

Recall that shades of green are obtained by mixing yellow and blue. It is a straight-forward task to verify that visible green in an image is rendered digitally with a mixture of red and blue channel intensities.

Colour Subimage

In this leaf image segment, the original mixture various shades of greens is replaced by a mixture of greys. The change from pixel color intensity to greyscale intensity can be seen in the sample pixel values in Fig. 2.6.   ∎

At pixel level, pixel modification can be carried out by replacing each pixel in a colour or greyscale image I with either the average of the colour channel values or maps of pixels values to real numbers using functions such as $ln(x)$, $exp(x)$ or with a weighted sum of the colour channel values. For example, for a greyscale image pixel $I_{gr}(x, y)$ at (x, y), the pixel intensity of I_{gr} at (x, y) is

$$I_{gr}(x, y) = \alpha I(x, y, r) + \beta I(x, y, g) + \gamma I(x, y, b),$$

where the weighting coefficients α, β, γ approximate the perceptual response of the human eye to the red, green and blue (r, g, b, respectively) colour band values.

There is a NTSC (National Transportation Safety Commission) television standard for greyscale image pixels such that

$$\alpha = 0.2989,\ \beta = 0.5870,\ \gamma = 0.1140$$

In Matlab, we write

$$I_{gr}(x, y) = \alpha. * I(x, y, 1) + \beta. * I(x, y, 2) + \gamma. * I(x, y, 3),$$

Problem 2.5 Given a colour image such as I = **rainbow.jpg**, do the following.

(**step**.1)   Choose your own values for the weighting coefficients α, β, γ.
(**step**.2)   Convert a column 30 pixels wide to greyscale intensities in image I. Call the new image $I gr$.
(**step**.3)   Display the resulting mixed colour-greyscale image $I gr$.
(**step**.4)   ☕ Use **cpselect** to compare the pixels in the 5 wide pixel column in I and $I gr$.   ∎

2.4 Algebraic Operations on Pixel Intensities

In this section, we consider various operations on image pixel intensities, resulting in changes in visual appearance of the image. Let g be a digital image (either colour or greyscale). Let $k \in [0, 255]$. Then new images i_1, i_2, i_3, i_4 are obtained using the image variable in simple algebraic expressions.

Image Algebraic Expressions I.

$$i_1 = g + g,$$
$$i_2 = (0.5)(g + g),$$
$$i_3 = (0.3)(g + g),$$
$$i_4 = \left(g \left(\frac{g}{2} \right) \right)(0.2).$$

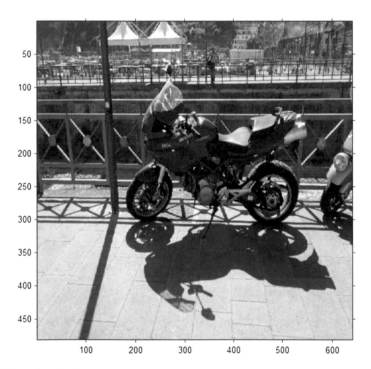

Fig. 2.7 Sample colour image

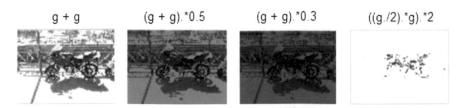

Fig. 2.8 Pixel value intensity changes induced by algebraic expressions I

Example 2.6 **Algebraic Operations I on Images**.
In algebraic operations I, notice that image g is added to itself. Various algebraic expressions can be put together to modify the pixel values in an image. MScript A.13 in Appendix A.2.3 implements algebraic expressions I on the colour image in Fig. 2.7 to obtain the resulting images shown in Fig. 2.8. For example, the implementation of $g + g$ in the leftmost image i_1 in Fig. 2.8 results in a brighter image (all the pixel intensities in the cycle image g have been doubled). ∎

Let h be a colour image and use the following algebraic expressions to change the pixel intensities in h.

Image Algebraic Expressions II .

$$i_5 = h + 30,$$
$$i_6 = h - (0.2)h,$$
$$i_7 = |h - (0.2)(h + h)|,$$
$$i_8 = (0.2)\,(h + (0.5)(h + h))\,.$$

Fig. 2.9 Sample colour image

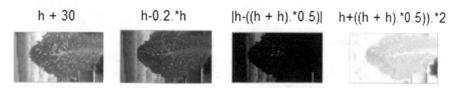

Fig. 2.10 Pixel value intensity changes induced by algebraic expressions II

Example 2.7 **Algebraic Operations II on Images**.
In algebraic operations II, notice that 30 is added to each of the intensities in image
h. MScript A.14 in Appendix A.2.3 implements algebraic expressions II on the leaf
colour image in Fig. 2.9 to obtain the resulting images shown in Fig. 2.10. For
example, the implementation of (0.2) $(h + (0.5)(h + h))$ in the rightmost image in
Fig. 2.10 results in a brighter image (all the pixel intensities in the cycle image are
sharply increased).   ■

Let img be a colour image and use the following algebraic expressions to change
the pixel intensities in img.

Image Algebraic Expressions III .

$i_9 = (0.8)img(:, :, 1)$ decrease red channel intensities,

$i_{10} = (0.9)img(:, :, 2)$ slightly decrease green channel intensities,

$i_{11} = (0.5)img(:, :, 2)$ sharply decrease green channel intensities,

$i_{12} = (16.5)img(:, :, 3)$ sharply increase blue channel intensities.

Fig. 2.11 Sample video frame image showing scene edges

Example 2.8 **Algebraic Operations III on Images**.
In algebraic operations III, notice that the pixel intensities in each colour channel are
decreased by varying amounts or increased by a huge amount (in the blue channel).
MScript A.15 in Appendix A.2.3 implements algebraic expressions III on the video
frame colour image in Fig. 2.11 to obtain the resulting images shown in Fig. 2.12. For

i9 (0.8).*red   i10 (0.9).*green   i11 (0.5).*green   i12 (16.5).*blue

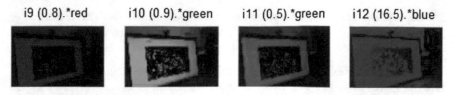

Fig. 2.12 Color channel pixel intensity changes induced by algebraic expressions III

example, the implementation of $(16.5)img(:, :, 3)$ in the rightmost image in Fig. 2.12 results in a brighter blue channel image (all the blue pixel intensities in the video frame image are sharply increased). ∎

2.13.1: Colour Array   2.13.2: Sq

Fig. 2.13 Sample Thai grocery shelves

Problem 2.9 Offline Video Frame Colour Channel Changes.
Use the approach to changing image channel intensities in MScript A.15 in Appendix AA.2.3 as a template for offline video processing, do the following.

1° ᧖ Using Matlab script A.8 in Appendix A.1.5 as a template for offline video processing, change the red channel intensities in each video frame image. **Hint**: Replace the lines of Voronoï tessellation code with lines to code in MScript A.15 to handle and display changes in the green channel of each video frame image.

2° Repeat Step 1 to change the green channel intensities in each video frame image.

3° Repeat Step 1 to change the blue channel intensities in each video frame image.
   ∎

Problem 2.10 Real-Time Video Frame Colour Channel Changes.
Use the approach to changing image channel intensities in MScript A.15 in Appendix A.2.3 as a template for real-time video processing, do the following.

1° 🐾 Using Matlab script A.9 in Appendix A.1.5 as a template for offline video processing, change the red channel intensities in each video frame image. **Hint**: Replace the lines of Voronoï tessellation code with lines to code in Matlab script A.8 to handle and display changes in the green channel of each video frame image in real-time.

2° Repeat Step 1 to change the green channel intensities in each video frame image in real-time.

3° Repeat Step 1 to change the blue channel intensities in each video frame image in real-time.   ∎

Distinct images g and h can be added, provided the images are approximately the same size. To combine pixel values in different images, it is necessary that the distinct images $I, \Im$ have the same dimensions. To get around this same-size images problem, choose any $n \times m$ image img, which is the larger of two images and just copy a second image onto an $n \times m$ array of 1s or 0s (call it $copy$). Then img and $copy$ can be combined in various ways.

Example 2.11 **Combining Pixel Intensities Across Separate Images**.
The images in Fig. 2.13 showing Thai grocery store shelves. These Thai shelf images are both approximately 1.5 MB. MScript A.16 in Appendix A.2.1 illustrates how to combine pixel intensities in pairs of different images. Two Thai grocery shelf images are combined in different ways is the first row of images in Fig. 2.14. The second row of images in Fig. 2.14 are result of algebraic operations on just one of the original images.   ∎

Problem 2.12 Choose three different pairs of colour images g, h and do the following.

1° 🚲 In Image Algebraic Expressions I, replace g, g with g, h and display the changed images using MScript A.16 in Appendix A.2.1.

2° Repeat Step 1 using the Image Algebraic Expressions II.

3° Repeat Step 1 using the Image Algebraic Expressions III.   ∎

There are many other possibilities besides the constructed images $I_1, \dots, I_{12}$ using the Algebraic Operations I, II and III. For example, one can determine largest red colour value in a selected image row r, using

$$[r, c] = max(g(row, :, 1)).$$

Using $g(r, c)$, new images can be constructed by modifying the red channel values using a maximum red channel value.

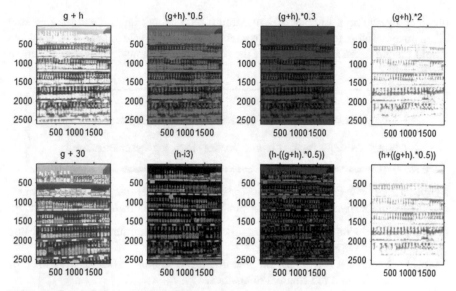

Fig. 2.14 Combining image pixel values using thai.m

Example 2.13 **Experiment with maximum pixel intensities.**
Sample results using MScript A.17 in Appendix refApCh2Sec:PixelValueChanges
are shown in Fig. 2.15. This is an external view of the modified red channel intensities

Fig. 2.15 External colour view of max-modified red channel intensities

Fig. 2.16 Internal greyscale view of max-modified red channel intensities

obtained by adding a fraction of a maximum red channel intensity in the first row of an image. Internally, a colour channel is just a greyscale image (not what we would imagine). An internal view of the modified red channel intensities is shown in Fig. 2.16. ∎

Internal View of Colour Image Channels

Internally, a colour image channel is viewed as a greyscale image.

Problem 2.14 🚲 Example 2.13 illustrates the addition of fractions of a maximum red channel intensity. For three colour images your own choosing, do the following.

(**step**.1)   Use the **min** instead of **max** function to find a minimum red channel value for an entire colour image.

(**step**.2)   Subtract a fraction of the maximum red channel intensity from each of the original red channel intensities.

(**step**.3)   Display the results both as colour images and greyscale images.   ∎

Problem 2.15 🚲 Repeat the steps in Problem 2.14 using a minimum colour channel intensity.   ∎

Finding Image Edges

The hardest thing of all is to find a black cat in a dark room, especially if there is no cat.
—Confucius [114].

2.5 Pixel Selection Illustrated with Edge Pixel Selection

One of the commonest forms pixel selection is in the form of edge pixels. The basic approach is to detect those pixels that are on edges in either in a greyscale image or in a colour channel.

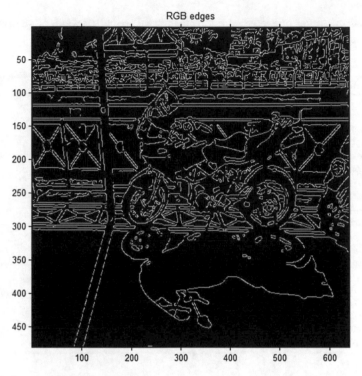

Fig. 2.17 Sample greyscale image edges

Briefly, to find edge pixels, we first find the gradient orientation (gradient angle) of each image pixel, i.e., angle of the tangent to each pixel. Let img be a 2D image and let $img(x, y)$ be a pixel at location (x, y). Then the **gradient angle** φ of pixel $img(x, y)$ is found in the following way.

$$G_x = \frac{\partial img(x, y)}{\partial x}.$$
$$G_y = \frac{\partial img(x, y)}{\partial y}.$$
$$\varphi = \tan^{-1}\frac{G_y}{G_x} = \tan^{-1}\left(\frac{\frac{\partial img(x,y)}{\partial y}}{\frac{\partial img(x,y)}{\partial x}}\right).$$

In Canny's approach to edge pixel detection [24], each image is filtered to remove noise, which has the visual effect of smoothing an image. After the gradient orientation for each pixel is found, then a double threshold for an hysteresis interval on orientation angles is introduced by Canny. The basic idea is to choose all pixels with gradient orientations that fall within the hysteresis interval. Edge pixels that fall within the selected hysteresis interval are called **strong edge pixels**. All edge pixels with gradient angles outside the hysteresis interval are called **weak edge pixels**. The weak edge pixels are ignored.

Algorithm 7: Colour Channel Edges Selection
___

   Input   : Read digital image img.
1 /* Capture colour image channel edges. */ ;
   Output: $img \longmapsto edgesR, edgesR, edgesR.$
2 /* Capture red channel pixel intensities. */ ;
3 $gR \longleftarrow img(:, :, 1)$;
4 /* Capture green channel pixel intensities. */ ;
5 $gG \longleftarrow img(:, :, 2)$;
6 /* Capture blue channel pixel intensities. */ ;
7 $gB \longleftarrow img(:, :, 3)$;
8 /* Capture blue channel pixel intensities. */ ;
9 $edge(gR,' canny') \longmapsto imgR$;
10 $edge(gG,' canny') \longmapsto imgG$;
11 $edge(gB,' canny') \longmapsto imgB$;
12 /* Map edge pixel intensities in each channel onto a black channel image bk. */ ;
13 $edgesR \longleftarrow cat(3, imgR, bk, bk)$;
14 $edgesG \longleftarrow cat(3, bk, imgG, bk)$;
15 $edgesG \longleftarrow cat(3, bk, bk, imgB)$;
16 /* Capture modified black image embossed with channel edges. */ ;
17 $Display\ edgesR, edgesR, edgesR$;

___

Before we separate out the edges from each colour image channel, we consider the conventional approach to separating greyscale image edges embossed as white pixels on a binary image.

Example 2.16 Figure 2.17 shows the result of finding the strong edge pixels in a greyscale image derived from a colour image using MScript A.18 in Appendix A.2.5. The basic approach is to start by converting a colour image to a greyscale image. If we ignore the location of each colour pixel, then a colour image is an example of a 3D image. Mathematically, each pixel p in location (x, y) in a colour image is described by a vector (x, y, r, g, b) in a 5-dimensional Euclidean space, where r, g, b are the colour channel brightness (intensity) values of pixel p. Traditionally, edge detection algorithms require a greyscale image, which is a 2D image in which each pixel intensity is visually a shade of grey ranging from pure white to pure black. After choosing the pixels in a colour channel, then any of the usual edge detection methods can be used on the single colour channel pixels. In this example, we use the edge detection method introduced by John Canny [24].

Here are some of the details.

Colour Subimage
In this cycle image segment, the combined RGB channel pixels are shown.

BW Subimage Edges
In this cycle BW image segment, white edge pixels on a binary subimage are shown.   ■

The steps to follow in edge pixel detection in each of the colour channels are given in Algorithm 7. Notice the parallel between the conventional approach to pixel edge detection and colour channel edge detection in Algorithm 7. In both cases, edge pixels (either in white or in colour) are embossed on a black image. Sample strong edge pixels for the red channel of a cycle image are shown in Fig. 2.18.

Example 2.17 Figure 2.19 shows the result of finding the strong edge pixels in the green channel of a colour image using MScript A.18 in Appendix A.2.5. The story starts by selecting all of the pixels in a colour image. Traditionally, edge detection algorithms require a greyscale image. The pixels in a single channel of a colour image have the appearance of a typical 2D greyscale image, except that pixel intensities are pixel colour brightness values in a single channel. After choosing the pixels in a colour channel, then any of the usual edge detection methods can be used on the single colour channel pixels. Here again, we use Canny's edge detection method.

$$\gg img = imread('carCycle.jpg'); \ \% \text{ select RGB image}$$
$$\gg gR = img(:, :, 1); \ \% \text{ select red channel pixels}$$
$$\gg imgR = edge(gR, 'canny'); \ \% \text{ select red channel pixels}$$

Here are some of the details.

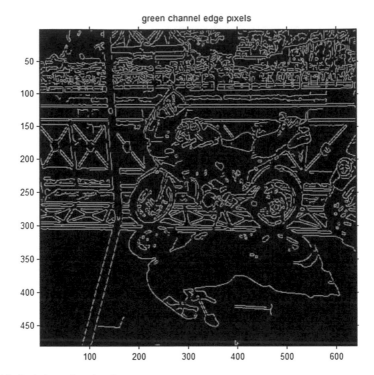

green channel edge pixels

Fig. 2.18 Red channel cycle edges

Colour Channel Subimage
   In this cycle image segment, only the green channel pixels are shown.

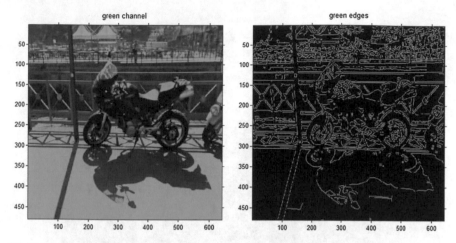

Fig. 2.19   Green channel edges

Colour Channel Subimage Edges

In this cycle image segment, the green channel edge pixels on the wheel subimage are shown.   ∎

It is possible to combine colour channel edge pixels on a black image.

Example 2.18   Figure 2.20 shows the result of combining the red channel and the green channel edge pixels again using MScript A.18 in Appendix A.2.5. This is accomplished in a straightforward fashion by concatenating the separate images, namely, $imgR$ (red channel edges), $imgG$ (green channel edges) and a (entirely black image).

Here are some of the details.

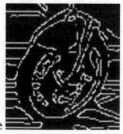

Binary Edges Subimage

In this cycle image segment, only the green channel pixels are shown.

technicolor RG edges

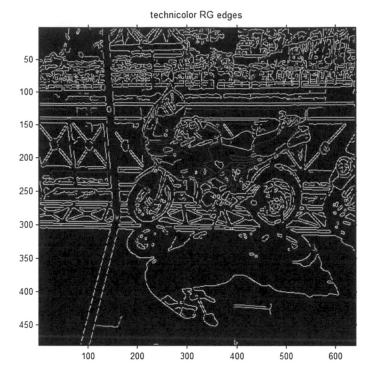

Fig. 2.20 Red green channel cycle edges

Colour Channel Subimage Edges

In this cycle image segment, the red channel and green channel edge pixels on the wheel subimage are shown. Notice that many yellow edges are included in the RG edges. The yellow edge pixels are at the higher (brighter) ends of the Canny hysteresis intervals used to identify strong edge pixels. An entirely different situation will arise, if we consider either RB or GB edge pixels (see Problem 2.19) (Fig. 2.21). ■

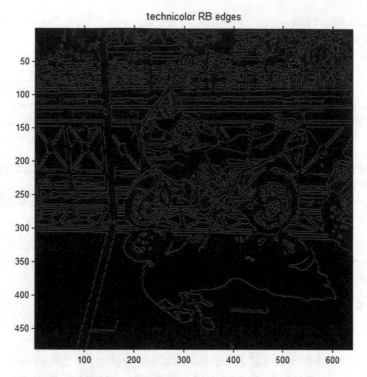

Fig. 2.21 Red blue channel cycle edges

Problem 2.19 Combined Color Channel Edge Pixels.
Extending the approach to combining colour edge pixels in Example 2.18, do the following:

1° 🚲 Display a combination of the red channel and blue channel edges on a black image. **Hint:** See how this is done in MScript A.18 in Appendix A.2.5.

2° 🚲 Display a combination of the green channel and blue channel edges on a black image.

3° ☕ Display a combination of the red, green, and blue channel edges on a black image. ■

Problem 2.20 Offline Video Frame Colour Channel Edges.
Use the approach to changing image channel intensities in MScript A.18 in Appendix A.2.5 as a template for offline video processing, do the following.

1° ☕ Using Matlab script A.8 in Appendix A.1.5 as a template for offline video processing, display the red channel edges in each video frame image. **Hint**: Replace the lines of Voronoï tessellation code with lines to code in MScript A.18 to handle and display the red channel edges in each video frame image.

2° Repeat Step 1 to handle and display the green channel edges in each video frame image.

3° Repeat Step 1 to handle and display the blue channel edges in each video frame image.

4° ☕ Using Matlab script A.8 in Appendix A.1.5 as a template for offline video processing, display the combined red and green channel edges in each video frame image. **Hint**: Replace the lines of Voronoï tessellation code with lines to code in MScript A.18 to handle and display the combined red and green channel edges in each video frame image.

5° Repeat Step 4 to handle and display the combined red and blue channel edges in each video frame image.

6° Repeat Step 4 to handle and display the combined green and blue channel edges in each video frame image.   ∎

Problem 2.21 Offline Video Frame Combined Colour Channel Edges.
Write a script to display offline the combined RGB channel edges in each video frame image. Do this for two different videos.   ∎

Problem 2.22 Real-Time Video Frame Colour Channel Edges.
Use the approach to changing image channel intensities in MScript A.18 in Appendix A.2.5 as a template for real-time video processing, do the following.

1° Using Matlab script A.9 in Appendix A.1.5 as a template for offline video processing, display the red channel edges in each video frame image. **Hint**. Replace the lines of Voronoï tessellation code with lines to code in Matlab script A.9 to handle and display the red channel edges in each video frame image in real-time.

2° Repeat Step 1 to handle and display the green channel edges in each video frame image.

3° Repeat Step 1 to handle and display the blue channel edges in each video frame image.

4° ☕ Using Matlab script A.8 in Appendix A.1.5 as a template for offline video processing, display the combined red and green channel edges in each video frame image. **Hint**: Replace the lines of Voronoï tessellation code with lines to code in MScript A.18 to handle and display the combined red and green channel edges in each video frame image in real-time.

5° Repeat Step 4 to handle and display the combined red and blue channel edges in each video frame image in real-time.

6° Repeat Step 4 to handle and display the combined green and blue channel edges in each video frame image in real-time.   ∎

Problem 2.23 Real-Time Video Frame Combined Colour Channel Edges.
Write a script to display in real-time the combined RGB channel edges in each video frame image. Do this for two different videos.   ∎

Algorithm 8: Log-Based Image Pixel Changes

Input : Read digital image img.
Output: $img \longmapsto log(img)$.

1 $gR \longleftarrow img(:, :, 1)$;
2 /* Capture red channel pixel intensities. */ ;
3 $gG \longleftarrow img(:, :, 2)$;
4 /* Capture green channel pixel intensities. */ ;
5 $gB \longleftarrow img(:, :, 3)$;
6 /* Capture blue channel pixel intensities. */ ;
7 $log(gR) \longmapsto imgR$;
8 $log(gG) \longmapsto imgG$;
9 $log(gB) \longmapsto imgB$;
10 /* Map log of pixel intensities in each channel to a modified channel image. */ ;
11 $captureModifiedImage \longleftarrow cat(3, imgR, imgG, imgB)$;
12 /* Capture modified channel intensities in a single image. */ ;
13 $Display\ captureModifiedImage$;

2.6 Function-Based Image Pixel Value Changes

This section briefly introduces an approach to modifying image pixel values using various functions. We illustrate this approach using the natural log of pixel values over an selected colour image channels. The steps to follow in modifying the each of the channel intensities resulting from the log of each colour channel pixel intensity are show in Algorithm 8.

Example 2.24 Figure 2.22 shows the result of a log-based modification of channel pixel intensities in a colour image using Algorithm MScript A.19 in Appendix A.2.6. Here are sample coding steps in the basic approach.

$\gg img = imread('carCycle.jpg')$; % select RGB image

$\gg gR = img(:, :, 1)$; % select red channel pixels

$\gg imgR = log(double(gR))$; % find log of red channel edge pixel intensities

$\gg sf = 0.2$; % scaling factor

$\gg imgR = (sf). * log(double(gR))$; % select lower edge pixel intensities

Here are some of the details.

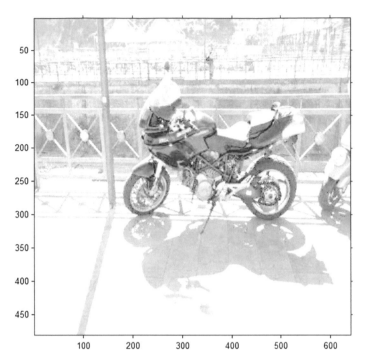

Fig. 2.22 Sample image after log-based pixel intensity changes

Colour Subimage
In this colour image segment, only the front wheel is shown.

Log-Based Colour Subimage
In this colour image segment, the combined log-modified channel intensities are shown. ■

Problem 2.25 Function-Based Colour Channel Intensity Modifications.
Select three colour images of your own choosing and do the following.

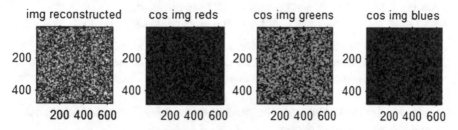

Fig. 2.23 Sample image after cosine-based pixel intensity changes

1° 🚲 Compute the cosine of each colour channel intensity and produce four images like ones in Fig. 2.23. **Hint**: Modify MScript A.19 in Appendix A.2.6 to get the desired result.

2° Repeat the preceding step for two different choices of the scaling factor to adjust the brightness of the modified images. For example, 0.2 is the scaling factor in MScript A.19 and 1.8 is the scaling factor used to obtain the results in Fig. 2.23.

Problem 2.26 Colour Channel Edge Information Content.
Select three colour images of your own choosing and do the following.

1° 👐 Compute the information content of each colour channel edge pixel intensity and produce four images like ones in Fig. 2.23. **Hint**: Find the total number of pixels in each image. Assume that the edge pixel intensities in the digital image img are random. In addition, let the probability $p(img(x, y)) = \frac{1}{x*y}$ for each image intensity $img(x, y)$ for a pixel with coordinates (x, y), $1 \leq x \leq m$, $1 \leq y \leq n$ in an $n \times m$ image.[1] Then, for each colour channel pixel intensity, compute the colour channel edge pixel **information content** $h(img(x, y, k))$, $k = 1, 2, 3$ of an edge pixel defined by

$$h(img(x, y, k)) := log_2 \left(\frac{1}{p(img(x, y, k))} \right) \text{ (colour channel pixel info. content).}$$

And, for each colour pixel edge intensity, compute the colour edge pixel **information content** $h(img(x, y))$, $1 \leq x \leq m$, $1 \leq y \leq n$ of an edge pixel defined by

$$h(img(x, y)) := log_2 \left(\frac{1}{p(img(x, y))} \right) \text{ (pixel information content).}$$

---

[1] Many other ways to compute the probability of a pixel intensity $img(x, y)$ are possible. There is a restriction:

$$\sum_{i=1}^{n*m} p_i(img(r, c)) = 1, 1 \leq r \leq m, 1 \leq c \leq n.$$

2° Repeat the preceding steps for two different choices of the scaling factor to adjust
the brightness of the modified images.   ∎

Problem 2.27 Colour Image Entropy and Its Modifications.
Select three colour images of your own choosing and do the following.

1° 🚲 Give a formula for the Shannon entropy of a $n \times m$ colour image img.
2° 🚲 Using the assumptions in Problem 2.26, write a Matlab or Mathematica script
to compute and display the Shannon entropy of three colour images of your own
choosing.
3° ☕ Modify the Matlab script in Step 2 to do the following with the three colour
images of your own choosing.

3(a) Change the color image pixel intensities so that the entropy of the image
increases.
3(b) Change the color image pixel intensities so that the entropy of the image
decreases.   ∎

2.7   Logical Operations on Images

The logical operations are not, and, or, and xor (exclusive or). This section introduces
the use of not, or, and xor (exclusive or) on image pixels. Later, it will be shown how
the **and** operation can be combined with what is known as thresholding to separate
the foreground from the background of images (see Sect. 2.8).

2.7.1 Complementing and Logical not of Pixel Intensities

For a greyscale image, the complement of the image makes dark areas lighter and
bright areas darker. For a binary image g, $not(g)$ changes background (black) values
to white and foreground (white) values to black. The $not(g)$ produces the same results
as $imcomplement(g)$.

Example 2.28 Mscript A.20 in Appendix A.2.7 illustrates changes in a greyscale
image in which the complement of each intensity is complemented and changes in a
binary image in which the logical **not** of each pixel intensity is computed. Figure 2.24
shows two modifications every intensity in a greyscale image:

1° Complement of each greyscale pixel intensity. Notice how the photographer's
coat is now mostly (not entirely) white and dull gray background areas become
very dark.
2° Addition of the maximum intensity to each greyscale pixel intensity. The result is
surprising, since it demonstrated the presence of blurred segments in the original
greyscale image.

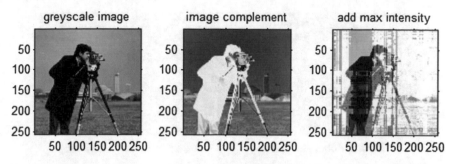

Fig. 2.24 Sample complement and increased greyscale pixel intensities

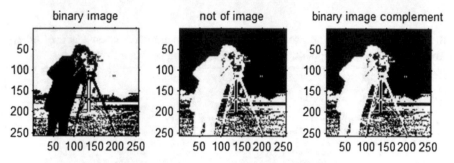

Fig. 2.25 Sample complement and logically negated binary pixel intensities

Figure 2.24 shows two modifications every intensity in a binary image:

3° Logical **not** of each greyscale pixel intensity. Notice how all black areas of the binary become white and all white areas become black.

4° Complement of each binary pixel intensity. This produces the same result as the complement of the binary image (Fig. 2.25).   ∎

Table 2.1 XOR

x	y	xor(x,y)
0	0	0
0	1	1
1	0	1
1	1	0

2.7.2 Xor Operation on Pairs of Binary Images

To see what the xor operation does, consider Table 2.1, where x, y are pixel intensities in a binary image. Table 2.1 is modelled after an exclusive or truth table. In Matlab,

2.26.1: Robots at start   2.26.2: Robots competing

Fig. 2.26 Sample colour images using `robots.m`

the exclusive or operation produces the following sample result on a pair of binary images. To see what happens, consider the following pair of colour images.

```
% constructing new images from old images using xor
% idea from Solomon and Breckon, 2011
clc, close all, clear all
                                           % What's happening?
g = imread('race1.jpg'); h = imread('race2.jpg'); % read images
gbw = im2bw(g); hbw = im2bw(h);            % convert to binary
check = xor(gbw,hbw);
subplot(1,3,1), imshow(gbw);               % display g
subplot(1,3,2), imshow(hbw);               % display h
subplot(1,3,3), imshow(check);             % display xor(gbw,hbw)
```

Listing 2.1 Matlab code `cars.m` to produce Fig. 2.26.

Fig. 2.27 Sample xor images `robots.m`

Next, a pair of .png colour images in Fig. 2.26 are converted to binary images (every pixel value is either 1 (white) or 0 (black) after applying the **im2bw** function to each image. Then the **xor** function is applied (see Listing 2.2) to the pair of binary images to obtain the result shown in Fig. 2.27.

```
% constructing new images from old images
close all
clear all
                                           % What's happening?
%g = imread('birds1.jpg'); h = imread('birds2.jpg'); % read png images
g = imread('race1.jpg'); h = imread('race2.jpg'); % read png images
gbw = im2bw(g,0.3); hbw = im2bw(h,0.3);    % convert to binary
check = xor(gbw,hbw);                       % xor binary
     intensities
```

```
figure,
subplot(1,3,1), imshow(gbw);                        % display gbw
subplot(1,3,2), imshow(hbw);                        % display hbw
subplot(1,3,3), imshow(check);                      % display xor(gbw,hbw)
```

Listing 2.2 Matlab code to produce Fig. 2.27.

For the sake of completeness, the same experiment is performed on a pair of .jpg colour images showing two different Thai grocery store displays. The interesting thing here is seeing how the xor operation on the displays reveals movements of similar items (bottles) from one display to the other (Fig. 2.28).

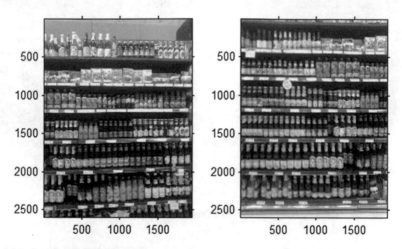

Fig. 2.28 Sample Thai Shelf images

```
% constructing new images from old images
clc, clear all, close all               % housekeeping
g = imread('P9.jpg'); h = imread('P7.jpg');         % read jpg images
%
gbw = im2bw(g); hbw = im2bw(h);                     % convert to binary
check = xor(gbw,hbw);                               % xor binary
     intensities
subplot(1,3,1), imshow(gbw);                        % display gbw
subplot(1,3,2), imshow(hbw);                        % display hbw
subplot(1,3,3), imshow(check);                      % display xor(gbw,hbw)
```

Listing 2.3 Matlab code xor2.m to produce Fig. 2.29.

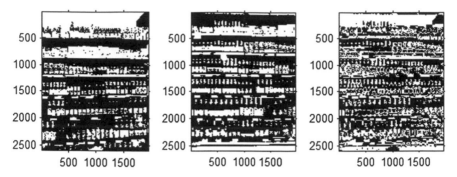

Fig. 2.29 Sample .png colour images

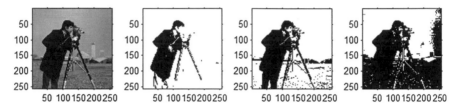

Fig. 2.30 Sample greyscale image thresholding

2.8 Separating Image Foreground From Background

Greyscale and colour images can be transformed into binary (black and white) images, where the pixels in the foreground of an image are black and pixels in the background of an image are white. The separation of image foreground from background is accomplished using a technique called **thresholding**. The thresholding method results in a binary image by changing each background pixel value to 0, if a pixel value is below a threshold, and to 1, if a foreground pixel value is greater than or equal to the threshold. Let $th \in (0, \infty]$ denote a threshold and let $\Im$ denote a greyscale image. Then

$$\Im(x, y) = \begin{cases} 1, & \text{if } \Im(x, y) > th, \\ 0, & \text{otherwise.} \end{cases}$$

```
% Thresholding on greyscale image
clc, clear all, close all      % housekeeping
g = imread('cameraman.tif');   % read greyscale image
h1 = im2bw(g,0.1);             % threshold = 0.1
h2 = im2bw(g,0.4);             % threshold = 0.5
h3 = im2bw(g,0.6);             % threshold = 0.5
subplot(1,4,1), imshow(g);     % display greyscale image
subplot(1,4,2), imshow(h1);    % display transformed image
subplot(1,4,3), imshow(h2);    % display transformed image
subplot(1,4,4), imshow(h3);    % display transformed image
```

Listing 2.4 Matlab script to produce Fig. 2.30 using ex_greyth.m.

Notice that $th = 0.5$ works best in separating the cameraman from the background (in fact, the background is no longer visible in Fig. 2.30 for $th = 0.5$). If there is interest in isolating the foreground of a greyscale image, it is necessary to experiment with different thresholds to obtain the best result. The code used to produce Fig. 2.30 is given in Listing 2.4.

Problem 2.29 Reversing Greyscale Pixel Separation Process.
Partially reverse the thresholding process for a greyscale. Wherever there is a white pixel in a thresholded image, change to white the pixel in the corresponding greyscale image. This reversal process will result in a greyscale where the foreground consists of pixels with varying intensities and the background of the greyscale image is entirely white. This reversal process will be important later, when feature extraction methods are used based on varying pixel intensities. ■

Separating the foreground from the background in colour images can either be done uniformly (treating all three colour channels alike) or finely by thresholding each colour channel individually. Sample results of the uniform separation approach are shown in Fig. 2.31 using the code Listing 2.5.

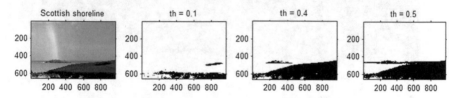

Fig. 2.31 Sample colour image thresholding

```
% Thresholding a colour image
                            % What's happening?
g = imread('rainbow.jpg');    % read colour image
% g = imread('penguins.jpg'); % read colour image
h1 = im2bw(g,0.1);            % threshold = 0.1
h2 = im2bw(g,0.4);            % threshold = 0.4
h3 = im2bw(g,0.5);            % threshold = 0.5
subplot(1,4,1), imshow(g);  title('Scottish shoreline');
subplot(1,4,2), imshow(h1); title('th = 0.1');
subplot(1,4,3), imshow(h2); title('th = 0.4');
subplot(1,4,4), imshow(h3); title('th = 0.5');
```

Listing 2.5 Matlab script to produce Fig. 2.31 using ex_2th.m.

Problem 2.30 Reversing Colour Pixel Separation Process.
Partially reverse the thresholding process for a colour image. Wherever there is a white pixel in a thresholded colour image, change to white the pixel in the corresponding colour image. This reversal process will result in a colour where the foreground consists of pixels with varying intensities for each colour channel for each pixel and the background of the colour image is entirely white. This reversal process will be important later, when feature extraction methods are used based on

varying pixel colour intensities. Common applications of this reversal process are in signature forgery detection and camouflage detection in paintings and in satellite images. ■

Fig. 2.32 Sample colour image

2.9 Conjunction of Thresholded Colour Channels

Another useful technique in separating the foreground from the background in colour images stems from an application of the logical **and** operation. The basic idea is to threshold the pixel intensities in each colour channel and then experiment with the conjunction of the resulting colour changes, either in pairs or the conjunction of all three thresholded colour channels. Let $\Im$ be a colour image, r, g, b colour channels in $\Im$, and let rth, gth, bth denote thresholds on the red, green, blue colour channels, respectively. Then

$$\gg rbw = im2bw(\Im(:, :, r), rth); \text{ thresholded red channel,}$$
$$\gg gbw = im2bw(\Im(:, :, g), gth); \text{ thresholded green channel,}$$
$$\gg bbw = im2bw(\Im(:, :, b), bth); \text{ thresholded blue channel.}$$

Then using the logical *and* operation, compute

> $arg = and(rbw, gbw)$; conjunction of r,g channels,

> $arb = and(gbw, bbw)$; conjunction of g,b channels,

> $agb = and(rbw, bbw)$; conjunction of r,b channels,

> $agb = and(and(rbw, gbw), bbw)$; conjunction of r,g,b channels.

The colour image in Fig. 2.32 is an example of macrophotography, showing a closeup of grasshoppers. **Macrophotography** is closeup photography. A **macro lens** is capable of reproduction ratios greater than 1:1. The onscreen reproduction of a 1:1 macroimage results in a photograph greater than a lifesize image. Reproduction ratios much greater than 1:1 is called **photomicroscopy**, usually accomplished with a stereo zoom digital microscope. An application of the conjunction form of thresholding on the macrophotograph of grasshoppers is shown in Fig. 2.33 using the sample code in Listing 2.6. Notice the best separation of the foreground from background is achieved with a conjunction of the thresholded red and blue channels. This is not always the case (see Problem 2.31).

> $agb = and(rbw, bbw)$; conjunction of r,b channels.

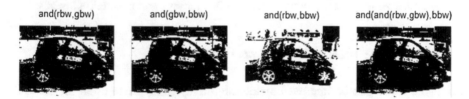

Fig. 2.33 Sample colour image thresholding with conjunction

```
% Thresholding colour channels
clc, clear all, close all      % housekeeping
g = imread('carPoste.jpg');         % read colour image
rth = 0.2989; gth = 0.587; bth = 0.114; % NTSC weights
r = g(:,:,1); gr = g(:,:,2);  b = g(:,:,3); % channels
rbw = im2bw(r,rth);                 % threshold r
gbw = im2bw(gr,gth);                % threshold g
bbw = im2bw(b,bth);                 % threshold b
o1 = and(rbw,gbw); o2 = and(gbw,bbw); o3 = and(rbw,bbw);
o4 = and(and(rbw,gbw),bbw);
subplot(1,4,1),imshow(o1), title('and(rbw,gbw)');
subplot(1,4,2),imshow(o2), title('and(gbw,bbw)');
subplot(1,4,3),imshow(o3), title('and(rbw,bbw)');
subplot(1,4,4),imshow(o4), title('and(and(rbw,gbw),bbw)');
```

Listing 2.6 Matlab script to produce Fig. 2.33 using ex_2th2.m.

Problem 2.31 Threshold and Conjunction Separation Process.

Use a combination of thresholding and conjunction on the colour channels for several different colour images, starting with **rainbow.jpg** (Scottish rainbow) and **seq4a.jpg** (hand). Do the following.

(**and**.1)   Vary the weights for the thresholded rgb channels,

(**and**.2)   Point out which conjunction of thresholded channels gives the best results. The result will be best when there are more details in the foreground.

(**and**.3)   For a particular colour image, explain why a particular conjunction of colour channels works best.

(**and**.4)   Besides the **rainbow.jpg** (Scottish rainbow) and **seq4a.jpg** (hand) images, find a third colour image (your choice), where the conjunction of all thresholded colour channels works best.   ∎

Problem 2.32   Reversing Threshold and Conjunction Separation Process.
Partially reverse the thresholding-conjunction process for a colour image. Wherever there is a white pixel in a binary image resulting from a conjunction of a combination of thresholded colour channels, change to white the pixel in the corresponding colour image. This reversal process will result in a colour where the foreground consists of pixels with varying intensities for each colour channel for each foreground pixel and the background of the colour image will be entirely white.   ∎

The reversal process from the solutions of Problems 2.30 and 2.32 will also be important later, when feature extraction methods are used based on varying pixel colour intensities. Common applications of this reversal process are in signature forgery detection and camouflage detection in paintings and in satellite images.

2.10   Improving Contrast in an Image

Image contrast can be improved by altering the dynamic range of an image. The **dynamic range** of an image equals the difference between the smallest and largest image pixel values. Transforms can be defined by altering the relation between the

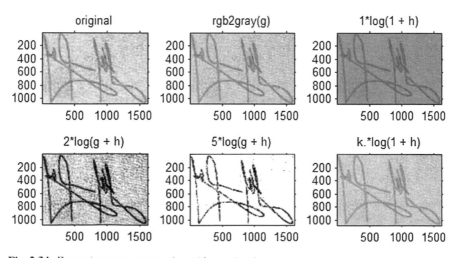

Fig. 2.34 Dynamic range compression with `eg_log1.m`

dynamic range and the greyscale (colour) image pixel values. For example, an image dynamic range can be altered by replacing each pixel value with its logarithm. Let $\Im$ denote an image. Then alter the pixel value at (x, y) using

$$\Im(x, y) = k \log_e(1 + (e^\sigma - 1)\Im(x, y)), \tag{2.1}$$

where (assuming 8 bit pixel values),

$$k = \frac{255}{\log_e(1 + max(\Im))}.$$

To simplify the implementation of Eq. (2.1), use the following technique to alter all pixel values in $\Im$.

$$\gg \Im = k. * log(1 + im2double(\Im))$$

Next observe that, since $\Im$ is a matrix, **max**($\Im$) returns a row vector containing the maximum pixel value from each column. To complete the implementation of k, use

$$\gg k = mean((255)./log(1 + max(\Im))).$$

The results of a number of different experiments in modifying the dynamic range are shown in Fig. 2.34, using the code in Listing 2.7.

```
% Compressing dynamic range of an image
clc, clear all, close all     % housekeeping
g = imread('sig.jpg');   % Read in image
subplot(2,3,1), imshow(g); title('original');
g = rgb2gray(g);
subplot(2,3,2), imshow(g); title('rgb2gray(g)');
g = im2double(g);             % pixel values -> double
h = im2double(g);             % pixel values -> double
k = (max(max(g)))./(log(1 + max(max(g))));
com1 = 1*log(1 + h);          % 1st compression
com2 = 2*log(g + h);          % 2nd compression
com3 = 5*log(g + h);          % 3rd compression
com4 = k.*log(1 + h);         % 4th compression
subplot(2,3,3), imshow(com1); title('1*log(1 + h)');
subplot(2,3,4), imshow(com2); title('2*log(g + h)');
subplot(2,3,5), imshow(com3); title('5*log(g + h)');
subplot(2,3,6), imshow(com4); title('k.*log(1 + h)');
```

Listing 2.7 Matlab script to produce Fig. 2.34 using eg_log1.m.

Notice that by increasing the value of the multiplier k, the overall brightness of the image increases.[2] The best result for the signature image is shown in the third image in row 2 of Fig. 2.34, where **5.*log(g + h)** is used on the image g. A less than satisfactory result is obtained using **k.*log(1 + im2double(g))**. The logarithmic transform in Eq. (2.1) induces a brightening of the foreground by spreading the foreground pixel values over a wider range and a compression of the background pixel range. The

---

[2]Many thanks to Patrik Dahlström for pointing out the corrections in eg_log1.m.

narrowing of the background pixel range provides a sharper contrast between the background and the foreground.

Problem 2.33 Let g denote either a greyscale or colour image. In Matlab, implement Eq. (2.1) using $(e^{\sigma} - 1)g(x, y)$ instead of **im2double**(g) and show sample images using several choices of σ. Use the cameraman image as well as the signature image to show the results for different values of σ. ■

Use the Matlab **whos** function to display the information about the current variables in the workspace, e.g., variables k and $com4$ in Listing 2.7. Matlab constructs the double data type in terms of the definition for double precision in IEEE Standard 754, i.e., double precision values require 64 bits (for Matlab, double is the default data type for numbers). The **im2double(g)** function converts pixel intensities in image g to type double.

2.11 Gamma Transform

An alternative to the logarithm approach in compressing the dynamic range of intensities in an image, is the gamma (*raise to a power*) transform. Let I denote a digital image, $I(x, y)$ a pixel located at (x, y), $k \in \mathbb{N}$ (natural number $1, \ldots, \infty$), and $\gamma \in \mathbb{R}^{+}$ (positive reals). Basically, each pixel value is raised to a power using

$$I(x, y) = k\,(I(x, y))^{\gamma}\,.$$

The constant k provides a means of scaling the transformed pixel values. Here are rules-of-thumb for the choice of γ.

(**rule**.1) $\gamma > 1$: Increase contrast between high-value pixel values at the expense of low-valued pixels.

(**rule**.2) $\gamma < 1$: Decrease contrast between high-value pixel values at the expense of high-valued pixels.

```
% Gamma transform
clc, clear all, close all              % housekeeping
g = imread('P9.jpg');                  % read image
% h = imread('P7.jpg ');               % read image
g = im2double(g);
g1 = 2*(g.^(0.5)); g2 = 2*(g.^(1.5)); g3 = 2*(g.^(3.5));
subplot(1,4,1), imshow(g);             % display g
title('Thai shelves');
subplot(1,4,2), imshow(g1);            % gamma = 0.5
title('gamma = 0.5');
subplot(1,4,3), imshow(g2);            % gamma = 1.5
title('gamma = 1.5');
subplot(1,4,4), imshow(g3);            % gamma = 3.5
title('gamma = 3.5');
```

Listing 2.8 Matlab script to produce Fig. 2.35 using myGamma.m.

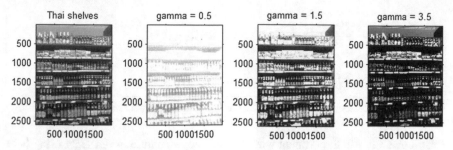

Fig. 2.35 Gamma transformation of a Thai colour image

2.12   Gamma Correction

There is a nonlinear relationship between input voltage and output intensity in monitor displays. This problem can be corrected by preprocessing image intensities with an inverse gamma transform (also called inverse power law transform) using

$$g_{out} = \left(g_{in}^{\frac{1}{\gamma}}\right)^{\gamma+k},$$

where g_{in} is the input image and g_{out} is the output image after gamma correction. Gamma correction can be carried out using the Matlab **imadjust** function as shown in `gamma_adjust.m` with sample results shown in Fig. 2.36. Unlike the results in Fig. 2.35 with the gamma transform, the best result in Fig. 2.36 is obtained with a lower γ value, namely, $\gamma = 1.5$ in Fig. 2.36 as opposed to $\gamma = 3.5$ in Fig. 2.35.

```
% Gamma correction transform
clc, clear all, close all      % housekeeping
g = imread('P9.jpg');              % Thai shelves image
%g = imread('sig.jpg');             % Currency signature
g = im2double(g);
g1 = imadjust(g,[0  1],[0  1],0.5);   % in/our range [0,1]
g2 = imadjust(g,[0  1],[0  1],1.5);   % in/our range [0,1]
g3 = imadjust(g,[0  1],[0  1],3.8);   % in/our range [0,1]
subplot(1,4,1), imshow(g);          % display g
```

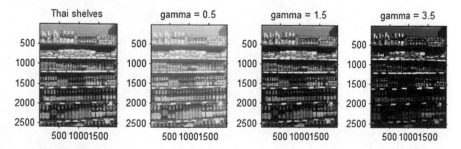

Fig. 2.36 Gamma correction of a Thai colour image

```
title('Thai shelves');
subplot(1,4,2), imshow(g1);           % gamma = 0.5
title('gamma = 0.5');
subplot(1,4,3), imshow(g2);           % gamma = 1.5
title('gamma = 1.5');
subplot(1,4,4), imshow(g3);           % gamma = 3.5
title('gamma = 3.5');
```

Listing 2.9 Matlab script to produce Fig. 2.36 using `gamma_adjust.m`.

Problem 2.34 ⮞ Experiment with the currency signature in Fig. 2.34 using both the gamma transform and inverse gamma transform. Which value of γ gives the best result in each case? The best result will be the transformed image that has the clearest signature.

Chapter 3
Visualising Pixel Intensity Distributions

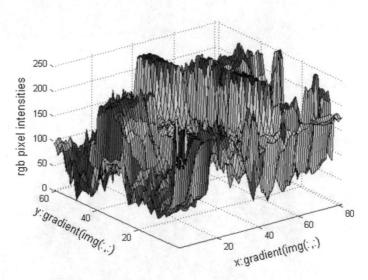

Fig. 3.1 3D view of combined rgb pixel intensities in Fig. 3.4

This chapter introduces various ways to visualize pixel intensity distributions (see, e.g., Fig. 3.1). Also included here are pointers to sources of generating points useful in image tessellations and triangulations. In other words, image structure visualizations carries with it tacit insights about image geometry.

The basic approach here is to provide 2D and 3D views of pixel intensities in cropped digital images. By cropping a colour image, it is possible to obtain different views of either the combined pixel colour values or the individual colour colour channel pixel values within the same image. The importance of image cropping cannot be overestimated. **Image cropping** extracts a subimage from an image. This makes it possible to concentrate on that part of a natural scene or laboratory sample that is considered interesting, relevant, deserving a closer look. Pixel intensities are

© Springer International Publishing AG 2017
J.F. Peters, *Foundations of Computer Vision*, Intelligent Systems
Reference Library 124, DOI 10.1007/978-3-319-52483-2_3

Fig. 3.2 Sample RGB image for the Salerno train station

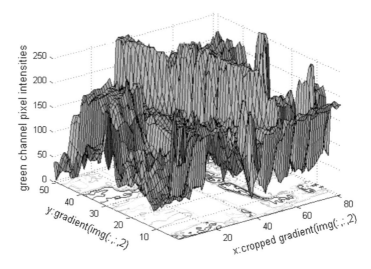

Fig. 3.3 3D View of *green* ● channel pixel Intensities with contours for Fig. 3.2

yet another source of generating points (sites) used to tessellate an image, resulting in image meshes that reveal image geometry and image objects from different perspectives.

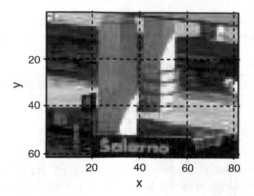

Fig. 3.4 Subimage from cropped image in Fig. 3.2

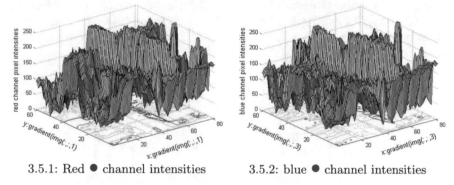

3.5.1: Red ● channel intensities   3.5.2: blue ● channel intensities

Fig. 3.5 *Red* and *blue* channel values in Fig. 3.2

Example 3.1 Matlab script A.22 in Appendix A.3 is used to do the following:

1° Crop an rgb image to obtain a subimage. For example, the tiny image in Fig. 3.4 is the result of cropping the larger image in Fig. 3.2.

2° Produce a 3D mesh showing the combined rgb pixel values. The result for cropped image is shown in Fig. 3.1.

3° Produce a 3D mesh with contours for the red ● channel values. The result for the red channel values for the pixels in the cropped image is shown in Fig. 3.5.1.

4° Produce a 3D mesh with contours for the green channel values. The result for the green ● channel values for the pixels in the cropped image is shown in Fig. 3.3. The green channel values in a colour image often tend to have the greatest number of changes between the minimum and maximum values. Hence, the green channel is good place to look for non-uniformity in the selection of generating points (sites) use in a Voronoï tessellation of an image. To see this, consider the difference between the 3D meshes and their contours, starting with 3D mesh for the green channel in Fig. 3.3, compared with the red channel values in Fig. 3.5.1 and blue channel values in Fig. 3.5.2.

5° Produce a 3D mesh with contours for the blue channel values. The result for the
blue ● channel values for the pixels in the cropped image is shown in Fig. 3.5.2.
■

N.B.: Notice that histogram construction requires an intensity image.

3.1   Histograms and Plots

There are a number of ways to visualize the distribution of pixel intensities in a digital
image. A good way to get started is to visualize the distribution of pixel intensities
in an image.

Example 3.2 **Sample Greyscale Histogram**. Sample pixel intensity counts for each
greyscale pixel in Fig. 3.2 are shown in Fig. 3.6. To experiment with image pixel
intensity counts, see script A.21 in Appendix A.3. For the details, see Sect. 3.1.1
given next.   ■

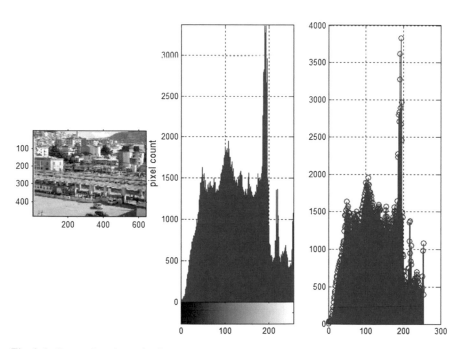

Fig. 3.6 Greyscale train station image histogram

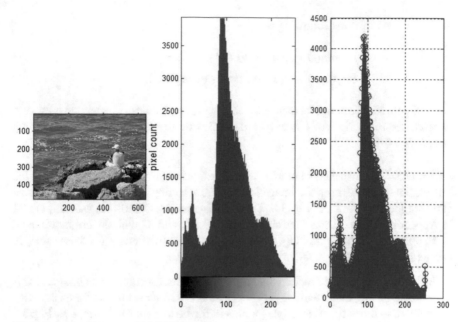

Fig. 3.7 Image histogram with 256 bins

3.1.1 Histogram

An image histogram plots the relative frequency of occurrence of image pixel intensity values against intensity values. Histograms are constructed using binning, since it is usually not possible to include individual pixel intensity values in a histogram. An **image intensity bin** (also called an **image intensity bucket**) is a set of pixel intensities within a specified range. Typically, a histogram for an intensity image contains 256 bins, one pixel intensity per bin. Each intensity image histogram displays the size (cardinality) of each pixel intensity bin. Image histograms are constructed using a technique called binning. **Image binning** is a method of assigning each pixel intensity to a bin containing matching intensities. Here is another example.

Example 3.3 **Sample histogram bins**.
The intensity (greyscale) image *img* in Fig. 3.7 has a wide range pixel counts for each of the 256 intensities in the image. Thanks to the white T-shirt of the fisherman, there are over 200 highest intensity pixels (intensity = 1). Again, notice that there over 500 pixels with 0 intensity (black pixels). Let histogram bins cover the range of intensities (one bin for each of the 256 intensities) be represented by

$$0, 1, 2, 3, \ldots, i, i+1, \ldots, 253, 254, 255 \ (\textbf{imageintensitybins}).$$

And let $img(x, y)$ be a pixel intensity at location (x, y). Let $0 \leq i \leq 255$ represent the intensity of bin i. Then all pixels with intensities matching the intensity of $img(x, y)$

are binned in bin(i) the following way:

$$bin(i) = \{img(x, y) : img(x, y) = i\} \text{ so that}$$

$$img(x, y) \in [i, i + 1) \text{ (} \mathbf{i}^{th} \textbf{ intensity bin}).$$

A sample Matlab script for exploring binning for both colour images and greyscale images, see see script A.21 in Appendix A.3. For an expanded study of binning, see [21, §3.4.1]. ■

Example 3.4 To inspect the numbers of intensities in a subimage, crop the a selected image. For example, crop the image in Fig. 3.7, selecting only the fisherman's head and shoulders as shown in Fig. 3.8. Then, for intensities 80, 81, 82, use script A.21 in Appendix A.3 to compare the size of bins 80, 81, and 82 with the original image:

In other words, the cardinality of the bins in the cropped image decreases sharply compared with the bins in the original image. ■

In Matlab, the **imhist** function displays a histogram for a greyscale image. If $\Im$ is a greyscale image, the default display for **imhist** is 255 bins, one bin for each image intensity. Use **imhist($\Im$,n)** to display *n* bins in the histogram for $\Im$ (see, *e.g.*, Fig. 3.6 a sample greyscale histogram for the rgb plant image). Use

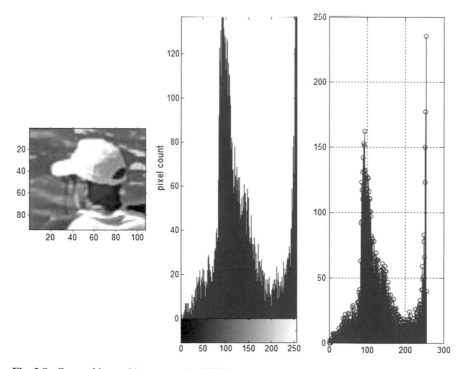

Fig. 3.8 Cropped image histogram with 256 bins

$$\gg [counts, x] = imhist(\Im); \qquad (3.1)$$

to store the relative frequency values in *counts* for histogram with horizontal axis values stored in x. See, also, the **histeq** function introduced in Sect. 3.6.

3.1.2 Stem Plot

A stem plot is a 2D plot that display function values as lollipops ⸮ (a stem with a circle end). Use the **stem** function to produce a stem plot. In the case of the relative frequency of pixel intensities, a stem plot provides a visual alternative to a histogram (the vertical lines in a histogram are replaced with ⸮s (see, *e.g.*, Fig. 3.12). A stem plot is derived from a histogram using the information gathered in (3.1) to obtain **stem(x,counts)**. See, *e.g.*, the stem plot in Fig. 3.12. Notice, also, that it is possible to produce a 3D stem plot, using the **stem3** function.

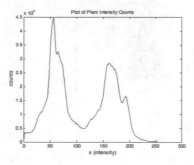

$$\gg plot(x, counts);$$

3.1.3 Plot

Relative to the vectors *Counts* and x extracted from a histogram, the **plot** function produces a 2D plot of the relative frequency counts (see, e.g., Table 3.1).

3.1.4 Surface Plot

Relative to matrices *gf* (filtered image) and *g* (double precision greyscale image), the **surf** and **surfc** functions produce 3D surface and 3D surface contour plots, respectively. For example, to obtain a surface plot with a contour plot beneath the surface using the rice image *g* and filtered image *gf* from Listing 3.4, try

Table 3.1   Two sets of bin counts

image			bin 80	bin 81	bin 82
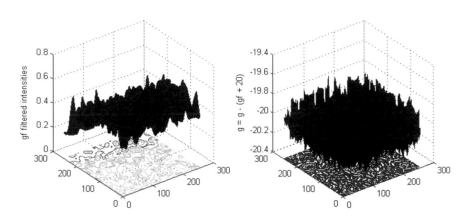					
			39	41	35
			2044	2315	2609

$\gg surfc(gf, g);$ **Filteredcontourplot**

$\gg surfc(gth, g);$ **Thresholdedimagecontourplot**

Notice evidence of the separation of image intensities in the first of the first of the contour plots and the evidence of the results of thresholding the rice image in the contour plot drawn beneath the surface in the second of the above surface plots. Using **surfc**, we obtain a visual perspective of the results produced by the script in Listing 3.4 in Fig. 3.16.

```
% Visualisation experiment

g = imread('rice.png');  % read greyscale image
g = im2double(g);
[x y] = meshgrid(max(g));
z = 20.*log(1 + g);
figure,surfc(x,y,z); zlabel('z = 20.*log(1 + g)');
```

Listing 3.1 Matlab code in mesh.m to produce Fig. 3.9.

3.1.5 Wireframe Surface Plot

The **meshgrid** combined with **surfc** produces a wireframe surface plot with a contour plot beneath the surface. This form of visualizing image intensities is shown in Fig. 3.9. In Listing 3.1, **meshgrid(g)** is an abbreviation for **meshgrid(g,g)**, transforming the domain specified by the image g into arrays x and y, which are then used to construct 3D wireframe plots.

3.1.6 Contour Plot

The **contour** function draws contour plot, with 3D surface values mapped to isolines, each with a different colour.

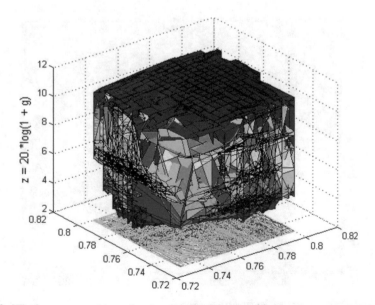

Fig. 3.9 Wireframe parametric surface for **rice.png** max intensities

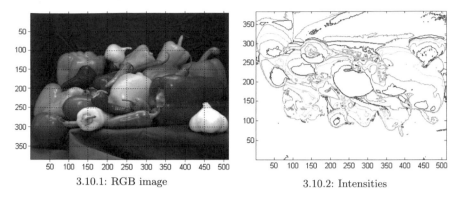

3.10.1: RGB image   3.10.2: Intensities

Fig. 3.10 *Red* channel isolines for peppers.png

3.2 Isolines

An **isoline** for a digital image connects points in a 2D plane all representing the
same intensity. The points in an isoline represent heights above the x-y plane. In
the case of the red channel in an rgb, the points in an isoline represent brightness
levels of the colour red (see, e.g., Fig. 3.11). Each isoline belongs to a surface, say
the line indicating the locations of, for example, intensity 100 value. For an isoline,
there is an implicit scalar field defined in 2D such as the value 100 at locations (0,0),
(25,75). The Matlab **clabel** function is can be used to insert intensity values in an

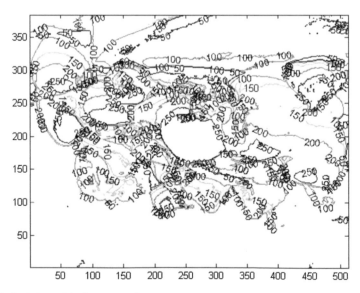

Fig. 3.11 *Red* channel isoline values for peppers.png

isoline. However, an attractive alternative to **clabel** is the combination of the **set** and **get** functions, which make it possible to control the range of values represented by isolines.[1] Sample isolines are shown in the contour plot in Figs. 3.10.2 and 3.1.5, produced by script in MScript A.23 Appendix A.3.3.

Problem 3.5 ☺ Experiment with surf, surfc, stem, stem3, plot, mesh, meshc, meshgrid, contour and, in each case, display various visualizations, using the mini-image represented by the array g in

$$\gg g = [12, -1, 55; 34, -1, 66; -123, 3, 56];$$

Repeat the same experiments with the **pout.png** greyscale image. Then demonstrate the use of each of the visualization functions with each of the colour channels in a sample colour image. ∎

Example 3.6 By way of illustration of the histogram and stem plot for an image intensity distribution, consider the distribution of intensities for **pout.tif** (a useful image in the Matlab library). A good overview of image histograms is given by M. Sonka, V. Hlavac and R. Boyle [185, §2.3.2]. The utility of a histogram can be seen in the fact that it is possible, for some images, to choose a threshhold value in logarithmically compressing the dynamic range of an image, where the threshold is an intensity in a valley between dominant peaks in the histogram. This is the case in Fig. 3.12, where an intensity of approximately 120 provides a good threshold.

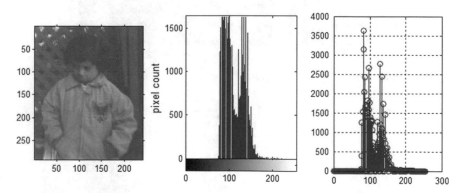

Fig. 3.12 Image pixel intensities distribution

---

[1] Instead of clabel, try a combination of set and get, *e.g.*,

$$\gg set(h,' ShowText',' on',' TextStep', get(h,' LevelStep'))$$

This approach to labelling contour lines will give control over the height labels that are displayed on the contour lines. For example, try $get(h,' LevelStep') * 2$ to inhibit the lower height labels.

```
% Histogram experiment
%% housekeeping
clc, clear all, close all
%%
% This section for colour images
I = imread('fishermanHead.jpg');
% I = imread('fisherman.jpg');
% I = imread('football.jpg');
I = rgb2gray(I);
%%
% This section for intensity images
%I = imread('pout.tif');
%%
% Construct histogram:
%
h = imhist(I);
[counts,x] = imhist(I);
counts
size(counts)
subplot(1,3,1), imshow(I);
subplot(1,3,2), imhist(I);
ylabel('pixel count');
subplot(1,3,3), stem(x,counts);
grid on
```

Listing 3.2 Matlab code in hist.m to produce Fig. 3.12.

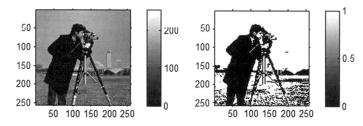

Fig. 3.13 Thresholded image via a histogram

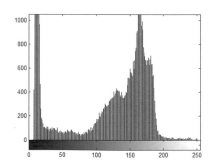

cameraman.tif histogram:

Problem 3.7 ☕ Experiment with threshold in ex_greyth.m using an image histogram to choose an effective threshold from the valley between peaks in relative frequencies distribution displayed in the histogram. Using the **cameraman.tif** image, obtain a result similar to that shown in Fig. 3.13 (call the modified Matlab

script **histh.m**). Display the original image, thresholded image, and image histogram, indicating the intensity you have chosen for the threshold. Also, display **histh.m**.
■

3.3 Colour Histograms

Also of interest is the distributions colour channel intensities in a colour image. After recording the color channel intensities for each pixel, superimposed stem plots for each colour serve to produce a colour histogram.

```
% Colour Image histogram
% algorithm from
% http://www.mathworks.com/matlabcentral/fileexchange/authors/100633
close all
clear all
                                % What's happening?
g = imread('rainbow-plant.jpg'); % read rgb  image
%g = imread('sitar.jpg');         % read greyscale image
nBins = 256;                      % bins for 256 intensities
rHist = imhist(g(:,:,1), nBins);  % save red intensities
gHist = imhist(g(:,:,2), nBins);  % save green intensities
bHist = imhist(g(:,:,3), nBins);  % save blue intensities
figure
subplot(1,2,1);imshow(g), axis on % display orig. image
subplot(1,2,2)                    % display histogram
h(1) = stem(1:256, rHist); hold on % red stem plot
h(2) = stem(1:256 + 1/3, gHist);  % green stem plot
h(3) = stem(1:256 + 2/3, bHist);  % blue stem plot
hold off
set(h, 'marker', 'none')          % set properties of bins
set(h(1), 'color', [1 0 0])
set(h(2), 'color', [0 1 0])
set(h(3), 'color', [0 0 1])
axis square                       % make axis box square
```

Listing 3.3 Matlab code in hist.m to produce Fig. 3.14.

Fig. 3.14 image pixel colour channel intensities distribution

The plot in Fig. 3.14 shows a histogram the presents the combined colour channel intensities for the plant image. By modifying the code in Listing 3.3, it is possible to display three separate histograms, one for each colour channel in the plant image (see Fig. 3.15).

3.4   Adaptive Thresholding

The limitations of global thresholding are overcome by using a different threshold for each pixel neighbourhood in an image. Adaptive thresholding focuses on local thresholds that are determined by pixel intensity values in the neighbourhood of each pixel. The form of thresholding is important because pixel intensities tend to be fairly uniform in pixel small neighbourhoods. Given an image $\Im$, neighbourhood filtered image $\Im_f$ and threshold $\Im_f + C$, modify each pixel value using

$$\Im \doteq \Im - (\Im_f + C).$$

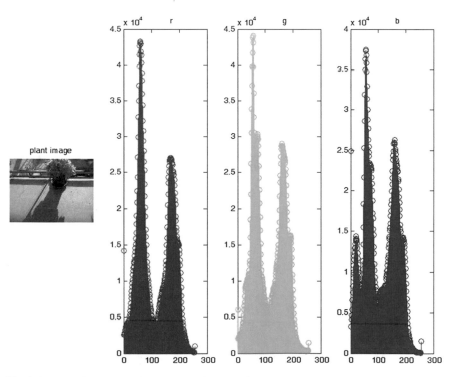

Fig. 3.15 Three image pixel colour channel distributions

A combination of the **imfilter** and **fspecial** filter functions can be used to compute filtered image neighbourhood values. First, decide on an effective $n \times n$ neighbourhood size and use the averaging filter **average** filter option for **fspecial**. Then use the **replicate** option to populate all image neighbourhoods with the average filter values for each neighbourhood. For example, choose $n = 9$ for a 9×9 neigbourhood and combine the two Matlab filters to obtain

$$\gg \Im_f = imfilter(\Im, fspecial('average', [99]), 'replicate');$$

```
% Histogram experiment
close all
clear all

g = imread('rainbow-plant.jpg');  % read rgb image
g = rgb2gray(g);
%g = imread('rice.png');  % read greyscale image
gf = imfilter(g,fspecial('average',[15 15]),'replicate');
gth = g - (gf + 20);
gbw = im2bw(gth,0);
subplot(1,4,1), imshow(gbw);
%set(gca,'xtick',[],'ytickMode','auto');
subplot(1,4,2), imhist(gf); title('avg filtered image');
grid on
glog = imfilter(g,fspecial('log',[15 15]),'replicate');
gth = g - (glog + 100);
gbw = im2bw(gth,0);
%glog = imfilter(g,fspecial('prewitt'));
%glog = imfilter(g,fspecial('sobel'));
%glog = imfilter(g,fspecial('laplacian'));
%glog = imfilter(g,fspecial('gaussian'));
%glog = imfilter(g,fspecial('unsharp'));
gbw = im2bw(gth,0);
subplot(1,4,3), imshow(gbw);
set(gca,'xtick',[],'ytickMode','auto');
subplot(1,4,4), imhist(glog); title('filtered image');
grid on
```

Listing 3.4 Matlab code in `adapt2.m` to produce Fig. 3.16.

Problem 3.8 Adaptive Thresholding.
In addition to the **average** and **log** neighbourhood filtering options used by the **fspecial** function, the following options are also available.

(**filter**.1) disk (circular averaging method),
(**filter**.2) gaussian (Gaussian lowpass filter),
(**filter**.3) laplacian (approximating 2D Laplacian operator),
(**filter**.4) motion (motion filter),
(**filter**.5) prewitt (Prewitt horizontal edge-emphasising filter) [160],
(**filter**.6) sobel (Sobel horizontal edge-emphasising filter) [180],
(**filter**.7) unsharp (unsharp contrast enhancement filter).

Try each of these filters as well as the **average** and **log** filters using the adaptive thresholding method on the following images: **pout.png** and **tooth819.tif**. In each case, show all of your results and indicate which filtering method works best using adaptive thresholding (see, e.g., Fig. 3.16). ■

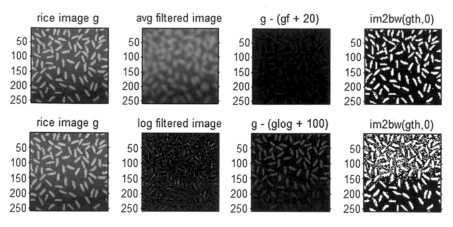

Fig. 3.16 Adaptive thresholding on rice image

3.5   Contrast Stretching

It is possible to stretch the dynamic range of an image, making image intensities occupy a larger dynamic range. As a result, there is increased contrast between dark and light areas of an image. This is accomplished using what is known as **contrast stretching**, another example of pixel intensity transformation. The transformation of each pixel value is carried out using the following method. Let

$$g = \text{input image,}$$
$$c, d = max(max(g)), min(min(g)), \text{ respectively,}$$
$$a, b = \text{new dynamic range for } g,$$
$$g(x, y) = (g(x, y) - c)\left(\frac{a - b}{c - d}\right) + a.$$

A combination of the **stretchlim** and **imadjust** functions can be used to carry out contrast stretching on an image. For example, the choice of the new dynamic range

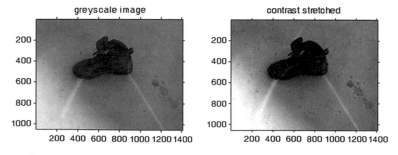

Fig. 3.17 Contrast stretching rainbow-on-shoe greyscale image

is the 10th and 90th percentile points in the cumulative distributions of pixel values.
This means that in the new dynamic range, 10% of the pixel values will be less than
the new min d and 90% of the new pixel values will be greater than the max c.

```
% Constrast−stretching experiment
clear all
close all

g = imread('rainbowshoe.jpg');   % read colour image
%g = imread('rainbow.jpg ');  % read colour image
% g = imread('tooth819.tif ');
% g = imread('tooth2.png ');
%g = imread('tooth.tif ');
%g = rgb2gray(g);
stretch = stretchlim(g,[0.03,0.97]);
h = imadjust(g,stretch,[]);
subplot(1,2,1), imshow(g);
title('rgb image');
%title('greyscale image ');
axis on
subplot(1,2,2), imshow(h);
title('contrast stretched');
axis on
```

Listing 3.5 Matlab code in contrast.m to produce Fig. 3.19.

Fig. 3.18 Contrast stretching rainbow-on-shoe rgb image

In the contrast-stretched image in Fig. 3.17, the shoe and the spots on the floor to
the right of the shoe are now more visible, *i.e.*, more distinguishable.[2]

Notice that contrast stretching is performed on an rgb image converted to
greyscale. It is possible to do contrast stretching directly on an rgb image (see,
e.g., Fig. 3.18). The changed distribution of relative frequencies of pixel values is
evident in the contrasting histograms in Fig. 3.20.

```
% Constrast−stretched dynamic ranges

g = imread('tooth819.tif');
stretch = stretchlim(g,[0.03,0.97]);
h = imadjust(g,stretch,[]);
```

---

[2]The shoe image in Fig. 3.17, showing refracted light from the windows overlooking on an upper
staircase landing in building E2, EITC, U of Manitoba, was captured by Chido Uchime with a cell
phone camera.

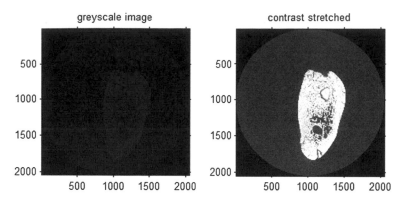

Fig. 3.19 Contrast stretching fossilised tooth image

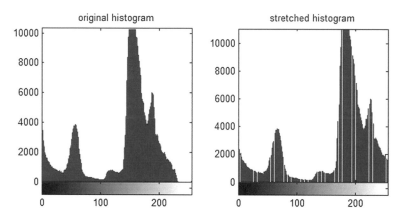

Fig. 3.20 Histogram for contrast-stretched fossil tooth image

```
subplot(1,2,1), imhist(g);
title('tooth histogram');
subplot(1,2,2), imhist(h);
title('new tooth histogram')
```

Listing 3.6 Matlab code in `histstretch.m` to produce Fig. 3.20.

The choice of the new dynamic range is image-dependent. Consider, for example, an image of a micro-slice of a 350,000 year old tooth fossil found in Serbia in 1990. This image and the corresponding contrast-stretched image is shown in Fig. 3.19. The features of the tooth image are barely visible in the original image. After choosing a new dynamic range equal to $[0.03, \ldots, 0.97]$, the features of the tooth are more sharply defined.

The contrast between the distribution of relative frequencies of pixel values in the original tooth image and contrast-stretched image can be seen by comparing the histograms in Fig. 3.21, especially for the high intensities.

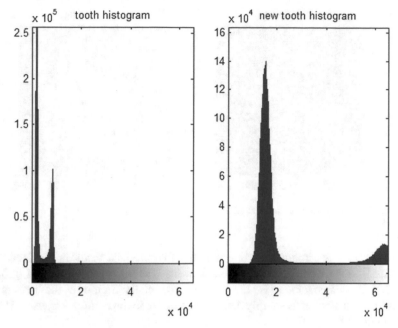

Fig. 3.21 Histogram for contrast-stretched tooth fossil image

Problem 3.9 ☕ Experiment with the tooth image **tooth819.tif** using contrast stretching. The challenge here is to find a contrast-stretched image that more sharply defines the parts of the tooth image. ∎

3.6 Histogram Matching

Contrast stretching can be generalised by extracting a target histogram distribution from an input image. The basic approach requires the user to specify a desired range for grey level intensities for histogram equalisation of an image. An example of how this is done is given in Listing 3.7 with the corresponding equalised image shown in Fig. 3.22.

```
% Histogram equalisation

g = imread('tooth819.tif'); % tooth image
ramp = 40:60;          % histogram distribution
h = histeq(g,ramp); % histogram equalisation
subplot(1,2,1), imshow(g);
title('tooth histogram');
subplot(1,2,2), imshow(h);
title('equalised image')
```

Listing 3.7 Matlab code in histeqs.m to produce Fig. 3.22.

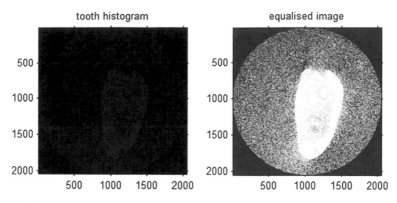

Fig. 3.22 Histogram equalisation of tooth image

After some experimentation, it was found that the best result is obtained with target histogram range equal to 40 : 60. This leads to the result shown in Fig. 3.23. Even with this narrow range for the target histogram, the regions of the tooth in the resulting image are not as sharply defined as they are in the contrast-stretched image of the tooth in Fig. 3.19.

Problem 3.10 Experiment with the tooth image **tooth819.tif** using histogram matching. The challenge here is to identify a target histogram that more sharply defines the parts of the tooth image. ■

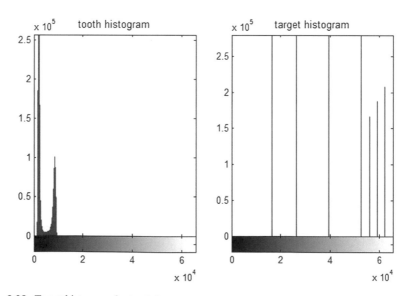

Fig. 3.23 Target histogram for tooth image

Chapter 4
Linear Filtering

This chapter introduces linear spatial filters. A **linear filter** is a time-invariant device (function, or method) that operates on a signal to modify the signal in some fashion. In our case, a linear filter is a function that has pixel (colour or non-colour) values as its input. In effect, a linear filter is a linear function on sets of pixel feature values such as colour, gradient orientation and gradient magnitude (especially, gradient magnitude which is a measure of edge pixel strength), which are either modified or exhibited in some useful fashion. For more about linear functions, see Sect. 5.1.

From an engineering perspective, one of the most famous as well as important papers on filtering is the 1953 paper by L.A. Zadeh [215]. Very relevant to the interests of computer vision are Zadeh's ideal and optimum filters. An **ideal filter** is a filter that yields a desired signal without any distortion or delay. A good example of an ideal filter is the Robinson shape filter from M. Robinson [167, Sect. 5.4, p. 159ff], useful in solving shape recognition problems in computer vision. Ideal filters are often not possible. So Zadeh introduced optimum filtering. An **optimum filter** is a filter that yields the best (close) approximation of the desired signal. Another classic paper that is important for computer visions J.F. Canny's edge filtering method introduced in [24] and elaborated in [25]. For a recent paper on scale-invariant filtering for edge detection in digital images, see S. Mahmoodi [118]. For more about linear filters in a general setting, see R.B. Holmes [84] and for linear filters in signal processing, see, especially, D.S. Broomhead, J.P. Huke and M.R. Muldoon [20, Sect. 3].

4.1 Importance of Image Filtering

Previously, the focus was on manipulating the dynamic range of images to improve, sharpen and increase the contrast of image features. In this chapter, the focus shifts from sharpening image contrast to image filtering, which is based on weighted sums of local neighbourhood pixel values. As a result, we obtain a means of removing

© Springer International Publishing AG 2017
J.F. Peters, *Foundations of Computer Vision*, Intelligent Systems
Reference Library 124, DOI 10.1007/978-3-319-52483-2_4

image noise, sharpening image features (enhancing image appearance), and achieve edge and corner detection. The study of image filtering methods has direct bearing on various approaches to image analysis, image classification and image retrieval methods. Indications of the importance of image filtering in image analysis and computer vision can be found in the following filtering approaches.

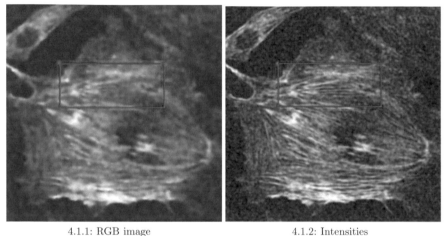

4.1.1: RGB image   4.1.2: Intensities

Fig. 4.1 Sample actin fibres image filtering

Fast elliptical filtering: Radially-uniform box splines are constructed via repeated convolution of a fixed number of box distributions by K.N. Chaudhury, A. Munoz-Barrutia and M. Unser in [28]. A sample result of the proposed filtering methods is shown in Fig. 4.1.

Gaussian Smoothing filtering: This method has two main steps given by S.S. Sya and A.S. Prihatmanto [189]: (1) a given raster image is normalized in the RGB colour so that the colour pixel intensities are in the range 0 to 225 and (2) the normalized RGB image is converted to HSV to obtain a threshold for hue, saturation and value in detecting a face that is being tracked by a Lumen social robot. This application of Gaussian filtering by Sya and Prihatmanto illustrates the high utility of the HSV colour space. Gaussian filtering is an example of non-linear filtering. For more about this, see Sect. 5.6 and for more about the Appendix B.8.

Nonlinear adaptive median filtering (AMF):   Nonlinear AMF is used by T.K. Thivakaran and R.M. Chandrasekaran in [193].

Nearness of open neighbourhoods of pixels to given pixels: This approach by S.A. Naimpally and J.F. Peters given in [128, 142, 151] and by others [73, 76, 137, 152, 162, 170]) introduces an approach to filtering an image that focuses on the proximity of an open neighbourhood of a pixel to pixels external to the neighbourhood. An **open neighbourhood of a pixel** is a set of pixels within a fixed distance of a pixel which does not include the pixels along the border of the neighbourhood. For more about open sets and neighbourhoods of points, see Appendix B.13 and B.14.

4.2 Filter Kernels

In linear spatial filters, we obtain filtered values of target pixels by means of linear combinations of pixel values in a $n \times m$ neighborhood. A **target pixel** is located at the centre of neighbourhood. A linear combination of neighbourhood pixel values is determined by a filter kernel or mask. A **filter kernel** is an array the same size as a neighbourhood, containing weights that are assigned to the pixels in the neighbourhood of a target pixel. A linear spatial filter *convolves* the kernel and neighbourhood pixel values to obtain a new target pixel value. Let w denote a $e \times 3$ kernel and let $g(x, y)$ be a target pixel in a 3×3 neighbourhood, then the new value of the target pixel is obtained as the sum of the dot products of pairs of row vectors. For a pair of $1 \times n$ vectors $\mathbf{a}, \mathbf{b}$ of the same size, the dot product is the sum of the products of the values in corresponding positions, i.e.,

$$A \cdot B = \sum_{i=1}^{n} (a_i)(b_i)$$

After selecting a 3×3 kernel, the value of a target pixel $g(x, y)$ in a 3×3 neighbourhood of a digital image g is computed as a sum of dot products, i.e.,

$$g(x, y) = \sum_{i=1}^{3} w(1, i)g(1, i) +$$

$$\sum_{i=1}^{3} w(2, i)g(2, i) +$$

$$\sum_{i=1}^{3} w(3, i)g(3, i).$$

For example, consider the following sample sum of dot products.

$$w = \begin{bmatrix} 1 & 0 & -1 \\ 2 & 0 & -2 \\ 1 & 0 & -1 \end{bmatrix}, n = \begin{bmatrix} 1 & 2 & 3 \\ 4 & 5 & 6 \\ 7 & 8 & 9 \end{bmatrix}, t = \sum_{i=1}^{3} w(i, :) \cdot n(i, :).$$

The kernel w is called a Sobel mask and is used in edge detection in images [58, Sect. 3.6.4] (see, also [Sect. 3][24, 54, 57, 120, 160]). In Matlab, the value of the target pixel $n(2, 2)$ for a given 3×3 array, can be computed using the **dot** (dot product) function. This is illustrated in Listing 4.1.

```
% Sample  target  pixel  value  using  a  3x3  filter  kernel

w = [1,0,−1;  2,0,−2;  1,0,−1];
n = [1,2,3;  4,5,6;  7,8,9];
p1 = dot(w(1,:),n(1,:))
p2 = dot(w(2,:),n(2,:))
p3 = dot(w(3,:),n(3,:))
t = (p1 + p2) + p3
```

Listing 4.1 Matlab code in `target.m` to produce a target value.

The steps in the convolution of a kernel with an image neighbourhood are summarised, next.

(**step**.1)   Define a $n \times n$ filter kernel k.

(**step**.2)   Slide the kernel onto a $n \times n$ neighbourhood n in an image g (the centre of the kernel lie on top of the neighbourhood target pixel).

(**step**.3)   Multiply the pixel values by the corresponding kernel weights. If $n(x, y)$ lies beneath $k(x, y)$, then compute $n(x, y)k(x, y)$. For the ith row $k(i,:)$ in k and ith row $n(i,:)$, compute the dot product $k(i, :) \cdot n(i, :)$. Then sum the dot products of the rows.

(**step**.4)   Replace the original target value with the new filtered value, namely, the total of the dot products from step 3.

4.3   Linear Filter Experiments

Using what is known as a function handle @, a kernel can be defined in Matlab in terms of an operation **op** such as **max, median, min** in the following manner.

$$\gg func = @(x)op(x(:));$$

The **nlfilter** (neighbourhood sliding filter) filters an image in terms of an $n \times n$ neighbourhood and kernel (e.g., the basic approach is to compute the value of a target pixel in a neighbourhood by replacing the original target pixel value with the median value of the kernel entries). For the cameraman, image consider the design of a median filter.

```
% Filter  experiment

g = imread('cameraman.tif');
subplot(1,4,1), imshow(g);
subplot(1,4,2), imhist(g);
func = @(x)median(x(:));    % set filter
%func = @(x)max(x(:));         % set filter
%func = @(x)(uint8(mean(x(:))));  % set filter
h = nlfilter(g,[3 3],func);
subplot(1,4,3), imshow(h);
title('nlfilter(g,[3 3],func)');
subplot(1,4,4), imhist(h);
```

Listing 4.2 Matlab code in `nbd.m` to produce Fig. 4.2.

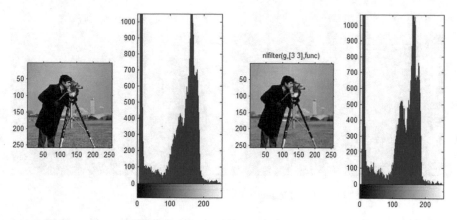

Fig. 4.2 Median filter of **cameraman.tif** with Listing 4.2

To see how the neighbourhood sliding filter works, try the following experiment shown in Listing 4.3.

```
% Experiment with function handle

g = [1,2,3; 4,5,6; 7,8,9]
func = @(x)max(x(:));        % set filter
func(g(3,:))
h = nlfilter(g,[3 3],func)
```

Listing 4.3 Matlab code in `sliding.m` to produce filtered image.

Problem 4.1 Using selected images, use the **max** and **mean** functions to define new versions of the **nlfilter** (see Listing 4.3 to see how this is done). In each case, use subplot to display the original image, histogram for the original image, filtered image, and histogram for the filtered image. ∎

4.4 Linear Convolution Filtering

The basic idea in this section is to use the **fspecial** function to construct various linear convolution filter kernels. By way of illustration, consider constructing a kernel that mimics the effect of motion blur. **Motion blur** is an apparent streaking of fast-moving objects (see, e.g., Fig. 4.3). This can be achieved either by taking pictures of fast moving objects while standing still or by continuous picture-taking while moving a camera. The motion blur effect can be varied with different choices of pixel length and counterclockwise angle of movement (the fspecial function is used set up a particular motion blur kernel). In Fig. 4.3, global motion blur is shown in a recent picture of a honey bee.

fspecial(motion,20,45)

Fig. 4.3 Lin. Convolution. filter of **honey bee** with Listing 4.4

```
% Linear convolution filtering
close all
clear all

g = imread('bee-polen.jpg');
%g = imread('kingfisher1.jpg');
subplot(1,2,1), imshow(g);
kernel = fspecial('motion',50,45); %len=20,CCangle=45
%kernel = fspecial('motion',30,45); %len=20,CCangle=45
h = imfilter(g,kernel,'symmetric');
subplot(1,2,2), imshow(h);
title('fspecial(motion,20,45)');
```

Listing 4.4 Matlab code in `convolve.m` to produce Fig. 4.3.

Problem 4.2 Apply motion blur filtering exclusively to the part of the image in Fig. 4.3 containing the honey bee. This will result in an image where only the honey bee is motion blurred.

Hint: Use a combination of the **roipoly** region-of-interest function and the **roifilt2** function (filter a region-of-interest (roi)). To see how this is done, try

> **help roifilt2**

The basic approach using the **roifilt2** function is shown in Listing 4.5 in terms of unsharp filtering one of the coins shown in Fig. 4.4. To take advantage of roi filtering, you will need to adapt the approach in Listing 4.4 in terms of mirror filtering a region-of-interest. ■

```
% Sample roi filtering

I = imread('eight.tif');
c = [222 272 300 270 221 194];
r = [21 21 75 121 121 75];
BW = roipoly(I,c,r);
H = fspecial('unsharp');
J = roifilt2(H,I,BW);
subplot(1,2,1), imshow(I); title('roi = upper right coin');
subplot(1,2,2), imshow(I); title('filtered roi = upper right coin');
```

Listing 4.5 Matlab code in `coins.m` to produce Fig. 4.4.

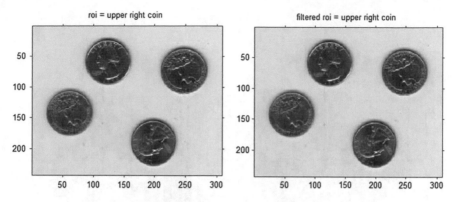

Fig. 4.4 Region-of interest filtering with Listing 4.5

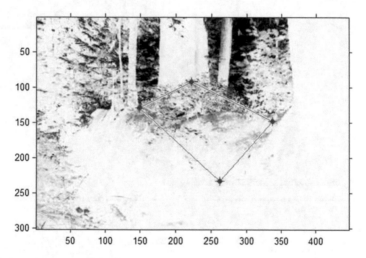

Fig. 4.5 Region-of interest selection with **roipoly**

4.5 Selecting a Region-of-Interest

With the **roipoly** or **impoly** functions, it is possible to select a polygon-shaped region-of-interest within an image, interactively. A sample roi selected using **roipoly** is shown in Fig. 4.5. This tool makes it possible to use the cursor to select the vertices of an roi. This function returns a binary image that can be used as a kernel in kernel filtering. A sample use of this tool is given in Listing 4.6. After you select an roi by clicking on the vertices of a polygon in an image, then the roi is an image in its own right. Sample histogram and bar3 (3D bar graph) for a selected roi are given in Fig. 4.6.

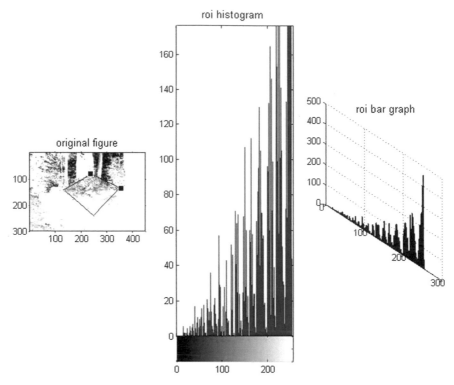

Fig. 4.6 Region-of interest selection with Listing 4.6

```
% How to use roipoly
clear all
close all
g = imread('rainbow-plant.jpg'); h = rgb2gray(g);
%g = imread('forest.tif');
%g = imread('kingfisher1.jpg');
%g = imread('bee-polen.jpg');
%g = imread('eight.tif');
%g = rgb2gray(g);
%c = [212 206 231 269 288 280 262 232 212]; % column from roitool
%r = [53 96 112 107 74 49 36 36 53]; % row from roitool
%c = [222 272 300 270 221 194]; % column from roitool
%r = [21 21 75 121 121 75]; % row from roitool
%[BW,r,c] = impoly(g);
% manually select r, c vectors, double-clicking after selection:
[BW,r,c] = roipoly(h);
B = roipoly(h,r,c);          % interactive roi selection tool
%p = imhist(g(B));
%npix = sum(B(:));
%figure,
subplot(1,3,1),imshow(g); title('original figure');
%subplot(1,3,2),imhist(g(B)); title('roi histogram');
subplot(1,3,2),bar3(h,0.25,'detached'), colormap([1 0 0;0 1 0;0 0 1]);
title('bar3(B,detached)');
subplot(1,3,3),bar(B,'stacked'),axis square; title('bar(B,stacked)');

%subplot(1,3,3),bar3(npix,'grouped'); title('bar3 graph');
%subplot(1,3,3),bar3(npix,'stacked'); title('bar3 graph');
```

Listing 4.6 Matlab code in `roitool.m` to produce Fig. 4.6.

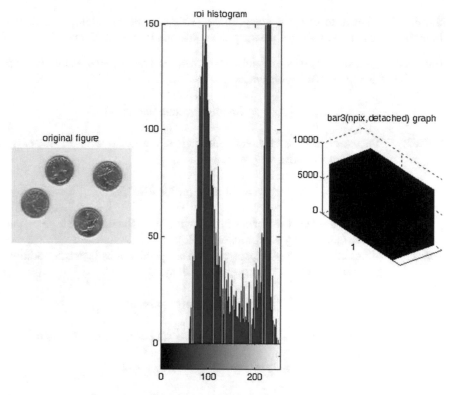

Fig. 4.7 Region-of interest selection with Listing 4.6

Problem 4.3 To solve the problem of finding the vectors c, r for the roi for an image g such as the one used in Listing 4.5, try

$$\gg [B,c,r] = roipoly(g)$$

Then rewrite the code in Listing 4.5 using **roipoly** to obtain the vectors c, r, instead of manually inserting the c, r vectors to define the desired roi. Show what happens when you select a roi containing the lower right hand coin in Fig. 4.4. A sample use of **roipoly** in terms of the **eight.tif** image is shown in Fig. 4.7. ∎

4.6 Adding Noise to Image

One of the principal applications of filtering in image enhancement is noise removal. By way of demonstrating the basic approach to removing noise from images, this section illustrates how noise can be added and then removed from an image. The

imnoise function is to create a noisy image. This is done by adding one of the following types of an image g and using mean filtering to remove the noise.

(**noise**.1)   **'gaussian'**: adds white noise with mean m (default = 0) and variance v (default = 0.01), with syntax

$$\gg g = \text{imnoise(g,'gaussian',m,v)}$$

(**noise**.2)   **'localvar'**: adds zero mean Gaussian white noise with an intensity-dependent variance, with syntax

$$\gg g = \text{imnoise(g,'localvar',V)}$$

where V is either a vector or a matrix with entries having double precision values (see Listing 4.8 for an example).

(**noise**.3)   **'poisson'**: generates Poisson noise from pixel values instead of adding artificial noise to the pixel values, with syntax

$$\gg g = \text{imnoise(g,'poisson')}$$

(**noise**.4)   **'salt & pepper'**: adds what looks like pepper noise to an image, with syntax

$$\gg g = \text{imnoise(g,'salt\&pepper', d)}$$

where d is the noise density (increasing the value of d increases the density of the pepper-effect.

(**noise**.5)   **'speckle'**: adds multiplicative noise to an image, with syntax

$$\gg g = \text{imnoise(g,'speckle', v)}$$

using the equation

$$j = g + ng,$$

where n is uniformly distributed random noise with mean 0 and variance v (default value of v is 0.04).

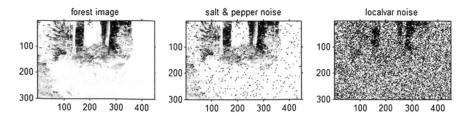

Fig. 4.8 Image with added noise with Listing 4.7

```
% Adding noise to an image

g = imread('forest.tif');
subplot(1,3,1), imshow(g); title('forest image');
nsp = imnoise(g,'salt & pepper',0.05);    %slight peppering
% nsp = imnoise(g,'salt & pepper',0.15); %increased pepper
subplot(1,3,2), imshow(nsp); title('salt & pepper noise');
g = im2double(g);
v = g(:,:);
np = imnoise(g,'localvar',v);
subplot(1,3,3), imshow(np); title('localvar noise');
```

Listing 4.7 Matlab code in noise.m to produce Fig. 4.8.

Problem 4.4 Do the following:

(**noisy**.1)   Show how to add noise to the **forest.tif** image using **gaussian** form of noise. Display the resulting image with gaussian noise.

(**noisy**.2)   Show how to add noise to the **forest.tif** image using **poisson** form of noise. Display the resulting image with poisson noise.

(**noisy**.3)   Show how to add noise to the **forest.tif** image using **speckle** form of noise. Display the resulting image with speckle noise.   ■

4.7 Mean Filtering

A mean filter is the simplest of the linear filters. This form of filtering gives equal weight to all pixels in an $n \times m$ neighbourhood, where a weight w is defined by

$$w = \frac{1}{nm}.$$

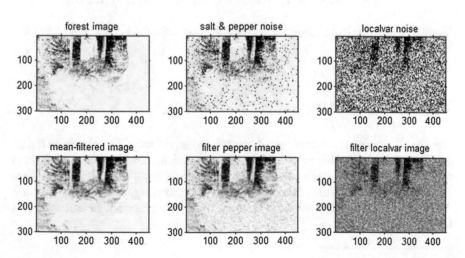

Fig. 4.9 Mean filtering an image with Listing 4.8

For example, in a 3×3 neighbourhood, $w = \frac{1}{9}$. Each pixel p value in an image is replaced by the mean value of the pixel values from the $n \times m$ neighbourhood of p. The end result of mean filtering is a smoothing of an image. Two applications of mean filtering are noise suppression and preprocessing (smoothing) an image so that subsequent operations on an image are more effective. After mean filter kernel has been set up, then the **imfilter** function is used to carry out mean filtering an image (see Listing 4.8).

```
% Mean filtering an image

g = imread('forest.tif');
subplot(2,3,1), imshow(g); title('forest image');
nsp = imnoise(g,'salt & pepper',0.05);   %slight peppering
% nsp = imnoise(g,'salt & pepper',0.15); %increased pepper
subplot(2,3,2), imshow(nsp); title('salt & pepper noise');
g = im2double(g);
v = g(:,:);
np = imnoise(g,'localvar',v);
subplot(2,3,3), imshow(np); title('localvar noise');
kernel = ones(3,3)/9;
g1 = imfilter(g,kernel);
g2 = imfilter(nsp,kernel);
g3 = imfilter(np,kernel);
subplot(2,3,4), imshow(g1); title('mean-filtered image');
subplot(2,3,5), imshow(g2); title('filter pepper image');
subplot(2,3,6), imshow(g3); title('filter localvar image');
```

Listing 4.8 Matlab code in `meanfilter.m` to produce Fig. 4.9.

Problem 4.5 Find the best mean filter for noise removal from **salt & pepper** and **localvar** noisy forms of the bf forest.tif image.
Hint: Vary the mean filter kernel.   ∎

Problem 4.6 Define an image g with the following matrix:

$$\gg g = [1,2,3,4,5; 6,7,8,9,10; 11,12,13,14,15; 16,17,18,19,20];$$

Show how the g matrix changes after mean-filtering with the kernel defined in Listing 4.8.   ∎

4.8   Median Filtering

Median filtering is more effective than mean filtering. Each pixel p value in an image is replaced by the median value of from the $n \times m$ neighbourhood of p. This form of filtering preserves image edges, while eliminating noise spikes in image pixel values. Rather than set up a filter kernel as in mean filtering, the **medfilt2** function is used to carry out median filtering in terms of a $n \times m$ image neighbourhood[1] (see Listing 4.9).

---

[1] Usually, $n = m = 3$.

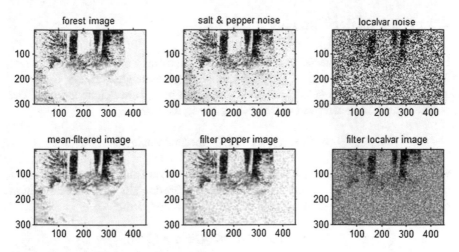

Fig. 4.10 Mean filtering an image with Listing 4.9

```
% Median filtering an image

g = imread('forest.tif');
subplot(2,3,1), imshow(g); title('forest image');
nsp = imnoise(g,'salt & pepper',0.05);   %slight peppering
% nsp = imnoise(g,'salt & pepper',0.15); %increased pepper
subplot(2,3,2), imshow(nsp); title('salt & pepper noise');
g = im2double(g);
v = g(:,:);
np = imnoise(g,'localvar',v);
subplot(2,3,3), imshow(np); title('localvar noise');
g1 = medfilt2(g,[3,3]);
g2 = medfilt2(nsp,[3,3]);
g3 = medfilt2(np,[3,3]);
subplot(2,3,4), imshow(g1); title('median-filtered image');
subplot(2,3,5), imshow(g2); title('filter pepper image');
subplot(2,3,6), imshow(g3); title('filter localvar image');
```

Listing 4.9 Matlab code in medianfilter.m to produce Fig. 4.10.

Problem 4.7 Find the best median filter for noise removal from **salt & pepper** and **localvar** noisy forms of the bf forest.tif image.
Hint: Vary the neighbourhood size. ■

Problem 4.8 Define an image g with the following matrix:

$$\gg g = [1,2,3,4,5; 6,7,8,9,10; 11,12,13,14,15; 16,17,18,19,20];$$

Show how the g matrix changes after median-filtering with the neighbourhood defined in Listing 4.9. ■

4.9 Rank Order Filtering

Median filtering is a special case of what is known as rank order filtering. A maximum order filter selects the maximum value in a given neighbourhood. Similarly, a minimum order filter selects the minimum value in a given neighbourhood. The **ordfilt2** function to carry out order filtering, using the syntax.

$$\gg \text{ filteredg} = \text{ordfilt2(g,order,domain)}$$

replaces each pixel value in image g by the orderth pixel value in an ordered set of neighbours specified by the nonzero pixel values in the domain. Using the maximum order filter on $g = $ **forest.tif** with a 5×5 neighbourhood, write

$$\gg \text{ maxfilter} = \text{ordfilt2(g,25,ones(5,5))}$$

To implement a minimum order filter on $g = $ **forest.tif** with a 5×5 neighbourhood, write

$$\gg \text{ minfilter} = \text{ordfilt2(g,1,ones(5,5))}$$

See Listing 4.9 for a sample maximum order filter with 5×5 neighbourhood.

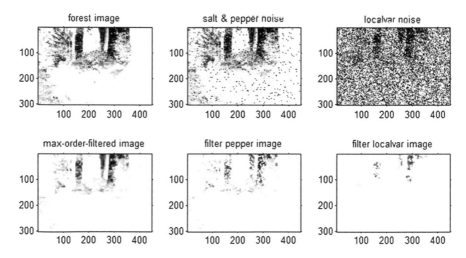

Fig. 4.11 Maximum order filtering an image with Listing 4.10

```
% Maximum order filtering an image

g = imread('forest.tif');
subplot(2,3,1), imshow(g); title('forest image');
nsp = imnoise(g,'salt & pepper',0.05);    %slight peppering
% nsp = imnoise(g,'salt & pepper',0.15); %increased pepper
subplot(2,3,2), imshow(nsp); title('salt & pepper noise');
```

```
g = im2double(g);
v = g(:,:);
np = imnoise(g,'localvar',v);
subplot(2,3,3), imshow(np); title('localvar noise');
g1 = ordfilt2(g,25,ones(5,5));
g2 = ordfilt2(nsp,25,ones(5,5));
g3 = ordfilt2(np,25,ones(5,5));
subplot(2,3,4), imshow(g1); title('max-order-filtered image');
subplot(2,3,5), imshow(g2); title('filter pepper image');
subplot(2,3,6), imshow(g3); title('filter localvar image');
```

Listing 4.10 Matlab code in `ordfilter.m` to produce Fig. 4.11.

Problem 4.9 Define an image g with the following matrix:

$$\gg g = [1,2,3,4,5; 6,7,8,9,10; 11,12,13,14,15; 16,17,18,19,20];$$

Show how the g matrix changes after maximum order filtering with a 3×3 neighbourhood rather than the 5×5 neighbourhood defined in Listing 4.8. ∎

Problem 4.10 Do the following:

(**ordfilt**.1) Find the best maximum order filter for noise removal from **salt & pepper** and **localvar** noisy forms of the bf forest.tif image.

(**ordfilt**.2) Find the best minimum order filter for noise removal from **salt & pepper** and **localvar** noisy forms of the bf forest.tif image. ∎

Problem 4.11 Do the following:

(**medfilt**.1) Using the **ordfilt2** function, give the formula for a median filter for any image $\Im$ in terms of a 5×5 and 9×9 neighbourhood.

(**medfilt**.2) Show the result of using **ordfilt2** median filter relative to a 5×5 neighbourhood for noise removal from **salt & pepper** and **localvar** noisy forms of the bf forest.tif image.

(**medfilt**.3) Show the result of using **ordfilt2** median filter relative to a 3×3 neighbourhood for noise removal from **salt & pepper** and **localvar** noisy forms of the **forest.tif** image. ∎

Problem 4.12 Use **roipoly** to select a polygon-shaped region (i.e., select a region-of-interest (**roi**)) of a noisy image. Then set up a Matlab script that performs median filtering on just the roi. The display the results of median filtering the roi for noise removal from **salt & pepper** and **localvar** noisy forms of the bf forest.tif image. ∎

4.10 Normal Distribution Filtering

Let x denote the pixel intensity of a digital image g, $\bar{x}$ the average image pixel intensity, and σ the standard deviation of the pixel intensities. The discrete form the of the normal distribution of the pixel intensities is Gaussian function $f : X \to \mathbb{R}$

defined by

$$f(x) = \frac{1}{\sigma\sqrt{2\pi}} e^{-\frac{(x-\bar{x})^2}{2\sigma^2}}.$$

To carry normal distribution filtering using the **fspecial** function, it is necessary to select an $n \times m$ kernel (usually, $n = m$) and standard deviation σ.

```
% Normal distribution filtering an image

g = imread('forest.tif');
subplot(2,3,1), imshow(g); title('forest image');
nsp = imnoise(g,'salt & pepper',0.05);    %slight peppering
% nsp = imnoise(g,'salt & pepper',0.15); %increased pepper
subplot(2,3,2), imshow(nsp); title('salt & pepper noise');
g = im2double(g);
v = g(:,:);
np = imnoise(g,'localvar',v);
subplot(2,3,3), imshow(np); title('localvar noise');
lowpass = fspecial('gaussian',[5 5], 2);
g1 = imfilter(g,lowpass);
g2 = imfilter(nsp,lowpass);
g3 = imfilter(np,lowpass);
subplot(2,3,4), imshow(g1); title('norm-filtered image');
subplot(2,3,5), imshow(g2); title('filter peppering');
subplot(2,3,6), imshow(g3); title('filter localvar noise');
```

Listing 4.11 Matlab code in gauss.m to produce Fig. 4.12.

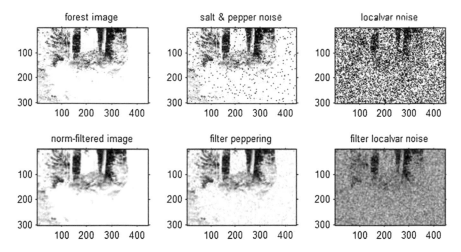

Fig. 4.12 Maximum order filtering an image with Listing 4.11

Problem 4.13 By experimenting with different values of n for a $n \times n$ kernel and σ, find a means of improving the normal distribution filtering of the **forest.tif** image and its salt-&-pepper and localvar noisy versions of the same image. ∎

Chapter 5
Edges, Lines, Corners, Gaussian Kernel and Voronoï Meshes

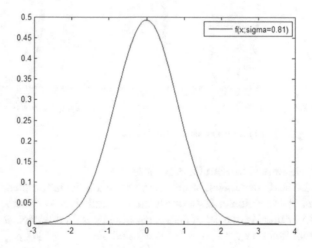

Fig. 5.1 Gaussian kernel $f(x; \sigma = 0.81)$ plot

This chapter focuses on the detection of edges, lines and corners in digital images. This chapter also introduces a number of non-linear filtering methods. A method is a **non-linear method**, provided the output of the method is not directly proportional to the input. For example, a method whose input is a real-valued variable x and whose output is x^α, $\alpha > 0$ (power of x) is non-linear.

© Springer International Publishing AG 2017
J.F. Peters, *Foundations of Computer Vision*, Intelligent Systems
Reference Library 124, DOI 10.1007/978-3-319-52483-2_5

5.1   Linear Function

Let α be a scalar such as $\alpha \in \mathbb{R}$. A function (mapping) $f : X \to Y$ is a **linear function**, provided, for $a, b \in X$, $f(a + b) = f(a) + f(b)$ (additivity property) and $f(\alpha b) = \alpha f(b)$ (homogeneity property). For example, the mapping $f(x) = x$ is linear, since $f(a + b) = a + b = f(a) + f(b)$ and $f(\alpha b) = \alpha b = \alpha f(b)$. In other words, the plot of a linear function is a straight line. By contrast, a **non-linear function** is a function that has non-linear output (a non-linear function does not satisfy the additivity and homogeneity properties of a linear function). In addition, the plot of a non-linear function is a curved line.

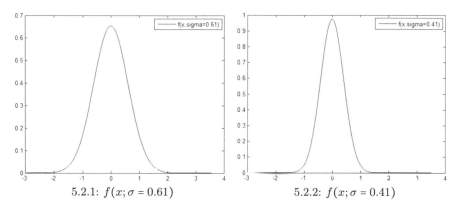

5.2.1: $f(x; \sigma = 0.61)$ 5.2.2: $f(x; \sigma = 0.41)$

Fig. 5.2   Varying widths of planar Gaussian kernel plots

Example 5.1 **Sample Gaussian Kernel Plots**.
Let $\sigma > 0$ be a scale parameter, which is the standard deviation (average distance from the mean of a set values). The expression σ^2 is called the **variance**. The average value or **mean** or **middle** of a set of data is denoted by μ. In this case, $\mu = 0$ The 1D Gaussian kernel function $f : \mathbb{R} \longrightarrow \mathbb{R}^2$ is defined by

$$f(x; \sigma) = \frac{1}{\sigma \sqrt{2\pi}} e^{-\frac{x^2}{2\sigma^2}} \text{ (\textbf{Gaussian kernel function}),}$$

is a non-linear function with a curved planar plot such as the ones shown in Fig. 5.2. In the definition of the 1D Gaussian kernel function $f(x; \sigma)$, x is a spatial parameter and σ is a scale parameter. Notice that as σ decreases (e.g., from $\sigma = 0.81$ in Fig. 5.1 to $\sigma = 0.61$ in Fig. 5.2.1 and then to $\sigma = 0.41$ in Fig. 5.2.2), the width of the Gaussian kernel plot shrinks. For this reason, σ is called a **width parameter**. For other experiments with the 1D Gaussian kernel, try the Matlab script A.24 in Appendix A.5.1. ∎

This Gaussian kernel is called a 1D (1 dimensional) kernel, since only a single spatial parameter is used to define the kernel, namely x. The name 1D Gaussian kernel comes from B.M. ter Haar Romeny in [65] .

Fig. 5.3 Corner-based Voronoï mesh on colour image

The corners in a digital image provide a good source of Voronoï mesh generators. A Voronoï mesh derived from image corners provides a segmentation of an image. Each segment in such a mesh is a convex polygon. Recall that the straight line segment between any pair points in a *convex polygon* belongs to the polygon. The motivation for considering this form of image segmentation is that mesh polygons provide a means of

1° **Image segmentation**. Voronoï meshes provide a straightforward means of partitioning an image into non-intersecting convex polygons that facilitate image and scene analysis as well as image understanding.   ∎

2° **Object recognition**. Object corners determine distinctive (recognizable) convex submeshes that can be recognized and compared.   ∎

3° **Pattern recognition**. The arrangement of corner-based convex image submeshes constitute image patterns that can be recognized and compared. See Sect. 5.13 for more about this.   ∎

Example 5.2 **Segmenting a Colour Image**. A sample segmentation of a colour image is shown in Fig. 5.3. In this image, a Voronoï mesh is derived from some of the corners (shown with ∗) in this image. Notice, for example, the rear wheel is mostly covered by a 7-sided convex polygon. For more about this, see Sect. 5.14.   ∎

```
% Edge detection filtering an image with logical not
clc, clear all, close all
g = imread('circuit.tif');
gz = edge(g,'zerocross');
subplot(1,3,1),imshow(g); title('circuit.tif');
```

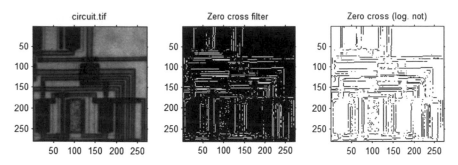

Fig. 5.4 Logical not versus non-logical not image with Listing 5.1

```
subplot(1,3,2),imshow(gz); title('Zero cross filter');
subplot(1,3,3),imshow(~gz); title('Zero cross (log. not)');
```

Listing 5.1 Matlab code in family mylogicalnot.m to produce Fig. 5.4.

5.2 Edge Detection

Quite a number of edge (and line) detection methods have been proposed. Prominent among these filtering methods are those proposed by L.G. Roberts [166], J.M.S. Prewitt [160], I. Sobel [180, 181] and the more recent Laplacian and Zero cross filtering methods. The Laplacian and Zero cross filters effect remarkable improvements over the earlier edge detection methods. This can be seen in Figs. 5.5 and 5.6.

```
% Edge detection filtering an image
clc, clear all, close all

%g = rgb2gray(imread('bee-polen.jpg'));
g = imread('circuit.tif');
gr = edge(g,'roberts');
gp = edge(g,'prewitt');
gs = edge(g,'sobel');
gl = edge(g,'log');
gz = edge(g,'zerocross');
subplot(2,3,1),imshow(g); title('circuit.tif');
subplot(2,3,2),imshow(~gr); title('Roberts filter');
subplot(2,3,3),imshow(~gp); title('Prewitt filter');
subplot(2,3,4),imshow(~gs); title('Sobel filter');
subplot(2,3,5),imshow(~gl); title('Laplacian filter');
subplot(2,3,6),imshow(~gz); title('Zero cross filter');
```

Listing 5.2 Matlab code in edges.m to produce Fig. 5.5.

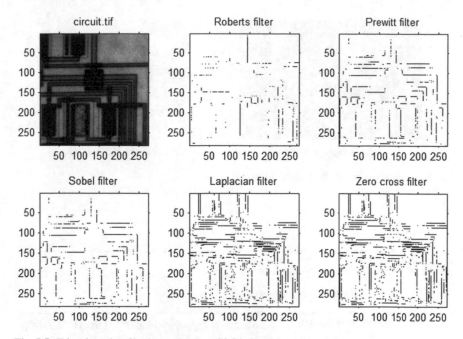

Fig. 5.5 Edge detection filtering an image with Listing 5.2

Fig. 5.6 Manitoba Dragonfly

Listing 5.2 illustrates the application of each of the common edge-filtering methods. Notice that the Matlab **logical not** operator. To experiment with logical not, try

5.7.1: Dragonfly Edges   5.7.2: Logical Not Edges

Fig. 5.7   Edges in a Manitoba Dragonfly image

```
% Sample logical not operation on an array
clc, clear all, close all

g = [1 1 1 1 0 0 0 0]
notg = ~g
```

Listing 5.3   from [11110000].]Use `logicalnot.m` to produce [00001111] from [11110000].

The approach in Script 5.3 can be used to reverse the appearance of each filtered image from white edges on black background to black edges on white background (see, e.g., Figs. 5.4 and 5.5, for edges extracted from Figs. 5.6 and 5.7).

The basic approach in edge detection filters is to **convolve** the $n \times n$ neighbourhood of each pixel in an image with an $n \times n$ mask (or filter kernel), where n is usually an odd integer. The term **convolve** means *fold (roll) together*. For a real-life example of convolving, see http://www.youtube.com/watch?v=7EYAUazLI9k.

For example, the Prewitt and Sobel edge filters are used to convolve each 3×3 image neighbourhood (also called an **8-neighbourhood**) with an edge filter. The notion of an **8-neighbourhood** of a pixel comes from A. Rosenfeld [170]. A **Rosenfeld 8-neighbourhood** is an square array of 8 pixels surrounding a center pixel. Prewitt and Sobel edge filters are a pair of 3×3 masks (one mask representing the pixel gradient in the x-direction and a second mask for the pixel gradient in the y-direction).

Matlab favours the horizontal direction, filtering an image with only the mask representing the gradient of a pixel in the x-direction. To see examples of masks, try

```
% Sample edge filter masks
clc, clear all, close all

mPrewitt = fspecial('prewitt')
mSobel = fspecial('sobel')
mLaplace = fspecial('laplacian')
```

Listing 5.4   `masks.m` to produce sample masks.

The masks available with the Matlab **fspecial** function favour the horizontal direction. For example, the Prewitt 3×3 mask is defined by

$$mPrewitt = \begin{bmatrix} 1 & 1 & 1 \\ 0 & 0 & 0 \\ -1 & -1 & -1 \end{bmatrix}.$$

The Laplacian edge filter $L(x, y)$ is a 2D isotropic[1] measure of the 2nd derivative of an image g with pixel intensities $g(x, y)$ defined by

$$L(x, y) = \frac{\partial^2 g}{\partial x^2} + \frac{\partial^2 g}{\partial y^2}.$$

A commonly used Laplacian mask is defined by the following 3×3 array.

$$Laplacian = \begin{bmatrix} 0 & -1 & 0 \\ -1 & 4 & -1 \\ 0 & -1 & 0 \end{bmatrix}.$$

For detailed explanations for the Laplacian, Laplacian of Gaussian, LoG, and Marr edge filters, see http://homepages.inf.ed.ac.uk/rbf/HIPR2/log.htm.

Problem 5.3 Using image enhancement methods from Chap. 3, preprocess the **dragonfly2.jpg** image and create a new image (call it **dragonfly2.jpg**). Find the best preprocessing method to do edge detection filtering to obtain an image similar to the one shown in Fig. 5.7. Display both the binary (black and white) and the (black on white or logical not) edge image as shown in Fig. 5.7. In addition, type

≫ **help edge**

and experiment with different choices of the **thresh** and **sigma** (standard deviation parameters for the Laplacian of the Gaussian (normal distribution) filtering method, using

≫ **gl = edge(g,′ log′, thresh, sigma)**

Hint: Use **im2double** on an input image. Also, edge detection methods operate on greyscale (not colour) images. ■

5.3 Double Precision Laplacian Filter

```
% Normal distribution filtering an image

g = imread('circuit.tif');
gr = edge(g,'roberts');
gp = edge(g,'prewitt');
gs = edge(g,'sobel');
subplot(2,3,1),imshow(g); title('circuit.tif');
```

---

[1]**Isotropic** means not direction sensitive, having the same magnitude or properties when measured in different directions.

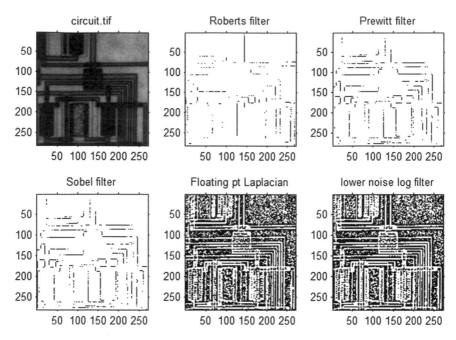

Fig. 5.8 Double precision Laplace filtering with Listing 5.5

```
subplot(2,3,2),imshow(~gr); title('Roberts filter');
subplot(2,3,3),imshow(~gp); title('Prewitt filter');
%
subplot(2,3,4),imshow(~gs); title('Sobel filter');
k = fspecial('laplacian'); % create laplacian filter
glap = imfilter(double(g),k,'symmetric'); % laplacian edges
glap = medfilt2(glap,[3 3]);
subplot(2,3,5),imshow(glap); title('Floating pt Laplacian');
%
k = fspecial('log'); % create laplacian filter
glog = imfilter(double(g),k,'symmetric'); % laplacian edges
glog = medfilt2(glog,[3 3]);
subplot(2,3,6),imshow(glog); title('lower noise log filter');
```

Listing 5.5 Matlab code in `laplace.m` to produce Fig. 5.8.

5.4 Enhancing Digital Image Edges

It has been observed by T. Lindeberg that the concept of an image edge is *only what we define it to be* [113, p. 118]. The earlier attempts at edge detection by Roberts, Prewitt and Sobel focused on the detection of points where the first order edge gradient is high. Starting in the mid-1960s, jumps in brightness values are the kinds of edges detected by Roberts [166]. Derivative approximations were computed by R.M. Haralick either directly from pixel values or from local least squares fit [69].

First order edge filters such as Roberts, Prewitt, and Sobel filters are commonly used as a step toward digital image segmentation. For images where edge-sharpening is important, then second order image filtering methods are used.

A common method of edge-sharpening as a step toward image enhancement is the second order Laplacian filter. For a pixel $g(x, y)$ in an image g, the non-discrete form of the Laplacian filter $\nabla^2 g(x, y)$ is defined by

$$\nabla^2 g(x, y) = \frac{\partial^2 g}{\partial x^2} + \frac{\partial^2 g}{\partial y^2}$$

For implementation purposes, the discrete form of the Laplacian filter $\nabla^2 g(x, y)$ is defined by

$$\nabla^2 g(x, y) = f(x + 1, y) + f(x - 1, y) - 4f(x, y) + f(x, y + 1) + f(x, y - 1)$$

The basic approach in a second order derivative approach to enhancing an image is to subtract a filtered image from the original image, i.e., in terms of a pixel value $g(x, y)$, compute

$$g(x, y) = g(x, y) - \nabla^2 g(x, y).$$

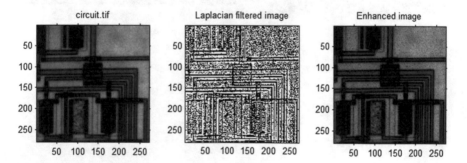

Fig. 5.9 Laplace image enhancement with Listing 5.6

```
% Laplacian edge-enhanced image

%A=imread('circuit.tif');
%g = rgb2gray(imread('Snap-04a.tif'));
g = imread('Snap-04a.tif');
k=fspecial('laplacian',1); %Generate Laplacian filter
h2=imfilter(g,k); %Filter image with Laplacian kernel
ge=imsubtract(g,h2); %Subtract Laplacian from original.
subplot(1,3,1), imshow(g); title('Snap-04a.tif fossil');
subplot(1,3,2), imagesc(~h2);
title('Laplacian filtered image'); axis image;
```

```
subplot(1,3,3), imshow(ge); title('Enhanced image');
```

Listing 5.6 Matlab code in enhance1.m to produce Fig. 5.9.

In Matlab, the second order Laplacian filter has an optional shape parameter α, which controls the shape of the Laplacian (e.g., see Listing 5.6, where $\alpha = 1$ (high incidence of edges)). The original image in Fig. 5.18 is a recent **Snap-04a.tif** image of an ostracod fossil from MYA (found in an ostracod colony trapped in amethyst crystal from Brasil). In this image, there is a very high incidence edges and ridges, handled with a high α value. Similarly, in the **circuit.tif** in Fig. 5.9, there are a high incidence of lines, again warranting a high α value to achieve image enhancement.

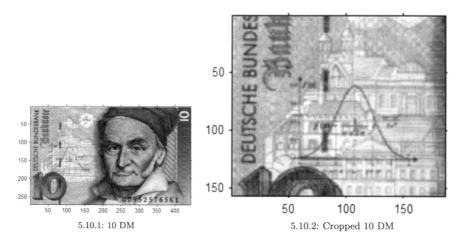

5.10.1: 10 DM   5.10.2: Cropped 10 DM

Fig. 5.10 1D Gaussian kernel experiments

5.5 Gaussian Kernel

It was Carl Friedrich Gauss (1777–1895) who introduced the kernel (or normal distribution) function named after him. Let x, y be linearly independent, random real-valued variables with a standard deviation σ and mean μ. The goal is to exhibit the distribution of either the x values by themselves or the combined x, y values around the origin with $\mu = 0$ for each experiment. The **width** $\sigma > 0$ of a set of x or x, y values is called the **standard deviation** (average distance from the middle of a set of data) and σ^2 is called the **variance**. Typically, the plot of a set of sample values with a normal distribution has a bell shaped curve (also called **normal curve** arranged around the middle of the values. The now famous example of the Gaussian kernel plot appears on 10 Deutsch mark (10 DM) note shown in Fig. 5.10.1. A cropped

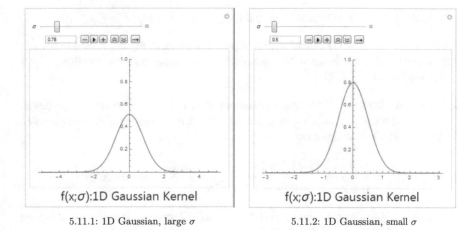

5.11.1: 1D Gaussian, large σ 5.11.2: 1D Gaussian, small σ

Fig. 5.11 1D Gaussian kernel experiments

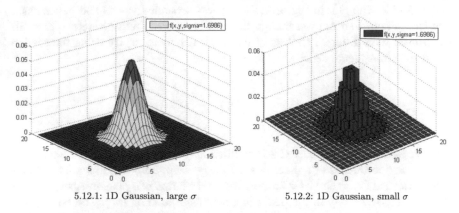

5.12.1: 1D Gaussian, large σ 5.12.2: 1D Gaussian, small σ

Fig. 5.12 2D Gaussian kernel experiments

version of the 10 DM images is shown in Fig. 5.10.2. A very good overview of the evolution of the Gaussian kernel is given by S. Stahl [186].

When all negative x or x, y values are represented are represented by their absolute values, then the Gaussian of the values is called a folded normal distribution (see, for example, F.C. Leone, L.S. Nelson and R.B. Nottingham [107]).

There are two forms of the Gaussian kernel to consider.

1D Gaussian kernel If we consider only sample values of x with standard deviation σ and mean $\mu = 0$, then the 1D Gaussian kernel function (denoted by $f(x; \sigma)$) is defined by

$$f(x; \sigma) = \frac{1}{\sigma\sqrt{2\pi}}e^{-\frac{(x-0)^2}{2\sigma^2}} = \frac{1}{\sigma\sqrt{2\pi}}e^{-\frac{x^2}{2\sigma^2}} \text{ (\textbf{1D Guassian kernel})}.$$

It is the plot of the 1D kernel that is displayed on the 10 DM in Fig. 5.10.2.

Example 5.4 Sample 1D Gaussian kernel plots are given in Fig. 5.11. To experiment with different choices of the width parameter σ, try using the Mathematica script 1 in Appendix A.5.2.   ■

2D Gaussian kernel   If we consider sample values of x and y with standard deviation σ and means $\mu_x = 0$, $\mu_y = 0$, then the 2D Gaussian kernel function (denoted by $f(x, y; \sigma)$) is defined by

$$f(x; \sigma) = \frac{1}{\sigma\sqrt{2\pi}}e^{-\frac{\left[(x-0)^2+(y-0)^2\right]}{2\sigma^2}} = \frac{1}{\sigma\sqrt{2\pi}}e^{-\frac{x^2 n+y^2}{2\sigma^2}} \text{ (2D Guassian kernel)}.$$

Example 5.5 Sample continuous and discrete 2D Gaussian kernel plots are given in Fig. 5.12. A **discrete plot** is derived from discrete values. By **discrete**, we mean that distinct, separated. In this example, discrete values are used to obtain the plot in Fig. 5.12.2. The plot in Fig. 5.12.1 is for less separated values and hence has a continuous appearance, even though the plot is derived from discrete values. To experiment with different choices of the width parameter σ, try using the Matlab script A.25 in Appendix A.5.3.   ■

5.6   Gaussian Filter

This section briefly introduces Gaussian filtering (smoothing) of digital images. Let x, y be the coordinates of a pixel in a 2D image Img, $Img(x, y)$ the intensity of a pixel located at (x, y) and let σ be the standard deviation of a pixel intensity relative to the average intensity of the pixels in a neighbourhood of Img. The assumption made here is that σ is the standard deviation of a probability distribution of the pixel intensities in an image neighbourhood. The Gaussian filter (smoothing) 2D function $G(x, y; \sigma)$ is defined by

$$G(x, y; \sigma) = \frac{1}{\sigma\sqrt{2\pi}}e^{-\frac{x^2+y^2}{2\sigma^2}} \text{ (Filtered value), or,}$$

$$G(x, y; \sigma) = e^{-\frac{x^2+y^2}{2\sigma^2}} \text{ (Simplified filtered value). Next,}$$

$Img(x, y) := G(x, y; \sigma)$ $(G(x, y; \sigma)$ **replaces pixel intensity** $Img(x, y))$.

The basic approach in Gaussian filtering an image is to assign each pixel intensity in a selected image neighbourhood with the filtered value $G(x, y, \sigma)$. M. Sonka, V. Hlavac and R. Boyle [184, Sect. 5.3.3, p. 139] observe that σ is proportional to the size of the neighbourhood on which the Gaussian filter operates (see, e.g., Fig. 5.14 for Gaussian filtering of the cropped train image in Fig. 5.13).

Fig. 5.13 Sample cropped CN train image

5.14.1: smooth   image   5.14.2: smooth   image   5.14.3: smooth   image
over 5×5 subimages, $\sigma =$ over 3×3 subimages, $\sigma =$ over 2×2 subimages, $\sigma =$
2   1.2   0.8

Fig. 5.14 Gaussian filtering a cropped image

Example 5.6 **Gausian Filtering in Smoothing an Image**.
To experiment with image smoothing using Gaussian filtering, try script A.26 in
Appendix A.5.4 (Figs. 5.15, 5.16 and 5.17).   ∎

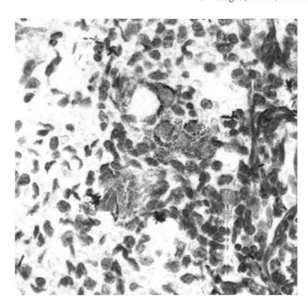

Fig. 5.15 Tissue sample image

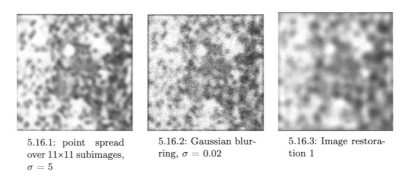

5.16.1: point spread over 11×11 subimages, $\sigma = 5$ 5.16.2: Gaussian blurring, $\sigma = 0.02$ 5.16.3: Image restoration 1

Fig. 5.16 Example 1: Restoring a noisy, blurred image

5.7   Gaussian Filter and Image Restoration

Example 5.7 **Gausian Filtering in Smoothing and Blurring Square Subimages**. To experiment with image restoration and Gaussian filtering, try script A.27 in Appendix A.5.5 (Fig. 5.18). ■

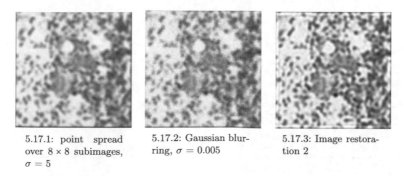

5.17.1: point spread over 8 × 8 subimages, $\sigma = 5$

5.17.2: Gaussian blurring, $\sigma = 0.005$

5.17.3: Image restoration 2

Fig. 5.17 Example 2: Restoring a noisy, blurred image

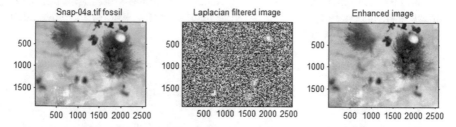

Fig. 5.18 Fossil image with Listing 5.6

5.8   Laplace of Gaussian Filter Image Enhancement

An alternative to a simple second order Laplacian filter, is the second order Laplacian of a Gaussian filter. This is implemented in Matlab using the **log** option with **fspecial** function.

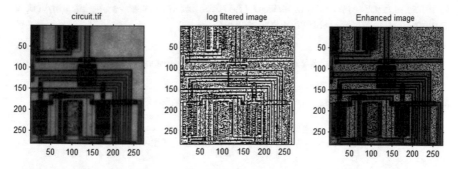

Fig. 5.19 2nd Order Laplace image enhancement with Listing 5.7

```
% Rotationally symmetric Laplacian of Gaussian enhanced image

g=imread('circuit.tif');
%g = rgb2gray(imread('Snap-04a.tif '));
%g = imread('Snap-04a.tif ');
k=fspecial('log',[3 3],0.2); %Generate Laplacian filter
h2=imfilter(g,k); %Filter image with Laplacian kernel
ge=imsubtract(g,h2); %Subtract Laplacian from original.
subplot(1,3,1), imshow(g); title('circuit.tif');
subplot(1,3,2), imagesc(~h2);
title('log filtered image'); axis image;
subplot(1,3,3), imshow(ge); title('Enhanced image');
```

Listing 5.7 Matlab code in `logsym.m` to produce Fig. 5.19.

Problem 5.8 Eliminate the salt-n-pepper effect of the second-order Laplacian image enhancement shown in Fig. 5.19. Show your results for the **circuit.tif** and one other image of your own choosing.

5.9   Zero-Cross Edge Filter Image Enhancement

In most cases, the most effective of the second order filter approaches to image enhancement stems from an application of the R. Haralick zero-crossing filtering method (see, e.g., the zero-crossing enhancement of the **circuit.tif** image in Fig. 5.20).

In a discrete matrix representation of a digital image, there are usually jumps in the brightness values, if the brightness values are different. To interpret jumps in brightness values relative to local extrema of derivatives, it is helpful to assume that pixel values come from a sampling of a real-valued function of a digital image g that is a bounded and connected subset of the plane $\mathbb{R}^2$. Then jumps in derivative values indicates points of high first derivative of g or to points of relative extrema in the second derivative of g [69, p. 58]. For this reason, Haralick viewed edge detection as *fitting a function to sample values*. The directional derivative of g at point (x, y) is defined in terms of a direction angle α by

$$g_\alpha'(x, y) = \frac{\partial g}{\partial x} \sin \alpha + \frac{\partial g}{\partial y} \cos \alpha,$$

and the second directional derivative of g at point (x, y) is then

$$g_\alpha''(x, y) = \frac{\partial^2 g}{\partial x^2} \sin^2 \alpha + \frac{2\partial^2 g}{\partial xy} \sin \alpha \cos \alpha + \frac{\partial^2 g}{\partial y^2} \cos^2 \alpha$$

Assuming that g is a cubic polynomial in x and y, then the gradient and gradient direction of g can be estimated in terms of α at the center of a neighbourhood used to estimate the value of g. In an $n \times n$ neighbourhood of g, the value of $g(x, y)$ is computed as a cubic in a linear combination of the form

$$g(x, y) = k_1 + k_2 x + k_3 y + k_4 x^2 + \cdots + k_{10} y^3.$$

The angle α is defined to be

$$\sin \alpha = \frac{k_2}{\sqrt{k_2^2 + k_3^2}},$$

$$\cos \alpha = \frac{k_3}{\sqrt{k_2^2 + k_3^2}}.$$

Then the second derivative of $g(x, y)$ in the direction α is approximated by

$$g_\alpha''(x, y) = 6[k_7 sin^3\alpha + k_8 sin^2\alpha$$
$$+ k_9 sin\alpha\, cos^2\alpha + k_{10} cos^3\alpha]\rho$$
$$+ 2[k_4 sin^2\alpha + k_5 sin\alpha\, cos\alpha + k_6 cos^2\alpha].$$

So when is a pixel marked as an edge pixel in the zero-crossing approach to edge detection? Haralick points to changes in the second and first derivatives as a zero-crossing indicator. That is, if, for some ρ, $|\rho| < \rho_0$, where ρ_0 is slightly smaller than the length of the side of a pixel, and

$$g_\alpha''(\rho) < 0 \text{ or } g_\alpha''(\rho) = 0, \text{ and } g_\alpha'(\rho) \neq 0,$$

then a negatively sloped zero crossing of the estimated second derivative has been found and the target neighbourhood pixel is marked as an edge pixel.

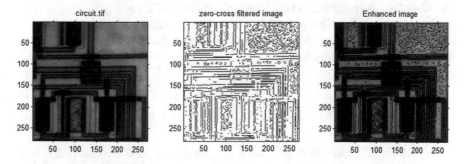

Fig. 5.20 Laplace image enhancement with Listing 5.8

```
% Zero−crossing image enhancement

%g=imread('circuit.tif');
g = rgb2gray(imread('Snap−04a.tif'));
%g = imread('Snap−04a.tif');
```

```
g = im2double(g);
h2=edge(g,'zerocross',0,'nothinning');
h2 = im2double(h2);
ge=imsubtract(g,h2); %Subtract Laplacian from original.
subplot(1,3,1), imshow(g); title('Snap-04a.tif');
subplot(1,3,2), imagesc(~h2);
title('zero-cross filtered image'); axis image;
subplot(1,3,3), imshow(ge); title('Enhanced image');
```

Listing 5.8 Matlab code in `zerox.m` to produce Fig. 5.20.

The Matlab **edge** function implementation has two optional parameters, namely, **thresh** and filter **h**. By choosing $h = 0$, the output image has closed contours and by choosing **no thinning** as the filtering method, the edges in the output image are not thinned. Notice that the edge-detection image in Fig. 5.20 is superior to the edge-detection image in Fig. 5.19 or in 5.9. Why? For some images such as the **Snap_04a.tif** image, the zero-crossing method does not work well. Evidence of this can be seen in Fig. 5.21.

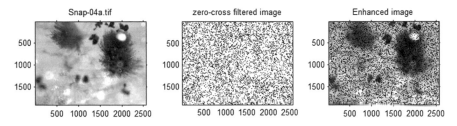

Fig. 5.21 Laplace image enhancement with Listing 5.8

Problem 5.9 Try other filters besides **nothinning** (used in Listing 5.8) and look for the best zero-crossing filter image enhancement of the **dragonfly2.jpg** and as well as one other image of your own choosing. For each of the two images, give both the binary and logical not edge image.

5.10 Anisotropy Versus Isotropy in Edge Detection

The term **isotropic** means having the same magnitude or properties when measured in different directions. The isotropic edge detection approach is direction-independent. Isotropic edge detection was proposed by D. Marr and E. Hildreth [120], an approach that offers simplicity and uniformity at the expense of smoothing across edges. Gaussian smoothing of edges was proposed by A.P. Witkin [212] by convolving an image with a Gaussian kernel. Let $I_o(x, y)$ denote an original image, $I(x, y, t)$ a derived image and $G(x, y, t)$ a Gaussian kernel with variance t. Then the original image is convolved with the Gaussian kernel in the following way.

$$t \in [0, \infty], \text{ continuum of scales } t \geq 0,$$

$$G(x, y; t) = \frac{1}{2\pi t} e^{-\frac{x^2+y^2}{2t}},$$

$$I(x, y, t) = I_o(x, y) * G(x, y; t),$$

where the convolution is performed only over the variables x, y and the scale parameter t after the semicolon specifies the scale level (t is the variance of the Gaussian filter $G(x, y; t)$). At $t = 0$, the scale space representation is the original image. An increasing number of image details are removed as t increases, i.e., image smoothing increases as t increases. Image details smaller than the $\sqrt{t}$ are removed from an image. The **fspecial** function is used to achieve Gaussian smoothing an image.

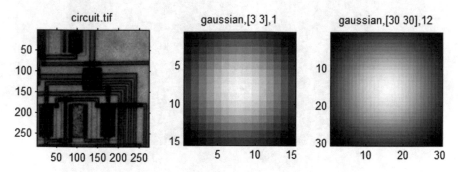

Fig. 5.22   Gaussian smoothing **circuit.tif** with Listing 5.9

```
% Gaussian image smoothing

g = imread('circuit.tif');
subplot(2,3,1),imshow(g); title('circuit.tif');
g1 = fspecial('gaussian',[15 15],6);
g2 = fspecial('gaussian',[30 30],12);
subplot(2,3,2),imagesc(g1); title('gaussian,[3 3],1');
axis image;
subplot(2,3,3),imagesc(g2); title('gaussian,[30 30],12');
axis image;
```

Listing 5.9   Matlab code in `iostropy.m` to produce Fig. 5.22.

An alternative to the isotropic edge detection is anisotropic diffusion, proposed by P. Pierona and J. Malik [139] (see, also, [152]). The term **anisotropic** means having different magnitude or properties when measured in different directions. In other words, the anisotropic approach to edge detection is direction-dependent (Fig. 5.23).

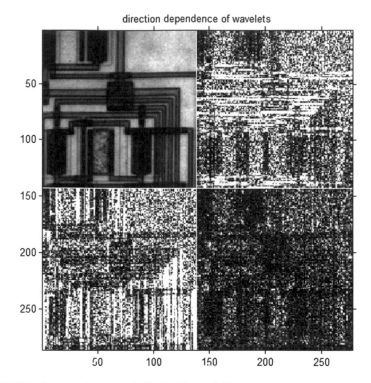

Fig. 5.23 Wavelet smoothing **circuit.tif** with Listing 5.10

```
% This function uses functions:
%     wavefast, wavecut, wavedisplay, waveback

g = imread('circuit.tif');
% Isolate edges of picture using the 2D wavelet transform
[c, s] = wavefast(g, 1, 'sym4');
figure,wavedisplay(c,s,-6);
title('direction dependence of wavelets');
% Zero the approximation coefficients
% [nc, y] = wavecut('a', c, s);
% Compute the absolute value of the inverse
% edges = abs(waveback(nc, s, 'sym4'));
% Display before and after images
% figure;
% subplot(1,2,1), imshow(g), title('Original Image');
% subplot(1,2,2), imshow(mat2gray(edges))
```

Listing 5.10 Matlab code in `directions.m` to produce Fig. 5.23.

Next, consider enhancing the **circuit.tif** image using the edges found using the wavelets to detect edges. A preliminary result of wavelet image enhancement is shown in Fig. 5.24. Two things can be observed. First, the wavelet form of edge detection is less effective than Haralick's zero crossing edge detection method. Second, at this very preliminary stage, it can be observed that the wavelet edge detection method does not result in satisfactory image enhancement. More work needs to be done before one can evaluate the image enhancement potential of the wavelet edge detection method (see Problem 5.10).

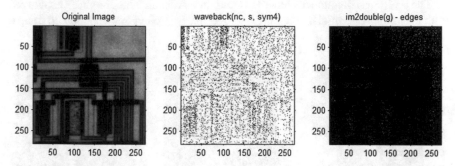

Fig. 5.24 Image Enhancement **circuit.tif** with Listing 5.11

```
% This function uses wavefast, wavecut, waveback

g = imread('circuit.tif');
% Isolate edges using 2D wavelet transform
[c, s] = wavefast(g, 1, 'sym4');
% Zero the approximation coefficients
[nc, y] = wavecut('a', c, s);
% Compute the absoluate value of the inverse
edges = abs(waveback(nc, s, 'sym4'));
% Display before and after images
figure;
subplot(1,3,1), imshow(g), title('Original Image');
subplot(1,3,2), imshow(edges);
title('waveback(nc, s, sym4)');
g = im2double(g); h = g - edges;
subplot(1,3,3), imshow(h);
title('im2double(g) - edges');
```

Listing 5.11 Matlab code in `directions2.m` to produce Fig. 5.24.

Problem 5.10 Experiment with enhancing images using the wavelet detection method with 3 other images besides the **circuit.tif** image. For example, use wavelets to detect edge and to perform image enhancement with the **Snap_4a.tif** and the **blocks.jpg** images.

5.11   Detecting Edges and Lines in Digital Images

This section briefly presents J.F. Canny's approach[2] to edge detection based on his M.Sc. thesis completed in 1983 at the MIT Artificial Intelligence Laboratory [24]. The term **edge direction** means the direction of the tangent to a contour that an edge

---

[2]See http://www.cs.berkeley.edu/~jfc/papers/grouped.html.

defines in 2D space. Canny introduced a mask to detect edge direction by convolving a linear edge detection function aligned normal to the edge direction of a projection with a projection function parallel to the edge direction.

The projection function of choice is a Gaussian. After an image has been convolved with a symmetric Gaussian, then the log function is applied to the smoothed image.

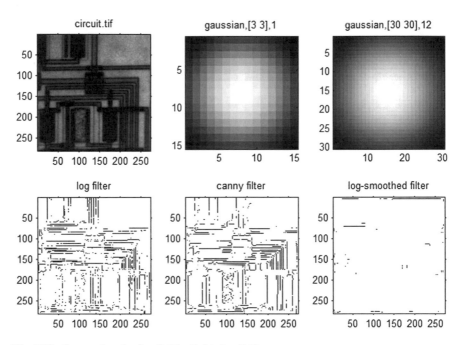

Fig. 5.25 Canny edges in **circuit.tif** with Listing 5.12

```
% Canny edge detection
clc, close all, clear all

g = imread('circuit.tif');
subplot(2,3,1),imshow(g); title('circuit.tif');
g1 = fspecial('gaussian',[15 15],6);
g2 = fspecial('gaussian',[30 30],12);
subplot(2,3,2),imagesc(g1); title('gaussian,[15 15],6');
axis image;
subplot(2,3,3),imagesc(g2); title('gaussian,[30 30],12');
axis image;
[bw,thresh] = edge(g,'log');
subplot(2,3,4),imshow(~bw,[]);title('log filter');
[bw,thresh] = edge(g,'canny');
subplot(2,3,5),imshow(~bw,[]);title('canny filter');
[bw,thresh] = edge(imfilter(g,g1),'log');
subplot(2,3,6),imshow(~bw,[]);title('log-smoothed filter');
```

Listing 5.12 Matlab code in logsmooth.m to produce Fig. 5.25.

In another round of experiments, the LoG (Laplacian of Gaussian) edge detection method is computed with Gaussian smoothing of **circuit.tif** using a 3×3 kernel with a standard deviation 1.5. This approach to edge detection does result in some improvement in Canny edge detection applied to the original image. This can be seen in the increased number of horizontal and vertical edges in log filtering $g0$. This is shown in Fig. 5.26. Also, it was found that increasing the size of the kernel decreases LoG filter performance (see Problem 5.11).

```
% Log of Gaussian edge detection

g = imread('circuit.tif');
%subplot(2,3,1),imshow(g); title('circuit.tif ');
g0 = fspecial('gaussian',[3  3],1.5);
subplot(2,3,1),imagesc(g1); title('g0=gaussian,[3  3],1.5');
axis image;
g1 = fspecial('gaussian',[15  15],7.5);
g2 = fspecial('gaussian',[31  31],15.5);
subplot(2,3,2),imagesc(g1); title('g1=gaussian,[15 15],7.5');
axis image;
subplot(2,3,3),imagesc(g2); title('g2=gaussian,[31 31],15.5');
axis image;
[bw,thresh] = edge(g,'log');
subplot(2,3,4),imshow(~bw,[]);title('log filter g');
[bw,thresh] = edge(g,'canny');
subplot(2,3,5),imshow(~bw,[]);title('canny filter g');
[bw,thresh] = edge(imfilter(g,g0),'log');
subplot(2,3,6),imshow(~bw,[]);title('log-smoothed filter g0');
```

Listing 5.13 Matlab code in `logsmooth2.m` to produce Fig. 5.26.

Problem 5.11 Try LoG filtering $g1$ and $g2$ Listing 5.13 as well as other Gaussian smoothing of the **dragonfly2.jpg** image and look for choices of kernel size and standard deviation that lead to an improvement over Canny filtering the original image. Notice that the LoG filter method has a **thresh** option (all edges not stronger than thresh are ignored) and a **sigma** option (standard deviation of the LoG filter (Laplacian of the Gaussian method). Experiment with these LoG optional parameters to obtain an improvement over the result in Fig. 5.26. In addition, notice that the Canny edge filter has an optional two element **thresh** parameter (the first element in the Canny thresh parameter is a low threshold and the second parameter is a high threshold). Experiment with the edge Canny thresh parameter to improve on the result given in Fig. 5.26.

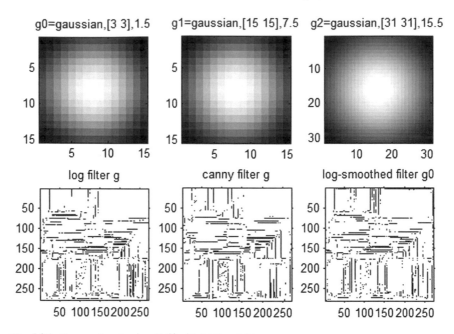

Fig. 5.26 Canny edges in **circuit.tif** with Listing 5.13

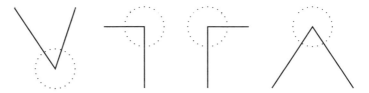

Fig. 5.27 Sample corners

5.12   Detecting Image Corners

This section introduces Harris–Stephens corner detection [71] (see, Fig. 5.28 for the results of finding corners in **circuit.tif**). A **corner** is defined to be the intersection of edges (i.e., a target pixel where there are two dominant and different edge directions in the neighbourhood of the target pixel). See, e.g., the corners inside the dotted circles in Fig. 5.27, where each corner is a juncture for a pair of edges with different edge directions. In conflict with corner detection are what are known as interest points. An **interest point** is an isolated point which is a local maximum or minimum intensity (a spike), line ending or point on a curve such as a ridge (concavity down) or valley (concavity up). If only corners are detected, then the detected points will include interest points. It is then necessary to do post processing to isolate real corners (separated from interest points). The details concerning this method will be given later. The corner detection results for **kingfisher1.jpg** are impressive, where only corner detection is perform only in a small region-of-interest in the image (see Fig. 5.29).

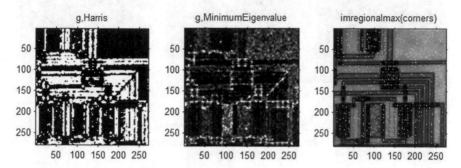

Fig. 5.28 Corners in **circuit.tif** with Listing 5.14

```
% Image corner detection

% g = imread('circuit.tif');
g = imread('kingfisher1.jpg');
g = g(10:250,300:600); % not used with circuit.tif
corners = cornermetric(g,'Harris'); % default
corners(corners <0) = 0;
cornersgray = mat2gray(corners);
figure,
subplot(1,3,1),imshow(~cornersgray);
title('g,Harris');
corners2 = cornermetric(g,'MinimumEigenvalue');
corners2 = mat2gray(corners2);
subplot(1,3,2),imshow(imadjust(corners2));
title('g,MinimumEigenvalue');
cornerpeaks = imregionalmax(corners);
results = find(cornerpeaks==true);
[r g b] = deal(g);
r(results) = 255;
g(results) = 255;
b(results) = 0;
RGB = cat(3,r,g,b);
subplot(1,3,3),imshow(RGB);
title('imregionalmax(corners)');
```

Listing 5.14 Matlab code in findcorners.m to produce Fig. 5.28.

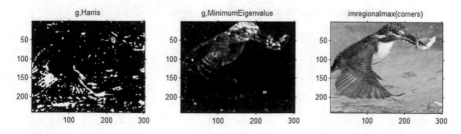

Fig. 5.29 Corners in **kingfisher1.jpg** with Listing 5.14

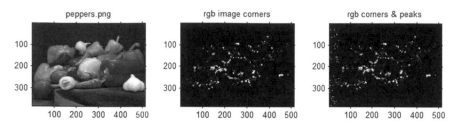

Fig. 5.30 Corners and peaks detected in a colour image

Problem 5.12 The corner and peak detection method implemented in Listing 5.14 is restricted to greyscale images (required by the **cornermetric** function). To see this, type

$$\gg \text{help cornermetric}$$

Give a matlab script called **cornerness.m** that makes it possible to use the cornermetric on colour images. Your adaptation of the cornermetric should produce (i) colour image showing the location of corners on the input colour image and (ii) colour image showing the location both the corners and peaks on the input colour image. Do this so that corners and peaks are visible on each input colour image. Demonstrate the use of your script on **peppers.png** and two other colour images that you select. For the peppers.png colour image, your cornerness.m script should produce output similar to the three images in Fig. 5.30, but instead of a black background, your script should display the locations of the corners and peaks on each input colour image.

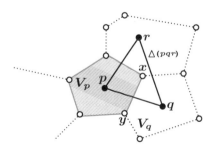

Fig. 5.31 Voronoï region V_p = Intersection of closed half-planes

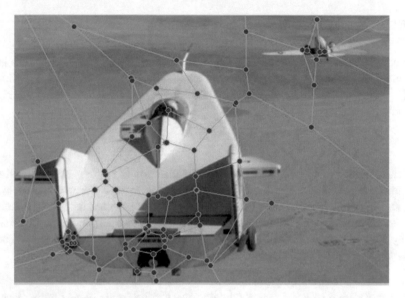

Fig. 5.32 Corner-based Voronoï mesh

5.13 Image Corner-Based Voronoï Meshes Revisited

This section revisits Voronoï meshes on digital images using image corners and carries forward the discussion on image geometry started in Sect. 1.22.

5.13.1 Voronoï Tessellation Details

A Voronoï *mesh* is also called a Voronoï tessellation. A Voronoï *tiling* (covering) of a digital image with convex polygons is called a Voronoï tessellation [202, 203]. This is different from the notion of 2D tessellation, which is a tiling of a plane region with regular polygons. Recall that a *regular polygon* is an n-sided polygon whose sides are all the same length. By contrast, the polygons in a Voronoï tiling are usually not regular.

The convex polygons in a Voronoï mesh are called Voronoï *regions*, based on Voronoï's method used to construct the polygons [40, Sect. I.1, p. 2] (see, also, [41, 143]).

5.13.2 Sites for Voronoï Polygons

Let $S \subset E$, a finite-dimensional normed linear space. The Euclidean plane is an example. Elements of S are called sites to distinguish them from other points in E [41, Sect. 2.2, p. 10]. Let $p \in S$. A *Voronoï region* of $p \in S$ (denoted V_p) is defined by

$$V_p = \left\{ x \in E : \|x - p\| \underset{\forall q \in S}{\leq} \|x - q\| \right\}.$$

The Voronoï region V_p depicted as the intersection of finitely many closed half planes in Fig. 5.31 is a variation of the representation of a Voronoï region in the monograph by H. Edelsbrunner [41, Sect. 2.1, p. 10], where each half plane is defined by its outward directed normal vector. The rays from p and perpendicular to the sides of V_p are comparable to the lines leading from the center of the convex polygon in G.L. Dirichlet's drawing [35, Sect. 3, p. 216].

Remark 5.13 **Voronoï Polygons**.
A Voronoï region of a site $p \in S$ contains every point in the plane that is closer to p than to any other site in S [52, Sect. 1.1, p. 99]. Let V_p, V_q be Voronoï polygons (see, e.g., Fig. 5.31). If $V_p \cap V_q$ is a line, ray or straight line segment, then it is called a Voronoï *edge*. If the intersection of three or more Voronoï regions is a point, that point is called a Voronoï *vertex*. ∎

A nonempty set A of a space X is a *convex set*, provided $\alpha A + (1 - \alpha)A \subset A$ for each $\alpha \in [0, 1]$ [12, Sect. 1.1, p. 4]. A *simple convex set* is a closed half plane (all points on or on one side of a line in R^2).

Lemma 5.14   ([41, Sect. 2.1, p. 9]) *The intersection of convex sets is convex.*

Proof Let $A, B \subset \mathbb{R}^2$ be convex sets and let $K = A \cap B$. For every pair points $x, y \in K$, the line segment $\overline{xy}$ connecting x and y belongs to K, since this property holds for all points in A and B. Hence, K is convex.

Lemma 5.15   ([143]) *A Voronoï region of a point is the intersection of closed half planes and each region is a convex polygon.*

Proof From the definition of a closed half-plane

$$H_{pq} = \left\{ x \in R^2 : \|x - p\| \underset{q \in S}{\leq} \|x - q\| \right\},$$

V_p is the intersection of closed half-planes H_{pq}, for all $q \in S - \{p\}$ [40], forming a polygon. From Lemma 5.14, V_p is a convex.

From an application point of view, Voronoï mesh segments a digital image. This is especially important in the case where the sites used to construct a mesh have some significance in the structure of a image. For example, by choosing the corners in an image as a set of sites, each Voronoï region of a site p that has the property that all points in the region are nearest p than to any other corner in the image. In effect, the points in a Voronoï region of a corner site p are symmetrically arranged around the particular corner p. This property holds true for each the Voronoï region in a corner mesh.

Fig. 5.33 Corners found with Matlab script A.28

5.14 Steps to Construct a Corner-Based Voronoï Mesh

The steps to construct a corner-based Voronoï mesh on a digital image are given next.

1° Select a digital image Im.
2° Select an upper bound n on the number of corners to detect in Im.
3° Find up to n corners in Im. The corners found form a set of sites.
4° Display the corners in Im. This display provides a handle for the next step. **N.B.:** At this point in a Matlab® script, use the **hold on** instruction. This hold-on step is not necessary in Mathematica® 10
5° Find the Voronoï region for each site. This step constructs a Voronoï mesh on Im. ∎

Example 5.16 **Constructing a Voronoï mesh on an Image**.
A sample Voronoï mesh is shown on the image in Fig. 5.32. To implement the Voronoï

5.34.1: Corners on full-size colour   5.34.2: Corners on cropped colour
image   image

Fig. 5.34 Image corners on full-size and cropped image

mesh construction steps in Matlab, use a combination of the **corner** function and
voronoi functions. Let X, Y be the x- and y-coordinates of the image corners found
using the **corner** function. Then use **voronoi**(X,Y) to find the x- and y-coordinates of
the vertices in each of the regions in a Voronoï mesh. Then the Matlab **plot** function
can be used to draw the Voronoï mesh on a selected digital image.   ■

Problem 5.17 For three digital images of your own choosing, construct a Voronoï
mesh on each image. Do this for the following upper bounds on the number of sites:
30, 50, 80, 130.   ■

5.15   Extreme Image Corners in Set of Mesh Generators

To include the extreme image corners in a set of mesh generators, used the following
steps.

1° $im :=$ greyscale image;
2° $[m, n] :=$ size of image im; % use size[im] in Matlab
3° let $C :=$ set of interior image corners;
4° let fc be the coordinates of the extreme image corners;
5° let $Cim := [C; fc]$; % Cim contains coords. of all im corners
6° superimpose Cim on image im;

Remark 5.18 **Superimposing corners on a full-size as well as on cropped image**.
A 480×640 colour image of a Salerno motorcycle is shown in Fig. A.49. Using the
Matlab script A.28, the corners are found in both the full image in Fig. A.51.1 and in
a cropped image in Fig. A.50.1. Notice that there are a number different methods that
can be used to crop an image (these cropping methods are explained in the comments
in script A.28.   ■

Example 5.19 A 480 × 640 colour image of an Italian Carabinieri auto is shown in Fig. 5.33. Using the Matlab script A.28 in Appendix A.5.6, the corners are found in both the full image in Fig. 5.34.1 and in a cropped image in Fig. 5.34.2. Notice that there are a number different methods that can be used to crop an image (these cropping methods are explained in the comments in script A.28.   ■

5.16   Voronoï Mesh on an Image with Extreme Corners

This section demonstrates the effectiveness of the inclusion of image corners in the set of sites (generators) in constructing a Voronoï mesh on a 2D digital image. To superimpose a Voronoï mesh on an image using the set of sites that includes the extreme image corners, do the following.

1° start with Cim from Step 5 in the image corner method;
2° let $X := Cim(:, 1)$, x-coordinates of the image corners;
3° let $Y := Cim(:, 2)$, y-coordinates of the image corners;
4° let $[vx, vy] := voronoi(X, Y)$, coordinates of the image corners;
5° superimpose the resulting Voronoï on image im;

Example 5.20 **Voronoï Mesh on Corner Sites**.
The corner-based Voronoï meshes shown in this section are obtained using the Matlab script reflst:VoronoiMeshOnImage. By including the extreme image corners in the

Fig. 5.35 Voronoï Mesh on image with extreme corners

Fig. 5.36 Voronoï Mesh on image without extreme corners

set of generating points (sites), we obtain a Voronoï mesh like the one shown in Fig. 5.35. Notice the convex polygons surrounding parts of the inside corners in Fig. 5.35 that result from including the extreme corners in the set of generators used to derived the image mesh (Figs. 5.36 and 5.37). ∎

```
% gradients: S, Garg, 2014, modified by J.F.P., 2015
% http://www.mathworks.com/matlabcentral/fileexchange/
% 46408-histogram-of-oriented-gradients--hog--code-using-matlab/
% content/hog_feature_vector.m
clear all; close all; clc;
im=imread('floorplan.jpg');
if size(im,3)==3
    im=rgb2gray(im);end
im=double(im);rows=size(im,1);cols=size(im,2);
Ix=im;Iy=im; % Basic Matrix assignments
for i=1:rows-2 % Gradients in X direction.
    Iy(i,:)=(im(i,:)-im(i+2,:));end
for i=1:cols-2 % Gradients in Y direction.
    Ix(:,i)=(im(:,i)-im(:,i+2));end
angle=atand(Ix./Iy); % edge gradient angles
angle=imadd(angle,90); % Angles in range (0,180)
magnitude=sqrt(Ix.^2 + Iy.^2);
imwrite(angle,'gradients.jpg');
imwrite(magnitude,'magnitudes.jpg');
subplot(2,2,1), imshow(imcomplement(uint8(angle))), title('edge
    gradients');
subplot(2,2,2), plot(Ix,angle),title('angles in [0,180]');
subplot(2,2,3), imshow(imcomplement(uint8(magnitude))),[0 255]),
title('x-,y-gradient magnitudes in situ');
subplot(2,2,4), plot(Ix,magnitude),title('x-,y-gradient magnitudes');
```

Listing 5.15 Matlab code in hog.m to produce Fig. 5.39.

5.37.1: Corners on full-size colour image

5.37.2: Corners on cropped colour image

Fig. 5.37 Image corners on full-size and cropped image

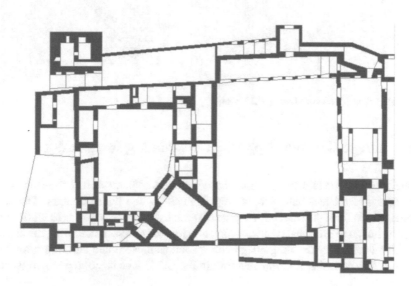

Fig. 5.38 Alhambra floorplan

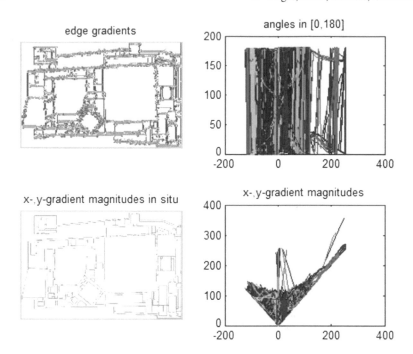

Fig. 5.39 Edges found with Listing 5.15 using Fig. 5.38

5.17   Image Gradient Approach to Isolating Image Edges

To arrive at a reasonable image corner-based segmentation mesh, it is often necessary to isolate image edges before an attempt is made to find image corners. The basic approach is to limit the search for image corners to parts of image edges without the noise by image regions that typically surround image corners. In addition, the corner-detection is aided by thinning image edges such as those found building floorplans (see, e.g., the floorplan for the Alhambra in Fig. 5.38). The basic steps to do this are as follows.

1° Find image gradients in the x- and y-directions (Gx, Gy). Notice that each pair of gradients defines a vector in the Euclidean plane for a 2D image.

2° Fine the Gradient magnitude $\|Gradx, Grady\| = \sqrt{Gx^2 + Gy^2}$ for each image gradient vector.   ∎

3° Let $magnitudes :=$ array of gradient magnitudes.   ∎

4° Convert the white edges surrounded by black regions to black edges surrounded by white regions. This can be done using either Matlab **imcomplement** or a combination of Mathematica 10 **ColorNegate** and **Binarize** to achieve a collection of crisp black edges on white.   ∎

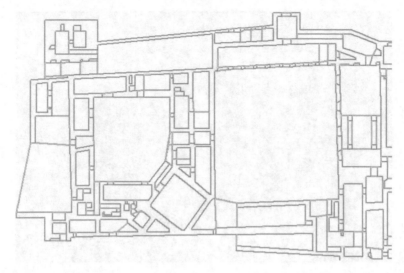

Fig. 5.40 Edges found with Listing 5.15

```
% gradients: S. Garg, 2014, modified by J.F.P., 2015
clear all; close all; clc;
% im=imread('floorplan.jpg');
im=imread('redcar.jpg');
if size(im,3)==3
    im=rgb2gray(im);end
im=double(im);rows=size(im,1);cols=size(im,2);
Ix=im;Iy=im; %Basic Matrix assignments
for i=1:rows-2 % Gradients in X direction.
    Iy(i,:)=(im(i,:)-im(i+2,:));end
for i=1:cols-2 % Gradients in Y direction.
    Ix(:,i)=(im(:,i)-im(:,i+2));end
angle=atand(Ix./Iy); % edge pixel gradients in degrees
angle=imadd(angle,90); %Angles in range (0,180)
magnitude=sqrt(Ix.^2 + Iy.^2);
imwrite(angle,'gradients.jpg');
imwrite(magnitude,'magnitudes.jpg');
figure,imshow(uint8(angle));
figure,imshow(imcomplement(uint8(magnitude)));
% figure,plot(Ix,angle);
% figure,plot(Ix,magnitude);
```

Listing 5.16 Matlab code in hog.m to produce Fig. 5.40.

Example 5.21 **Edge Thinning Using Image Gradient Magnitudes**. A sample thinning of the thick lines in the Alhambra floorplan image is shown in Fig. 5.40. In this
image, each of the thick floorplan borders has been reduced to thin line segments. The
result is a collection of thinly bordered large-scale convex polygons. The Alhambra
floorplan gradient angles are displayed in Fig. 5.41. ∎

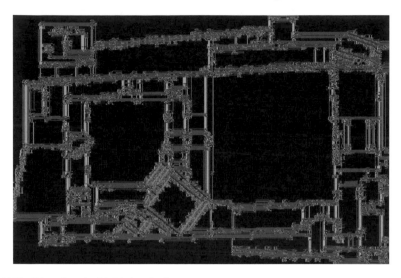

Fig. 5.41   Edges found with Listing 5.16

5.18   Corners, Edges and Voronoï Mesh

The results from Example 5.21 provide a basis for finding a minimum number of image corners, leading to the construction of an effective Voronoï Mesh. In section, we again consider the Alhambra floorplan image.

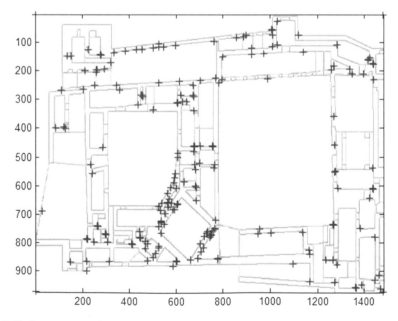

Fig. 5.42   Image corners found with Listing A.28

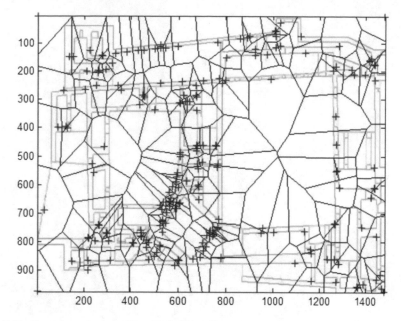

Fig. 5.43 Voronoï mesh on the thinned Alhambra floorplan

The Matlab script A.28 applied to the Alhambra floorplan (limited to thinned edges) produces the result shown in Figs. 5.42 and 5.43.

Problem 5.22 Voronoï Mesh on Corners in Image Edges.
Using the Alhambra floorplan image and three additional images from your own colour image archive (not images found on the web), superimpose a corner-based Voronoï Mesh on the thinned edges of each image. Notice that this approach differs from the approach given in Sect. 5.14, which does not consider image edges.

Hint: Choose images containing lots of straight edges such as images containing houses or buildings (Fig. 5.43). ∎

Chapter 6
Delaunay Mesh Segmentation

Fig. 6.1 Region centroid-based delaunay mesh on colour image

This chapter introduces segmentation of digital images using Delaunay meshes. An image is segmented by separating the image into almost disjoint regions. The interiors of image segments do not overlap. Each segment contains points that belong only to the segment. Adjacent segments have a common border. The common border of adjacent segments means (1) all points in the interior of a segment belong only to

© Springer International Publishing AG 2017

J.F. Peters, *Foundations of Computer Vision*, Intelligent Systems
Reference Library 124, DOI 10.1007/978-3-319-52483-2_6

the segment, (2) segments do not partition an image into disjoint regions, since each pair of adjacent segments in an image segmentation have a common border. In this chapter, an image is segmented into triangular segments in a mesh using an approach to planar triangulation introduced by Delaunay. A Delaunay mesh is the result of what is known as a triangulation.

Example 6.1 A sample Delaunay mesh covering of a colour image is shown in Fig. 6.1. This mesh is constructed with a set of regional centroids (used as generating points) found in the image. For more about this, see Sect. 6.4.   ∎

6.1 Delaunay Triangulation Generates a Triangular Mesh

Delaunay triangulations, introduced by B.N Delone [Delaunay] [33], represent pieces of a continuous space. This representation supports numerical algorithms used to compute properties such as the density of a space. A *triangulation* is a collection of triangles, including the edges and vertices of the triangles in the collection. A 2D *Delaunay triangulation* of a set of sites (generators) $S \subset \mathbb{R}^2$ is a triangulation of the points in S. Let $p, q \in S$. A straight edge connecting p and q is a *Delaunay edge* if and only if the Voronoï region of p [41, 143] and Voronoï region of q intersect along a common line segment [40, Sect. I.1, p. 3]. For example, in Fig. 1.3, $V_p \cap V_q = \overline{xy}$. Hence, $\overline{pq}$ is a Delaunay edge in Fig. 1.3.

Fig. 6.2 $p, q \in$ $S, \overline{pq} =$ delaunay edge

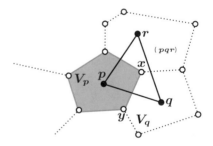

A triangle with vertices $p, q, r \in S$ is a *Delaunay triangle* (denoted $\triangle(pqr)$ in Fig. 1.3), provided the edges in the triangle are Delaunay edges. A *Delaunay mesh* on a plane surface is a collection Delaunay triangles that cover the surface. In other words, every point belongs to a triangle in a surface mesh. Here are the steps to generate a corner-based Delaunay mesh (see, e.g., Fig. A.5 using Matlab® A.4 in Appendix A.1.3).

1° Find the set of corners S in an image. Include in S the extreme NS and EW image corners.

2° Connect each pair of nearest corners $x, y \in S$ with a straight edge $\overline{xy}$. A Delaunay triangle results from connecting with straight edges corners x, y, r that are nearest each other.

3° Repeat step 2° until all pairs of corners are connected.   ■

Every planar convex polygon has a nonempty interior so that there are uncountably infinite number points between any pair of points in the polygon.

Theorem 6.2 *A planar Delaunay triangle is not a convex polygon.*

Problem 6.3 Prove Theorem 6.2. Give an example a Delaunay triangle in an image. ■

Delaunay Wedge

A planar *Delaunay wedge* is a Delaunay triangle with an interior that contains an uncountably infinite number of points. The interior of a Delaunay triangle is that part of the triangle between the edges. It is assumed that every Delaunay triangle connecting generating points in an image defines a Delaunay edge.   ■

Let S be a set of mesh generating points. Recall that a closed half-plane H_{ps}

$$H_{ps} = \left\{ x \in R^2 : \|x - p\| \underset{s \in S}{\leq} \|x - s\| \right\},$$

A *Delaunay wedge* with vertices $p, q, r \in S$ (denoted $W_{p,q,r}$) is defined by

$$V_{p,q,r} = \bigcap H_{ps} : \text{ for all } s \in \{q, r\}.$$

That is, a Delaunay wedge is the intersection of closed half planes H_{ps}, for all $s \in \{q, r\} - p$.

Theorem 6.4 *A planar Delaunay wedge is a convex polygon.*

Proof Immediate from Lemma 5.15, since a Delaunay wedge is the intersection of closed half planes in spanning a Delaunay triangle $\triangle(pqr)$, stretching from vertex p to the opposite edge $\overline{qr}$.

Problem 6.5 Give an example of a Delaunay wedge in an image.   ■

6.2 Triangle Circumcircles

For simplicity, let E be the Euclidean space $\mathbb{R}^2$. For a Delaunay triangle $\triangle(pqr)$, a *circumcircle* passes through the vertices p, q, r of the triangle (see Fig. 6.3 for an example). The center of a circumcircle u is the Voronoï vertex at the intersection of three Voronoï regions, i.e., $u = V_p \cap V_q \cap V_r$. The circumcircle radius $\rho = \|u - p\| = \|u - q\| = \|u - r\|$ [40, Sect. I.1, p. 4], which is the case in Fig. 6.3.

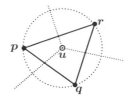

Fig. 6.3 Circumcircle

Lemma 6.6 *Let circumcircle* $\bigcirc(pqr)$ *pass through the vertices of a Delaunay triangle* $\triangle(pqr)$, *then the following statements are equivalent.*

1° *The center* u *of* $\bigcirc(pqr)$ *is a vertex common to Voronoï regions* V_p, V_q, V_r.
2° $u = cl\, V_p \cap cl\, V_q \cap clV_r.$
3° $V_p\, \delta\, V_q\, \delta\, V_r.$

Proof $1^o \Leftrightarrow 2^o \Leftrightarrow 3^o$.

Theorem 6.7 *A triangle* $\triangle(pqr)$ *is a Delaunay triangle if and only if the center of the circumcircle* $\bigcirc(pqr)$ *is the vertex common to three Voronoï regions.*

Proof The circle $\bigcirc(pqr)$ has center $u = clV_p \cap clV_q \cap clV_r$ (Lemma 6.6) $\Leftrightarrow \bigcirc(pqr)$ center is the vertex common to three Voronoï regions $V_p, V_q, V_r \Leftrightarrow \overline{pq}, \overline{pr}, \overline{qr}$ are Delaunay edges $\Leftrightarrow \triangle(pqr)$ is a Delaunay triangle.

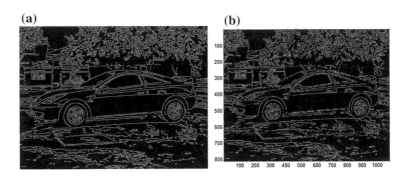

Fig. 6.4 a Image edges and **b** Image corners

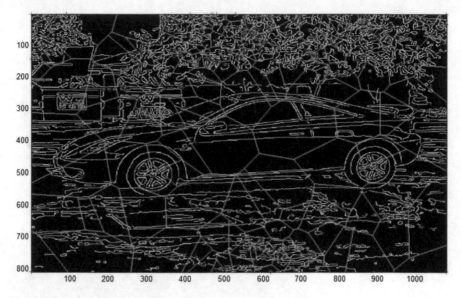

Fig. 6.5 Image mesh

6.3   Constructing a Corner-Based Delaunay Mesh on Image Edges

The steps to construct a corner-based Delaunay mesh on image edges are as follows.

1° Detect the edges in a given image Im.
   Example: Fig. 6.4a.   ∎
2° Find the set of corners S of the edges in the image Im. Include in S the extreme
   NS and EW image corners.
   Example: Fig. 6.4b.   ∎
3° Connect each pair of nearest corners $x, y \in S$ with a straight edge $\overline{xy}$. A Delaunay
   triangle results from connecting with straight edges between corners x, y, r that
   are nearest each other.   ∎
4° Repeat step 3° until all pairs of corners are connected. **N.B.** It is assumed that
   each triangular region of the mesh is a Delaunay wedge.
   Example: Fig. 6.5.   ∎

Problem 6.8 Give a Matlab script that constructs a corner-based Delaunay mesh on
image for three image of your own choosing. **N.B.**: Choose your images from a your
personal collection of images not taken from the web.   ∎

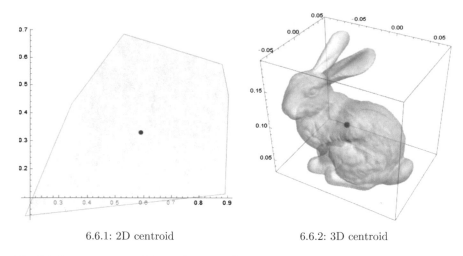

6.6.1: 2D centroid   6.6.2: 3D centroid

Fig. 6.6 2D convex region & 3D Wolfram Stanford Bunny centroids

6.4   Centroid-Based Delaunay Image Mesh

This section briefly introduces an alternative to the corner-based approach to constructing a Delaunay mesh on an image using geometric centroids. A **geometric centroid** is the center of mass of an image region. An **image region** is a bounded set of points in an image. For instance, let X be a set of points in a $n \times m$ rectangular 2D region containing points with coordinates (x_i, y_i), $i = 1, \ldots, n$ in the Euclidean plane. Then the coordinates x_c, y_c of the discrete form of the **centroid of a 2D region** are

$$x_c = \frac{1}{n} \sum_{i=1}^{n} x_i, \, y_c = \frac{1}{m} \sum_{i=1}^{m} y_i.$$

The coordinates x_c, y_c, z_c of the discrete form of the **centroid of a 3D region** in Euclidean space $\mathbb{R}^3$ are

$$x_c = \frac{1}{n} \sum_{i=1}^{n} x_i, \, y_c = \frac{1}{m} \sum_{i=1}^{m} y_i, \, z_c = \frac{1}{h} \sum_{i=1}^{h} z_i.$$

Example 6.9 **2D and 3D Image Region Centroids**.
In Fig. 6.6, the red dot ● indicates the location of a region centroid. Two examples are shown, namely, centroid ● in a 2D convex region in Fig. 6.6.1 and centroid ● in a 3D region occupied with the Wolfram Stanford Bunny in Fig. 6.6.2. To experiment with finding other region centroids, see MScript 2 and MScript 3 in Appendix A. See, also, Sect. 6.4.1.   ■

The basic approach is to use image region centroids as generating points in Delaunay mesh construction. Here are the steps to do this.

1° Find the region centroids in a given image Im.
2° Connect each pair of nearest centroids $x, y \in S$ with a straight edge $\overline{xy}$. A Delaunay triangle results from connecting with straight edges for centroids x, y, r that are nearest each other.
3° Repeat step 2° until all pairs of centroids are connected. **N.B.** It is also assumed that each triangular region of the mesh is a Delaunay wedge. ■

6.4.1 Finding Image Centroids

6.7.1: Salerno fisherman   6.7.2: Image region centroids

Fig. 6.7 Image region centroids

Example 6.10 **Region centroids on an image**.
A sample plot of the image region centroids are shown in Fig. 6.7.1 on the image in Fig. 6.7.2 using Matlab® script A.30 in Appendix A.6.2. For more about this, see Appendix B.3. ■

6.4.2 Finding Image Centroidal Delaunay Mesh

Example 6.11 **Region centroid-based Delaunay triangulation on an image**.
A sample plot of the image region centroid-based Delaunay mesh is shown in Fig. 6.8.2 (relative to the region centroids in Fig. A.57.1) using Matlab® script A.31 in Appendix A.6.3. For more about this, see Appendix B.3. ■

6.8.1: lifting body   6.8.2: Image region centroids

Fig. 6.8 Image region centroid-based Delaunay mesh

Maximal Nucleus Triangle Clusters.

Notice that clusters of small triangles define shapes of image objects such as the fisherman, fishing rod and prominent darker rocks in the picture in Fig. 6.7.1.

Also notice that every Delaunay triangle ▓▓▓▓▓ is the **nucleus** of a cluster of Delaunay triangles. Each image object shape is associated with a nucleus having a maximal number of adjacent triangles, forming a **Maximal Nucleus Triangle Cluster** (MNTC). An **object shape** is defined by an MNTC cluster. A triangle $\triangle A$ is **adjacent** to a nucleus triangle N, provided $\triangle A$ has either an edge or a vertex in common with N.

6.4.3   Finding Image Centroidal Voronoï Mesh

Example 6.12 **Region centroid-based Voronoï mesh on an image**.
A sample plot of the image region centroid-based Voronoï mesh is shown in Fig. 6.9.2 (relative to the region centroids in Fig. 6.9.1) using Matlab® script A.32 in Appendix A.6.4. For more about this, see Sect. 6.4.   ■

Maximal Nucleus [Polygon] Clusters.

Notice that clusters of Voronoï polygons with inscribed Delaunay Triangles define shapes of image objects such as the fisherman's head, fishing rod and prominent darker rocks in the picture in Fig. 6.7.1. Also notice that every Voronoï polygon is the **nucleus** of a cluster of Voronoï polygons. Each image object shape is associated with a nucleus having a maximal number of adjacent polygons, forming a **Maximal Nucleus Cluster** (MNC). An **object shape** is defined by an MNC cluster. A polygon ⊞A is **adjacent** to a nucleus triangle N, provided ⊞A has an edge in common with N. For more about MNCs in Voronoï meshes, see Appendix B.12.

6.9.1: lifting body   6.9.2: Image region Centroidal
                                   Voronoï Mesh

Fig. 6.9   Image region centroid-based Voronoï mesh

6.4.4   Finding Image Centroidal Voronoï Superimposed on a Delaunay Mesh

Example 6.13 **Region centroid-based Voronoï over Delaunay mesh on an image**. A sample plot of the image region centroid-based Voronoï over a Delaunay mesh is shown in Fig. 6.10.2 (relative to the region centroidal Delaunay mesh in Fig. 6.10.1) using Matlab® script A.33.   ∎

Maximal Nucleus [Polygon-Triangle] Clusters

Notice that clusters of Voronoï polygons with inscribed Delaunay triangle corners define shapes of image objects such as the fisherman's head, fishing rod and prominent darker rocks in the picture in Fig. 6.7.1. Also notice that every Voronoï polygon

is the **nucleus** of a cluster of Voronoï polygons. Each image object

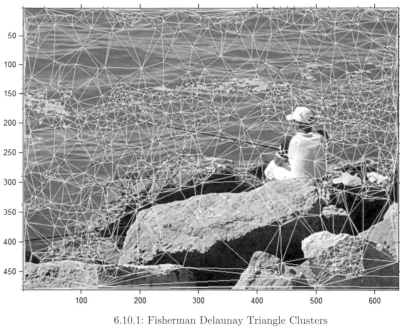

6.10.1: Fisherman Delaunay Triangle Clusters

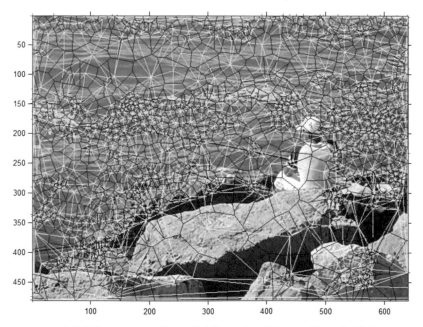

6.10.2: Image region Centroidal Voronoïon Delaunay Triangle Mesh

Fig. 6.10 Image region centroid-based Voronoï over Delaunay mesh

shape is associated with a nucleus having a maximal number of adjacent polygons with inscribed Delaunay Triangle corners, forming a **Maximal Nucleus [Polygon-Triangle] Cluster** (MNptC). An **object shape** is defined by an MNptC cluster. A polygon $\boxplus \triangle A$ is **adjacent** to a nucleus triangle N, provided $\boxplus \triangle A$ has an edge in common with N. For more about MNCs in Voronoï meshes, see Appendix B.12.

Problem 6.14 Give a Matlab script that false colours (your choice of colour) the maximal nucleus triangle of each MNTC in a centroid-based Delaunay mesh on an image for three images of your own choosing. False colour each triangle adjacent to the maximal nucleus triangle. **N.B.**: Choose your images from a your personal collection of images not taken from the web. In this problem, image centroids are used instead of corners as a source of generating points in constructing the Delaunay triangulation mesh. ∎

Problem 6.15 Give a Matlab script that false colours (your choice of colour) the maximal nucleus triangle of each MNC in a centroid-based Voronoï mesh on an image for three images of your own choosing. False colour each triangle adjacent to the maximal nucleus triangle. **N.B.**: Choose your images from a your personal collection of images not taken from the web. In this problem, image centroids are used instead of corners as a source of generating points in constructing the Voronoï mesh. ∎

Problem 6.16 Give a Matlab script that false colours (your choice of colour) the maximal nucleus triangle of each MNptC in a centroid-based Voronoï-Delaunay triangulation mesh on an image for three images of your own choosing. False colour each triangle adjacent to the maximal nucleus polygon with inscribed triangle corners. **N.B.**: Choose your images from a your personal collection of images not taken from the web. In this problem, image centroids are used instead of corners as a source of generating points in constructing the Voronoï mesh. ∎

Chapter 7
Video Processing. An Introduction to Real-Time and Offline Video Analysis

7.1.1: Initial Frame

7.1.2: Later Frame

Fig. 7.1 Voronoï tiling of video frames in tracking moving objects

© Springer International Publishing AG 2017
J.F. Peters, *Foundations of Computer Vision*, Intelligent Systems
Reference Library 124, DOI 10.1007/978-3-319-52483-2_7

Fig. 7.2 Perception angle

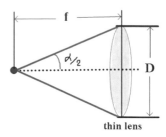

thin lens

This chapter introduces video processing with the focus on tracking changes in video frame images. Video frame changes can be detected in the changing shapes, locations and distribution of the polygons (regions) in Voronoï tilings of the frames (see, e.g., Fig. 7.1). The study of video frame changes can be done either in real-time or offline. Real-time video frame analysis is the preferred method, provided the analysis can be carried out in a reasonably short time for each frame. Otherwise, for more time-consuming analysis of video frame content, offline processing is used. From a computer vision perspective, scenes recorded by a video camera depend on the camera aperture angle and its view of a visual field, which is analogous to the human perception angle (see Fig. 7.2). For more about this, see Sect. 7.3.

Example 7.1 **Tiling Video Frames of Moving Toy Tractors in Real-Time**.
The tiling of an initial frame of a video[1] showing a race between rolling toy tractors is shown in Fig. 7.1.1. Image centroids are used as generating points of the Voronoï regions in this tiling. Each the frame centroids is represented by a *. The locations of the centroids as well as the numbers, locations and shapes of the tiling polygons change from one frame to the next one, reflecting changes in the positions of the tractors over time. For example, in Fig. 7.1.2, the number of polygons covering the larger of the two tractors has increased in number in a later frame in the same video. This is an example of video frame tiling carried out in real-time (during video capture). ■

7.1 Basics of Video Processing

This section briefly introduces some of the essentials of video processing, leading to object detection in videos. A good introduction to video processing is given by T.B. Moselund [125].

    The basic unit in a video is a frame. A **frame** is an individual digital image in a linear sequence of images.

---

[1] Many thanks to Braden Cross for this video frame.

Basis Steps in Video Analysis.

webcam $\longmapsto$ image acquisition $\longmapsto$ pre $-$ processing $\longmapsto$
frame structuring $\longmapsto$ classification

∎

7.1.1 Frame Point Processing

Every frame is a set of pixels susceptible to any of the standard image processing techniques such as false colouring, pixel selection (e.g., centroid, corner and edge pixels), pixel manipulation (e.g., RGB $\longmapsto$ greyscale), pixel (point) processing (e.g., adjusting color channel brightness), filtering (e.g., frame noise reduction, histogram equalization, thresholding) and segmentation (e.g., separation of pixels into non-overlapping regions).

7.1.2 Image Acquisition

Video processing begins with the image acquisition process. This process is markedly different from snapshots. **Image acquisition** is basically a two step process in which a single image is added to a sequence of images called frames.

Videos consume huge amounts of memory for their storage. Hence, image compression is a central concern in video image acquisition. The **MPEG** (Motion Picture Experts Group) standard was designed to compress video signals from 4 to 6 Mbps (megabits per second). MPEG-1 and MPEG-2 compression reduces spatial and temporal redundancies.

With the MPEG approach to compression, each frame is coded separated using JPEG (Joint Photographic Experts Group) lossy compression. JPEG uses piecewise uniform quantization. A **quantizer** is determined by an encoder that partitions an input set of signal values in classes and a decoder that specifies the set of output values. Let x be a signal value. This quantization process is modelled with a selector function $S_i(x)$ on a set R_i (a partition cell). A selector function $S_i(x)$ is an example of what is known as an **indicator function** 1_R of a partition cell, defined by

$$1_R(x) = \begin{cases} 1, & \text{if } x \in R \text{ (input signal } x \text{ belongs to partition cell } R), \\ 0, & \text{otherwise.} \end{cases}$$

A video selector function S_i for partition cell R_i is defined by the indicator function 1_{R_i} on cell R_i, i.e.,

$$S_i(x) = 1_{R_i}(x).$$

A good introduction to JPEG lossy compression is given by A. Gersho and R.M. Gray [56, Sect. 5.5, pp. 156–161].

Further compression of a video stream is accomplished by detecting redundance in consecutive frames are often almost the same. More advanced forms of the composition of audio-visual information is accomplished by the MPEG-4 standard. This standard views audio-visual data as objects that combine each object state with a set of methods that define object behaviour. For more about this as well as the MPEG-7 and MPEG-21 standards, see F. Camastra and A. Vinciarelli [23, Sect. 3.8.1, pp. 90–93].

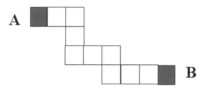

Fig. 7.3 Path-connected shapes A, B

7.1.3   Blobs

A **blob** (binary large object) is a set of path-connected pixels in a binary image. The notion of connectedness makes it possible to extend the notion of a blob to grey-blobs in greyscale and colour-blobs in colour images.

Polygons are connected, provided the polygons share one or more points. For example, a pair of Voronoï regions A and B that have a common edge are connected. Again, for example, a pair of Delaunay triangles that have a common vertex are connected. In that case, connected Voronoï polygons with a common edge containing n points are n-**adjacent**. Similarly, Delaunay triangles that have a common vertex containing n points are both connected and n-**adjacent**. A pair of Delaunay triangles with a common vertex are **1-adjacent**.

A sequence $p_1, \ldots, p_i, p_{i+1}, \ldots, p_n$ of n pixels or voxels is a **path**, provided p_i, p_{i+1} are adjacent (no pixels in between p_i and p_{i+1}). Pixels p and q are **path-connected**, provided there is a path with p and q as endpoints. Similarly, image shapes A and B (any polygons) are path-connected, provided there is a sequence $S_1, \ldots, S_i, S_{i+1}, \ldots, S_n$ of n adjacent shapes with $A = S_0$ and $B = S_n$.

Example 7.2 **Path-Connected Shapes**.
Shapes *A* and *B* in Fig. 7.3 are connected, since there is a path (sequence containing pairwise adjacent shapes) between *A* and *B*. ∎

For more about connectedness from digital image perspective, see R. Klette and A. Rosenfeld [94, Sect. 2.2.1, pp. 46–50].

Example 7.3 **Path-Connected Voronoï Nucleus Clusters**.
Polygons *A* and *B* in a Voronoï nucleus cluster are connected, since there is always a path (sequence containing pairwise adjacent polygons) between *A* and *B*. ∎

From a path-connectedness perspective, a **grey-blob** in a greyscale image is path-connected set of greyscale pixels. In fact, every collection of path connected shapes in a greyscale image are grey-blobs. Similarly, a **colour-blob** in a colour image is path-connected set of colour pixels. And every collection of path connected shapes in a colour image are colour-blobs. This means that one can always find blobs in video frame images. For more about video image blobs, see T.B. Moselund [125, Chap. 7, pp. 103–115].

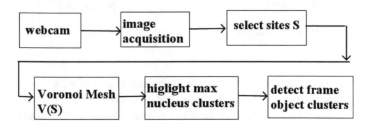

Fig. 7.4 Video object detection steps

7.1.4 Frame Tiling and Frame Geometry

Either in real-time or offline, every video frame can be tiled (tessellated) with a Voronoï diagram or tiling each frame using Delaunay's triangulation method, i.e., connect sites of neighbouring Voronoï regions with straight edges to form multiple triangles covering a video frame. A natural outcome of either form of frame tiling is mesh clustering and object recognition. The fundamentally important step in his form of video processing is the selection of sites (generating points) used to construct either frame Voronoï regions or Delaunay triangles. After the selection of frame generating points, a frame can be tiled. Frame tiling takes along the path that leads to video object detection (see Fig. 7.4 for the steps leading to frame object detection).

7.2   Voronoï Tiling of Video Frames

Recall that a Voronoï tiling of a plane surface is a covering of the surface with non-overlapping Voronoï regions. Each 2D Voronoï regions of a generating point is an n-sided polygon (briefly, **ngon**). In effect, a planar **Voronoï tiling** is a covering of a surface with non-overlapping ngons. Video frame tilings have considerable practical value, since the contour of the outer polygons surrounding a frame object can be measured and compared.

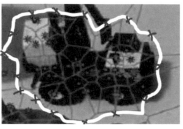
7.5.1: Contour 1

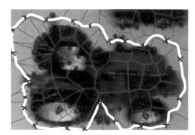
7.5.2: Contour 2

Fig. 7.5   Centroid-based contours video frame objects

Example 7.4 **Voronoï tiling of a Video Frame and Object Contours**.
The video frames in Fig. 7.1 are examples of Voronoï tilings using centroids as generating points. Let × represent a location of a centroid in a Voronoï region. Then the contour of a frame object is identified with a line that connects the centroids of the Voronoï regions surrounding the object. Two examples of centroid-based contours of frame objects are shown in Fig. 7.5.   ■

The contour of a frame object defines its shape. Shapes are similar, provided the shapes are, in some sense, close to each other (see Sect. 7.4 for an approach to measuring the similarity between shapes).

7.3   Detection of Shapes in Video Frames

The detection of personal spaces in the motion of people in video sequences is aided by constructing Voronoï tilings (also called Voronoï diagrams) on each video frame. A *personal space* is defined by a comfort distance between persons in motion. Let d be a distance (in meters) between persons. Four types of comfort distances between persons have been identified by E. Hall [66], namely,

Intimate: $0 \leq d \leq 0.5$ m   (Friendship distance).
Personal: $0.5 \leq d \leq 1.25$ m   (Conversational distance).

Social: $1.25 \leq d \leq 3.5$ m   (Impersonal distance).
Public: $d \geq 3.5$ m   (Public Speaking distance).

Based on the notion of comfort distance between persons, an approach to the study of motion of persons in a succession of video frames is suggested by J.C.S. Jacques and others [86]. In this study, a perceived personal space (PPS) is introduced relative to the changing distances between persons in a succession of Voronoï-tiling of video frames. Let f_v be a video frame, R_c the radius of a circular sector with angle α around a person at point c in a frame tiling, $\frac{\alpha\pi}{360°}$ the area of the personal sector. Then the the personal space $PPS(f_v)$ of a frame f_v [86, Sect. 3.2, p. 326] is defined by

$$PPS(V_f) \geq \frac{\alpha\pi}{360°} R_c^2 \ m \ \text{(Video frame perceived personal distance).}$$

Then PPS is defined to be the area of the region formed by the intersection of a person's visual field and corresponding Voronoï polygon [86]. The region of attention focus of a person's visual field is estimated to be a circular sector with an approximate aperture angle of 40°. Let f be the focal length and D the diameter of an aperture. The aperture angle [122] of a lens (e.g., human eye) is the **apparent angle** α of the lens aperture as seen from the focal point, defined by

$$\alpha = 2\tan^{-1}\left(\frac{D}{2f}\right) \text{ (aperture angle).}$$

Fig. 7.6 Contour distances

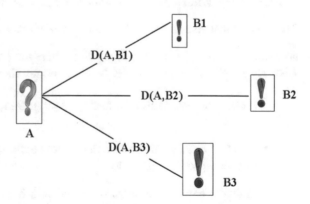

A form of clustering in Voronoï tilings on greyscale video frames is introduced by E.A.B. Over, I.T.C. Hooge and C.J. Erkelens [136]. The basic approach was to identify a cluster whose center is a Voronoï region of a point s and whose boundary region is defined by that part of the tiling occupied by all points between surround generating points and s. The homogeneity of the intensities of a polygon is used as a means of identifying each cluster center.

7.4   Measuring Shape Similarity and the Voronoï Visual Hull of an Object

Visual hulls of frame objects are introduced by K. Grauman, G. Shakhnarovich and T. Darrell [60]. In a Voronoï tiling of a video frame, a **visual hull** (VH) of an object a collection of polygons that surround an object. Each class of objects will have a characteristic VH. For example, the VH of a vehicle wheel will have a round shape containing small polygons surrounding a tiny polygon on the while hub. For a person who is standing, the VH will have a rectangular shape, covering the the silhouette of the person. Let A be the set of points on the contour of a sample object and let B be the contour of a known object. The Hausdorff distance [75, Sect. 22, p. 128] between a point x and a set A (denoted by $D(x, A)$) is defined by

$$D(x, A) = \min \left\{ \|x - a\| : a \in A \right\} \quad \textbf{(Hausdorff point} - \textbf{set distance).}$$

The similarity distance $D(A, B)$ between the two contours A and B, represented by a set of uniformly sampled points in A and B [60, Sect. 2, p. 29], is defined by

$$D(A, B) = \max \left\{ \max_{a \in A} D(a, B), \max_{b \in B} D(b, A) \right\} \quad \textbf{(Similarity Distance).}$$

Example 7.5 **Similarity Distance Between Visual Hulls of Objects.**

The contour distances $D(A, B_i)$ between question mark shape ❓ A in Fig. 7.7.1 is compared with the distance between points along the contours of three different ❗ **Aha!** shapes $B1, B2, B3$ in Fig. 7.6. In this example, check the value of

$$maxContourDistance := \max \left\{ D(A, B1), D(A, B2), D(A, B3) \right\}.$$

The shapes $B1, B2, B3$ would be considered close to the shape ❗, provided maxContourDistance is close to zero.   ∎

The similarity between the shapes of visual hulls in a sequence of video frames is measured relative to a distance between known visual hull shapes. Let A, B_i be known shapes with distance $D(A, B_i)$. And let S_0 be a known shape that is compared with shape S_j, $1 \leq j \leq n$ in a sequence of n video frames. The distance $D(A, B_i)$ between known shapes A, B_i is compared with the summation of the distances between a base shape S_0 and a succession of shapes S_i in a sequence of video frames. A and S_0 have **similar shapes**, provided the sum of the differences between n shape contours S_0 and S_j is close to (approximately equal to) the distance $D(A, B_i)$, i.e.,

$$D(A, B_i) \approx \sum_{j=1}^{n} \left(\sum_{\substack{a \in S_0 \\ b \in S_j}} \|a - b\| \right) \quad \textbf{(Similar Frame Shapes)}.$$

Let $\varepsilon > 0$ be a small number. Shapes A and S_0 are considered close, provided

$$shapeDiff(S_0, S_j) := \sum_{j=1}^{n} \left(\sum_{\substack{a \in S_0 \\ b \in S_j}} \|a - b\| \right).$$

$$\left| D(A, B_i) - shapeDiff(S_0, S_j) \right| \leq \varepsilon.$$

Example 7.6 **Comparing Shapes**.
Let A be a known shape and let B_i be a shape similar to A. Let $S_0 := A$, i.e., let S_0 be the same as the known shape A. Then, for shapes $S_1, \ldots, S_j, \ldots, S_n$ and some small $\varepsilon > 0$, check if

$$\left| D(A, B_i) - shapeDiff(A, S_j) \right| \leq \varepsilon \text{ for } 1 \leq j \leq n. \qquad \blacksquare$$

In measuring the similarity of shapes in a sequence of video frames, the basic approach is to compute the sum of the normed distances $\|a - b\|$ for points a, b along the contours of the shapes S_0 and S_j. This means that we start by remembering the known shape S_0 and compare this known shape with shape S_j found in each video frame.

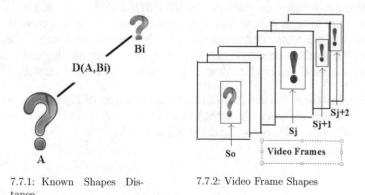

7.7.1: Known Shapes Distance

7.7.2: Video Frame Shapes

Fig. 7.7 Hunt for similar video frame shapes

The hunt for similar shapes in video frames reduces to a comparison of the distance between known shapes and similarity distances between a succession of frame shapes.

Example 7.7 **Hunt for Similar Video Frame Shapes**.
The distance between known similar shapes is represented in Fig. 7.7.1. A sequence of video frames containing shapes $S_0, \ldots, S_j, S_{j+1}, S_{j+2}, 0 \leq j \leq n$ is represented in Fig. 7.7.2.

The similarity distance $D(A, B_i)$ between question mark shapes **?** in Fig. 7.7.1 is compared with the distance between points along the contours of frame shapes containing a mixture of **?** and **! Aha!** shapes in a video frame sequence in Fig. 7.7.2. This comparison fails, since the similarity distance between **?** and **!** is usually not close for small ε values. Strangely enough, the **?** shape can be deformed (mapped) into the **!** shape. To see how this is done, see [144, Sect. 5.3].   ■

7.5   Maximal Nucleus Clusters

Notice that every polygon in a Voronoï tessellation of a surface is the centre (nucleus) of a cluster containing all polygons that are adjacent to the nucleus. A Voronoï *mesh nucleus* is any Voronoï region that is the centre of a collection of Voronoï regions adjacent to the nucleus.

Definition 7.8   Maximum Nuclear Cluster (MNC) [147]. A cluster of connected polygons with nucleus N is *maximal*, provided N has the highest number of adjacent polygons in a tessellated surface (denoted by (denoted by $\max \mathfrak{C} N$)). Similarly, a descriptive nucleus cluster is maximal, provided N has the highest number of polygons in a tessellated surface descriptively near N, (denoted by $\max \mathfrak{C}_\Phi N$).   ■

Maximal nucleus clusters (MNCs) serve as indicators of high object concentration in a tessellated image. A method that can be used to find MNCs in Voronoï tessellations of digital images is given in Algorithm 9.

Algorithm 9: Construct Maximal Nucleus Cluster

Input : Digital images *img*.
Output: MNCs on image *img*.
1 $img \longmapsto TitledImg$/*(Voronoï tessellation)*/;
2 *Choose a Voronoï region in TitledImg:* *;
3 $ngon \longleftarrow TiledImg$;
4 $NoOfSides \longleftarrow ngon$;
5 /* Count no. of sides in *ngon* & remove it from *TitledImg*. */;
6 $TiledImg := TiledImg \setminus ngon$;
7 $ContinueSearch := True$;
8 **while** *(TiledImg $\neq \emptyset$ and ContinueSearch)* **do**
9     $ngonNew \longleftarrow TiledImg$;
10     $TiledImg := TiledImg \setminus ngonNew$;
11     $NewNoOfSides \longleftarrow ngonNew$;
12   **if** *(NewNoOfSides > NoOfSides)* **then**
13 | $ngon := ngonNew$;
14   **else**
15 | /* Otherwise ignore *ngonNew*: */
16   **if** *(TiledImg $= \emptyset$)* **then**
17         $ContinueSearch := False$;
18         $max\mathfrak{C}N := ngon$;
19   /* MNC found; Discontinue search */;

Example 7.9 Let X be the collection of Voronoï regions in a tessellation of a subset of the Euclidean plane shown in Fig. 7.8 with nuclei $N_1, N_2, N_3 \in X$. In addition, let 2^X be the family of all subsets of Voronoï regions in X containing maximal nucleus clusters $\mathfrak{C}N_1, \mathfrak{C}N_2, \mathfrak{C}N_3 \in 2^X$ in the tessellation. Then, for example, int$\mathfrak{C}N_2 \cap$ int$\mathfrak{C}N_3 \neq \emptyset$, since $\mathfrak{C}N_2, \mathfrak{C}N_3$ share Voronoï regions. Hence, $\mathfrak{C}N_2$ overlaps $\mathfrak{C}N_3 \neq \emptyset$ (see [147]). Notice that there Voronoï regions surrounding nucleus $N1$ that share an edge (are adjacent to) Voronoï regions surrounding nucleus Ns. For this reason, we say that $\mathfrak{C} N_1$ adjacent $\mathfrak{C} N_2$. Either adjacent or overlapping MNCs have shapes determined by their perimeters, which surround regions-of-interest in a tessellated image. This observation leads to a useful method in the detection of objects-of-interest in video frame images. ∎

In short, a **maximal nucleus cluster** (**MNC**) is a cluster that is a collection of Voronoï regions in which the nucleus (center of the cluster) has the highest number of adjacent polygons. It is possible for a Voronoï tessellation of a digital image to have more than one MNC. Each cluster polygon is a Voronoï region of a generating point. One way to find an approximate contour of a cluster is to connect each neighbouring pair of adjacent polygon generating points with a straight edge.

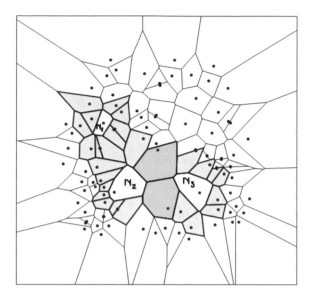

Fig. 7.8 $\mathfrak{C} N_1$ adjacent $\mathfrak{C} N_2$ and $\mathfrak{C} N_2$ overlaps $\mathfrak{C} N_3$

Example 7.10 **Sample Maximal Nuclei Cluster Contours**.
A map of the milkyway[2] is shown in Fig. 7.9.1. The contours of two maximal nucleus clusters are shown in Fig. 7.9.2. The orange • in Fig. 7.11 indicate the locations of pixels, each with a different gradient orientation angle, and which serve as mesh generators. The contours are found by connecting neighbouring pair of generating points of the Voronoï polygons that are adjacent to the cluster nucleus (see, e.g., Fig. 7.10).   ∎

There are two basic types of cluster contours useful in identifying object shapes in tessellated digital images.

1° **Fine Cluster Contour**. In polygons adjacent to a cluster nucleus, a **fine cluster contour** is a path in which each neighbouring pair of generating points is connected by a straight edge. In other words, a **file cluster contour** is a straight edge-connected path containing adjacent straight edges. Straight edges are adjacent, provided the edges have a common end-point.

2° **Coarse Cluster Contour**. In polygons that surround those polygons adjacent to a cluster nucleus, a **coarse cluster contour** is a path in which each neighbouring pair of generating points is connected by a straight edge.

---

[2]http://www.atlasoftheuniverse.com/milkyway2.jpg.

Fig. 7.9 Contours of
maximal nucleus clusters on
a milkyway image

7.9.1: milkyway

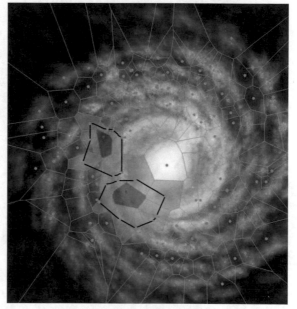

7.9.2: Cluster contours

Example 7.11 **Sample Coarse and Fine luster Contours.**
Two maximal nucleus clusters $\mathbb{C}N_1$, $\mathbb{C}N_2$ are shown in Fig. 7.10. Each of these clusters is surrounded by both fine- and coarse-contours. ■

Fig. 7.10 Straight-edge path-connected cluster contours on clusters $\mathbb{C}N_1$, $\mathbb{C}N_2$

7.6 Problems

Problem 7.12 Gradient orientation-based Voronoï mesh.
Capture two .mp4 files with 100 frames in each file and do the following in real-time during video capture.

1° For each frame image, find up to 100 pixels with different gradient orientations and locations.

2° Let S be the set of pixels with different gradient orientations found in Step 1.

3° For each frame, construct the Voronoï mesh $V(S)$, using S as the set of generating points.

4° Display two sample frames with mesh $V(S)$ superimposed on it.

5° Repeat Steps 1–4 for up to 300 pixel gradient orientations.

Problem 7.13 RGB-based Voronoï mesh.
Capture two .mp4 files with 100 frames in each file and do the following in real-time during video capture.

1° For each frame image, find up to 100 pixels with different colour intensities and locations.

2° Let S be the set of pixels with different colour intensities found in Step 1.

3° For each frame, construct the Voronoï mesh $V(S)$, using S as the set of generating points.

4° Display two sample frames with mesh $V(S)$ superimposed on it.

5° Repeat Steps 1–4 for up to 300 pixel colour intensities.

Problem 7.14 HSV-based Voronoï mesh.
Capture two .mp4 files with 100 frames in each file and do the following in real-time during video capture.

1° Convert each RGB frame to the HSV colour space.

2° For each frame image, find up to 100 pixels with different hue-values and locations, i.e., each pixel found will have a hue and value that is different from the other pixels in *img*.

3° Let S be the set of pixels with different hues and values found in Step 2.

4° For each frame, construct the Voronoï mesh $V(S)$, using S as the set of generating points.

5° Display two sample frames with mesh $V(S)$ superimposed on it.

6° Repeat Steps 2–5 for up to 300 pixel hue-value combinations.

Problem 7.15 Gradient orientation & green channel-based Voronoï mesh.
Capture two .mp4 files with 100 frames in each file and do the following in real-time during video capture.

1° For each frame image, find up to 100 pixels with different green channel colour intensity and gradient orientation combinations and locations, i.e., each pixel found will have a green intensity and gradient orientation that is different from the other pixels in each frame image.

2° Let S be the set of pixels with different hues and values found in Step 1.

3° For each frame, construct the Voronoï mesh $V(S)$, using S as the set of generating points.

4° Display two sample frames with mesh $V(S)$ superimposed on it.

5° Repeat Steps 1–4 for up to 300 pixel green intensity-gradient orientation combinations.

Problem 7.16 Fine Cluster Contours.
Capture three .mp4 files with 100 frames in each file and do the following in real-time during video capture.

1° capture video frames.

2° Select 100 corners in each frame image.

3° Tile (tessellate) each frame with a Voronoï diagram.

4° Recall that each Voronoï polygon is the **nucleus of a cluster**, which is a collection of polygons adjacent to a central polygon called the cluster nucleus. The focus of this step is on maximal nucleus clusters, i.e., a **maximal nucleus cluster** is a nucleus cluster in which the Voronoï nucleus polygon has a maximal number of adjacent polygons. In each frame, identify the Voronoï nuclei polygons with the maximum number of adjacent polygons.

5° False-colour with green each maximal nucleus polygon.

6° False-colour with yellow each of the polygon adjacent to the cluster nucleus.

7° **Fine Cluster Contours**. In each frame, identify the fine-contours of each maximal nucleus clusters. For the adjacent polygons surrounding each maximal polygon, connect each pair of *neighbouring* corners with a straight edge. For a sample pair of fine-contours on maximal nucleus clusters in a Voronoï tessellation of a milkyway image, see Example 7.10.

8° Repeat Step 1 for 300 corners.   ∎

Problem 7.17   Coarse Cluster Contours.
Instead of identifying the fine cluster contours, repeat the steps in Problem 7.16 and find the coarse cluster contours for the maximal nucleus clusters in each video frame. For two coarse cluster contours in the same image, see Example 7.11. **Important**: In your solution this problem, comment on which is more effective in identifying the shapes of frame object, fine cluster contours or coarse cluster contours.   ∎

Problem 7.18   Centroid-based Voronoï nucleus clustering.
Do Problem 7.16 for frame image centroids instead of frame corners to tessellate each video frame.   ∎

Problem 7.19   Gradient orientation-based Voronoï nucleus clustering.
Do Problem 7.16 for frame pixel gradient orientations instead of frame corners to tessellate each video frame.   ∎

Problem 7.20   Gradient orientation & green channel-based Voronoï nucleus clustering.
Do Problem 7.16 for frame pixel gradient orientations and green channel intensities instead of frame corners to tessellate each video frame.   ∎

Problem 7.21   Corner & green channel-based Voronoï nucleus clustering.
Do Problem 7.16 for frame corners and green channel intensities instead of just frame corners to tessellate each video frame. That is, for each frame image, find up to 100 corners with different green channel colour intensity combinations and locations, i.e., each corner found will have a green intensity that is different from the other corners in each frame image.   ∎

Problem 7.22   Corner & red channel-based Voronoï nucleus clustering.
Do Problem 7.16 for frame corners and red channel intensities instead of just frame corners to tessellate each video frame. That is, for each frame image, find up to 100 corners with different red channel colour intensity combinations and locations, i.e., each corner found will have a red intensity that is different from the other corners in each frame image.   ∎

Problem 7.23   Corner & blue channel-based Voronoï nucleus clustering.
Do Problem 7.16 for frame corners and blue channel intensities instead of just frame corners to tessellate each video frame. That is, for each frame image, find up to 100 corners with different blue channel colour intensity combinations and locations, i.e., each corner found will have a red intensity that is different from the other corners in each frame image.   ∎

Problem 7.24 Corner & RGB-based Voronoï nucleus clustering.
Do Problem 7.16 for frame corners and RGB colour intensities instead of just frame corners to tessellate each video frame. That is, for each frame image, find up to 100 corners with different RGB colour intensity combinations and locations, i.e., each corner found will have a RGB intensity that is different from the other corners in each frame image. ■

Problem 7.25 Frame Object Detection.
Do the following.

1° Select a digital image. Repeat the steps in Problem 7.16 to tile and find the contours of maximal nucleus clusters (MNCs) in the selected image. Choose and store (in a variable **Target**) a selected contour of a maximal nucleus cluster (MNC) in the tiled image.
Note: The Target contour will be used in the next steps to find contours in video frames that are similar to the Target contour.
Hint: Choose a target image containing objects that are similar to one or more of the objects in video frames made in the following steps. Also notice that *similarity* in this problem means **approximately the same**.

2° Repeat the steps in Problem 7.16 to capture and tile the frames in three .mp4 files in real-time.

3° Offline, do the following using the frames in the .mp4 files from Step 2.

(a) Let $S_1, \ldots, S_j, \ldots, S_n$ be MNC fine contours (shapes) in n frames in the selected .mp4 file.
(b) Select small number $\varepsilon > 0$. **Note**: This is the contour similarity threshold.
(c) Compute the similarity distance $D(Target, S_j)$, for $1 \leq j \leq n$, i.e., for each MNC contour found.
(d) Do the following for each video frame. If $D(Target, S_j) \leq \varepsilon$, then false colour the nucleus of the MNC for the S_j contour. In a separate .jpg file, save frame containing a contour of an MNC that has a shape similar to the Target contour.
(e) Record your findings in a table of comparisons between the known Target contour and at most 5 examples of contours of MNCs found in the captured .mp4 files:
Table:

Frame	Image	Image	Distance
j	Target	S_j	$D(Target, S_j)$

Hint: Use a Snipping Tool to capture the Target and S_j images for this Table.

4° Comment on the similarity between the Target vs. the video frame S_j contours (shapes) found.
5° Comment on which choice of the number ε that works best.
6° Repeat Steps 1 to 5 for 5 different Targets and 5 different videos of different scenes. ■

Problem 7.26 Gradient orientation-based fine cluster contour similarities.
Do Problem 7.25 for generating points that are frame pixel gradient orientations instead of frame corners to tessellate each video frame. That is, in each frame, choose generating points that are pixel gradient orientations instead of corners to test the similarity between a known shape and the shapes in captured .mp4 files.   ∎

Problem 7.27 RGB pixel intensities-based fine cluster contour similarities.
Do Problem 7.25 for frame RGB pixel intensities instead of frame corners to tessellate each video frame. That is, choose generating points that are RGB pixel intensities instead of corners to test the similarity between a known shape and the shapes in captured .mp4 files.   ∎

Problem 7.28 Green channel pixel intensities-based fine cluster contour similarities.
Do Problem 7.25 for generating points that are frame green channel pixel intensities instead of frame corners to tessellate each video frame. That is, in each frame, choose generating points that are frame green channel pixel intensities instead of corners to test the similarity between a known shape and the shapes in captured .mp4 files.   ∎

Problem 7.29 Corner and Green channel pixel intensities-based fine cluster contour similarities.
Do Problem 7.25 for generating points that are corners with different green channel pixel intensities to tessellate each video frame. That is, in each frame, choose generating points that are corners with different green channel pixel intensities instead of corners to test the similarity between a known shape and the shapes in captured .mp4 files.   ∎

Problem 7.30 Repeat the steps in Problem 7.25 by doing Problem 7.17 in Step 7.25.2. That is, measure the similarity between shapes by measuring the difference between the coarse cluster contour of a known image object *Target* and the coarse cluster contours that identify the shapes of objects in each video frame.   ∎

Problem 7.31 This problem focuses on coarse cluster contours (call them coarse perimeters). Consider three levels of coarse perimeters:

S1P: Level 1   coarse perimeter (our our starting point–call it supra 1 perimeter or briefly S1P).
S2P: Level 2   coarse perimeter (supra 2 perimeter or briefly S2P) that contains a supra 1 perimeter.
S3P: Level 3   coarse perimeter (supra 3 perimeter or briefly S3P) that contains a S2P and S1P.

Level 3 is unlikely.
The occurrence of S2P containing S1P is the promising case in terms of object recognition. Do the following:

1° Detect when a S1P is contained in a S2P. Announce this in the work space along with the lengths of the S1P and S2P perimeters. Put a tiny circle label (1) on the S1P and a (2) on S2P.

2° Detect when S3P contains S2P. Announce this in the work space along with the lengths of the S1P and S2P perimeters. Put a tiny circle label (2) on S2P, (3) on S3P.

3° Detect when S3P contains S2P and S2P contains S1P. Announce this in the work space along with the lengths of the S1P, S2P, S3P perimeters. Put a tiny circle label (1) on the S1P and a (2) on S2P and a (3) on S3P.

4° Detect when S2P does not contain S1P and S3P does not contain S1P. Announce this in the work space along with the lengths of the S1P, S2P, S3P perimeters. Put a tiny circle label (1) on the S1P and a (2) on S2P.

5° Produce a new figure that suppresses (ignores) MNCs on the border of an image and displays S1P (case 1).

6° Produce a new figure that suppresses (ignores) MNCs on the border of an image and displays S1P, S2P (case 2). Include (1), (2) circle labels. Announce this in the work space along with the lengths of the S1P and S2P perimeters.

7° Produce a new figure that suppresses (ignores) MNCs on the border of an image and displays S1P, S2P, S3P (case 3). Announce this in the work space along with the lengths of the S1P, S2P, S3P perimeters. Put a tiny circle label (1) on the S1P and a (2) on S2P and a (3) on S3P.

| **Suggestion by Drew Barclay** : | Select SnP contained within S(n+1)P so that the line segments making up each of the contours do not ever intersect. In addition, the minimum and maximum X/Y values have greater absolute values for S(n+1)P.

| **For this problem** |: Try object recognition in a traffic video to see which of the above cases works best.

| **Hint** : | Crop frame 1 of a video and use that crop for each of the video frames. Let k equal the number of SURF keypoints selected. Try $k = 89$ and $k = 377$ to see some very interesting coarse perimeters. ∎

An **edgelet** is a set of edge pixels. An edgelet has been found by S. Belongie, J. Malik and J. Puzicha [13] to be useful in the study of object shapes. In the context of maximal nucleus cluster (MNC) contours, a **contour edgelet** is a set of edge pixels in the contour of an MNC, restricted to just the mesh generating points that are the endpoints of the edges in a MNC contour. Let be the the the ith edgelet e_i be defined by

$$e_i = \{g \in MNC : g \text{ is a mesh generator of a MNC polygon}\}.$$

Let $|e_i|$ be the number of mesh generators in a mesh contour edgelet and let $Pr(e_i)$ (probability of the occurrence of edgelet e_i) be defined by

$$Pr(e_i) = \frac{1}{|e_i|} = \frac{1}{\text{size of } e_i} \text{ (MNC Contour Edgelet Probability)}.$$

In a sequence of tessellated video frames, let m_i be the frequency of occurrence of edgelets with the same number of mesh generators as edgelet e_1. For edgelets $e_1, e_2, \cdots, e_i, \cdots, e_k$ in k video frames, let $m_1, m_2, \cdots, m_i, \cdots, m_k$ be the frequencies of occurrence of the k edgelets. The histogram for the edgelet frequencies defines the shape context of the edgelets. The underlying assumption is that if a pair of edgelets e_i, e_j each has the same number of mesh generators, then e_i, e_j will have have similar shapes.

Problem 7.32 Let V be a video containing Voronoï tessellated frames. Do the following:

1. Crop each video frame (select only one or more the central rectangular regions, depending on the size of the rectangles used to partition a video frame). Work with the central rectangular region for the next steps.
2. Tessellate each frame using SURF points as mesh generators. Experiment with the number of SURF points to use as mesh generators, starting with 10 SURF points.
3. Find the MNCs in each video frame.

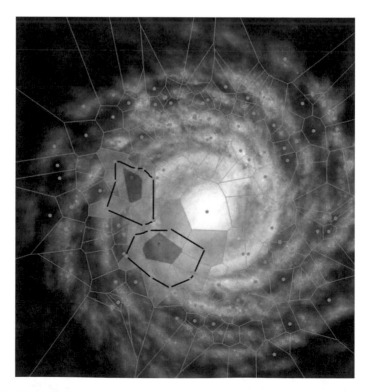

Fig. 7.11 Cluster contours

4. Find the edgelets $e_1, e_2, \cdots, e_i, \cdots, e_k$ in k video frames. Display each of the edgelets found in two of the video frames.

5. Determine $|e_i|$, the number of mesh generators in mesh contour edgelet e_i.

6. Display the edgelet shapes (a) by themselves and (b) superimposed on an MNC. Hint: extract an MNC from a video frame and display the MNC by itself with superimposed edgelet. See, e.g., Fig. 7.12.

7. Find the frequencies $m_1, m_2, \cdots, m_i, \cdots, m_k$ of occurrence of the k edgelets. That is, for each edgelet e_i, determine the number of edgelets that have the same size $|e_i|$ as e_i. For example, if three edgelets have size $|e_i|$ for edgelet e_1, then $m_1 := 3$.

8. Compute $Pr(e_i)$, the probability of the occurrence of edgelet e_i, for each of the edgelets in each of the video frames.

9. Give Matlab script to display a histogram for the frequencies $m_1, m_2, \cdots, m_i, \cdots, m_k$.

10. Give Matlab script to display a compass plot for frequencies $m_1, m_2, \cdots, m_i, \cdots, m_k$.

Fig. 7.12 Edgelet in a cluster contour

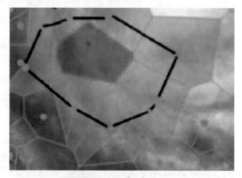

7.12.1: edgelet

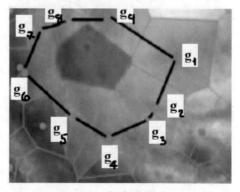

7.12.2: edgeletLabels

11. Give Matlab script to display a log-polar plot for edgelet frequencies. **Hint**: the basic approach is to bin the edgelet frequencies of a video frame into a polar histogram. For examples, see O. Tekdas and N. Karnad [192, Fig. 3.1, p. 8].
12. Give Matlab script to display a plot $Pr(e_i)$ against e_i.
13. Give Matlab script to display a 3D contour plot $Pr(e_i)$ against e_i and m_i. **Hint**: For a sample solution, see Matlab script A.34 in Appendix A.7.1.   ■

Let N be the sample size used to study video frame edgelets. For instance, if we shoot a video with 150 frames, then $N := 150$. For this work, N equals the number of frames containing tessellated video images. The chi-squared distribution[3] χ_s^2 is a measure of the deviation of a sample s from the expectation for the sample s and is defined by

$$\chi_s^2 = \sum_{i=1}^{k} \frac{m_i - NPr(e_i)}{NPr(e_i)}.$$

Eisemann-Klose-Magnor Shape Cost Function
$$C_{shape}(e_i, e_j)$$

The Eisemann-Klose-Magnor cost $C_{shape}(e_i, e_j)$ between two shape contexts [44, p. 10] e_i, e_j is defined to be the χ_s^2 for the pair of shape contexts.

Problem 7.33 Let V be a video containing Voronoï tessellated frames. Do the following:

1. Repeat steps 1 to 8 in Problem 7.32.
2. Give Matlab script to compute χ_s^2 for a tessellated video frame.
3. Give Matlab script to plot χ_s^2 values for 10 sample videos.   ■

7.7 Shape Distance

The focus here is on computing what is known as the cost for the distance between MNC contour edgelets. The approach here is an extension of the basic notion of a cost function for distance introduced by M. Eisemann, F. Klose and M. Magnor [44, p. 10]. Let e_i, e_j be edgelets and let $a, b > 0$ be constants used to adjust the cost function $C_{dist}(e_i, e_j)$ defined by

$$C_{dist}(e_i, e_j) = \frac{a}{\left(1 + e^{-b\|e_i - e_j\|}\right)}.$$

---

[3] http://mathworld.wolfram.com/Chi-SquaredTest.html.

The selection of a and b is based on arriving at the maximal cost of the distance between e_i and e_j. For example, let $b := 1$ and let $a := D(e_i, e_j)$ (Čech distance between the pair edgelet point sets) defined by

$$D(e_i, e_j) = \min \left\{ \|x - y\| : x \in e_i, y \in e_j \right\}.$$

We are interested in defining a cost function of the distance between a target MNC contour edgelet e_{target} and an sample edgelet e_j in a video. Then, for example, $C_{dist}(e_{target}, e_j)$ is defined by

$$C_{dist}(e_{target}, e_j) = \left. \frac{a}{\left(1 + e^{-b\|e_{target} - e_j\|}\right)} \right|_{a = D(e_{target}, e_j), b = 1}.$$

Problem 7.34 ☕

Let V be a video containing Voronoï tessellated frames. Do the following:

1. Crop each video frame (select only one or more the central rectangular regions, depending on the size of the rectangles used to partition a video frame). Work with the central rectangular region for the next steps.
2. Tessellate each frame using SURF points as mesh generators. Experiment with the number of SURF points to use as mesh generators, starting with 10 SURF points.
3. Select an edgelet e_{target} that is the set of generators that are the endpoints of the edges along a fine contour of a target object shape.
4. Select an edgelet e_j from a sample video. The selected edgelet should be extracted from a video frame containing a MNC contour that is similar to the known target shape. In other words, to select e_j, verify that

$$\left| D(e_{target}, e_j) - shapeDiff(e_{target}, e_j) \right| \le \varepsilon.$$

See Sect. 7.5 in this book for the methods used to compute $D(e_{target}, e_j)$ and $shapeDiff(e_{target}, e_j)$. It may be necessary to pad e_{target} or e_j with zeros, if one of these edgelets has fewer edge pixels than the other edgelet.
Hint: Check the number of pixels in both edgelets, before attempting to compute the distance between e_{target} and e_j.
5. Give Matlab script to compute the cost distance function $C_{dist}(e_{target}, e_j)$.
6. Repeat steps 1 to 5 for other choices of edgelets e_j and the same target. Also, experiment with other choices of a, b in the cost distance function.
7. Repeat steps 1 to 5 for other choices of edgelets e_j in 10 different videos and a different target. Also, experiment with other choices of a, b in the cost distance function.
8. Give Matlab script to display a 3D contour plot $C_{dist}(e_{target}, e_j)$ against a and b for the 10 selected videos.
9. Comment on the choices of a and b in computing $C_{dist}(e_{target}, e_j)$. ∎

7.8   Weight Function for Edgelets

In general, for edgelets e_i, e_j, the Eisemann-Klose-Magnor weight function $C\left(e_i, e_j\right)$ is defined by

$$C\left(e_i, e_j\right) = C_{dist}\left(e_i, e_j\right) + C_{shape}\left(e_i, e_j\right).$$

For this work, this cost function is specialized in terms of a pair edgelets e_{target}, e_j, giving rise to $C\left(e_{target}, e_j\right)$ defined by

$$C\left(e_{target}, e_j\right) = C_{dist}\left(e_{target}, e_j\right) + C_{shape}\left(e_{target}, e_j\right).$$

Problem 7.35 Let V be a video containing Voronoï tessellated frames. Do the following:

1. Repeat Steps 1 to 5 in Problem 7.34.
2. In Problem 7.34.5, include in your Matlab script the computation of the shape cost $C_{shape}\left(e_{target}, e_j\right)$.
3. In Step 2 of this Problem, compute the overall cost $C\left(e_{target}, e_j\right)$.
4. Repeat Steps 7.35.1 to 7.35.3 for 10 different videos and different targets.
5. Give Matlab script to display a 2D plot $C\left(e_{target}, e_j\right)$ against videos $1, \cdots, 10$.
6. Give Matlab script to display a 3D contour plot $C\left(e_{target}, e_j\right)$ against $C_{dist}\left(e_{target}, e_j\right)$ and $C_{shape}\left(e_{target}, e_j\right)$ for the 10 selected videos.
7. Comment on the results of the 2D and 3D plots obtained.   ∎

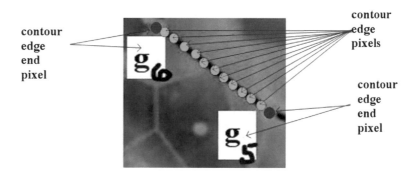

Fig. 7.13 Contour edge pixels from the edgelet in Fig. 7.12.2

7.9   Maximum Edgelets

The first new element in object recognition in tessellated images is the introduction of *maximum* MNC contour edgelets (denoted by **maxe_l**) that are edgelets containing all contour edge pixels, i.e.,

max$|e_i|$ = no. of contour edge pixels , **not** just straight edge endpoints.

Example 7.36 Let e_i denote the edgelet Fig. 7.12.2, which is the ith edgelet in a collection of tessellated video frames. Edgelet e_i contains 9 mesh generating points. In other words, e_i is not maximal. To obtain maxe_i, identify all of the edge pixels along each contour straight edge. For example, maxe_i would include the endpoints g_5, g_6 as well as the interior pixels in the contour straight edge $\overline{g_5 g_6}$ shown in Fig. 7.13.   ∎

Problem 7.37 Let V be a video containing Voronoï tessellated frames. Write a Matlab script to do the following:

1. Crop each video frame (select only one or more the central rectangular regions, depending on the size of the rectangles used to partition a video frame). Work with the central rectangular region for the next steps.
2. Tessellate each frame using SURF points as mesh generators. Experiment with the number of SURF points to use as mesh generators, starting with 10 SURF points.
3. Find the MNCs in each tessellated frame.
4. Display the MNC in the tessellated frame. Highlight the nucleus in red and the polygons surround the nucleus in yellow, all with with 50% opacity (adjust the opacity so that the subimage underlying the MNC can be clearly see).
5. For the selected MNC, determine the maximum fine contour edgelet for the MNC (call it maxe_i).
6. Display (plot) the points in maxe_i by itself.
7. Display (plot) the points (in red) in maxe_i superimposed on the image MNC.
8. Repeat Steps 1 to 7 for 10 different videos.
9. Comment on the results obtained.   ∎

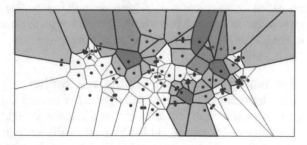

Fig. 7.14 Four sample MNCs in a Voronoï mesh

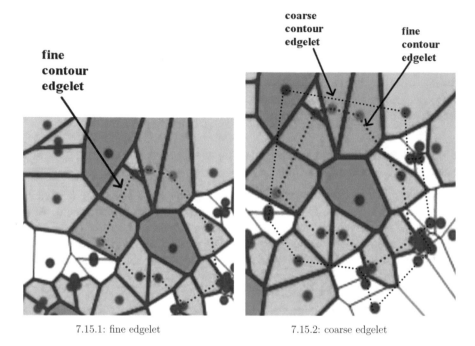

fine contour edgelet

coarse contour edgelet

fine contour edgelet

7.15.1: fine edgelet   7.15.2: coarse edgelet

Fig. 7.15 Fine and coarse MNC contour edgelets

7.9.1   Coarse Contour Edgelets

The second new element is the inclusion of coarse contour edgelets in the study of object shapes in tessellated digital images. Until now, the focus has been on fine contour edgelets define by the straight edges connecting the generating points for all polygons adjacent to the nucleus in a mesh MNC. Now we want to consider edgelets defined by the straight edges connecting the generating points for the polygons surround the fine contour polygons.

Example 7.38 Let $\mathbf{maxe}_{fine}$ denote a maximum fine contour edgelet and let $\mathbf{maxe}_{coarse}$ denote a maximum coarse contour edgelet. For example, in Fig. 7.15.1, the dotted lines $\cdots\cdots$ represent the endpoints and interior straight edge pixels in a fine MNC contour edgelet $\mathbf{maxe}_{fine}$ in the Voronoï mesh shown in Fig. 7.14. In Fig. 7.15.1, the dotted lines $\cdots\cdots$ represent the endpoints and interior straight edge pixels in a coarse MNC contour edgelet $\mathbf{maxe}_{coarse}$ in the Voronoï mesh shown in Fig. 7.14. ■

Problem 7.39 Let V be a video containing Voronoï tessellated frames. Write a Matlab script to do the following:

1. Crop each video frame (select only one or more the central rectangular regions, depending on the size of the rectangles used to partition a video frame). Work with the central rectangular region for the next steps.
2. Tessellate each frame using SURF points as mesh generators. Experiment with the number of SURF points to use as mesh generators, starting with 10 SURF points.
3. Find the MNCs in each tessellated frame.
4. Display the MNC in the tessellated frame. Highlight the nucleus in red and the polygons surround the nucleus in yellow, all with with 50% opacity (adjust the opacity so that the subimage underlying the MNC can be clearly see).
5. For the selected MNC, determine the maximum fine contour edgelet for the MNC (call it maxe_i).
6. Display (plot) the points in maxe_i by itself.
7. Display (plot) the points (in red) in maxe_i superimposed on the image MNC.
8. Repeat Steps 1 to 7 for 10 different videos.
9. Comment on the results obtained.   ■

7.9.2   Connected Mesh Regions that are MNCs

Let $MNC1$, $MNC2$ be a pair of maximal nucleus clusters in a Voronoï tessellation. $MNC1$, $MNC2$ are connected, provided the MNCs have adjacent polygons or have at least one polygon in common (see, e.g., Fig. 7.16).

Fig. 7.16 Three connected MNCs in a Voronoï mesh

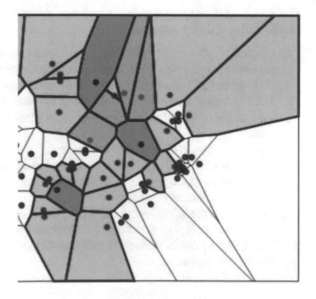

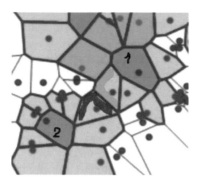

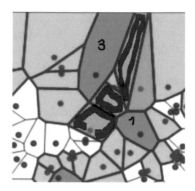

7.17.1: MNC1, MNC2 have adjacent Voronoï mesh polygons

7.17.2: MNC1, MNC3 share three Voronoï mesh polygons

Fig. 7.17 Two forms of connected MNCs

Example 7.40 Let *MNC*1, *MNC*2 be represented in Fig. 7.17.1. This pair of MNCs is connected, since they have a pair of adjacent Voronoï regions. It is also possible for a pair of MNCs to have one or more Voronoï regions in common. In Fig. 7.17.2, *MNC*1, *MNC*3 have three polygons in common. In both cases, these MNCs are considered strongly connected, since they share more than one pixel. ∎

Problem 7.41 Let *V* be a video containing Voronoï tessellated frames. Write a Matlab script to do the following:

1. Crop each video frame (select only one or more the central rectangular regions, depending on the size of the rectangles used to partition a video frame). Work with the central rectangular region for the next steps.
2. Tessellate each frame using SURF points as mesh generators. Experiment with the number of SURF points to use as mesh generators, starting with 10 SURF points.
3. Find the MNCs in each tessellated frame.
4. Determine if the MNCS are connected by adjacent polygons. If a pair of adjacent MNCs are found, display the MNCs and highlight in orange (or some other bright colour) the adjacent polygons.
5. Determine if the MNCS are connected by shared polygons, i.e., MNCs that have one or more polygons in common. If a pair of MNCs with shared polygons are found, display the MNCs and highlight in orange (or some other bright colour) the shared polygons.

6. Display each MNC in the tessellated frame. Highlight the nucleus in red and the polygons surround the nucleus in yellow, all with with 50% opacity (adjust the opacity so that the subimage underlying the MNC can be clearly see).
7. For the selected MNC, determine the maximum fine contour edgelet for the MNC (call it maxe_i).
8. Display (plot) the points in maxe_i by itself.
9. Display (plot) the points (in red) in maxe_i superimposed on the image MNC.
10. Repeat Steps 1 to 9 for 10 different videos.
11. Comment on the results obtained. ∎

Chapter 8
Lowe Keypoints, Maximal Nucleus Clusters, Contours and Shapes

Fig. 8.1 30 keypoints in Dirichlet tessellation of CN train video frame

This chapter carries forward the use of Voronoï meshes superimposed on digital images as a means revealing image geometry and the shapes that result from contour lines surrounding maximal nucleus clusters (MNCs) in a mesh. Recall that every polygon in a Voronoï mesh is a nucleus of a cluster of polygons. Here, the term **nucleus** refers to the fact that in each MNC, there is always a polygon that is a cluster center. For more about this, see Appendix B.12 (**MNC**) and Appendix B.13 (**nucleus**).

The focus of this chapter on an image geometry approach to image and scene analysis. To facilitate image and scene analysis, a digital image can be viewed as a set of points (pixels) susceptible to the whole spectrum of mathematical structures commonly found in geometry and in the topology of digital images.

© Springer International Publishing AG 2017

J.F. Peters, *Foundations of Computer Vision*, Intelligent Systems ʼ
Reference Library 124, DOI 10.1007/978-3-319-52483-2_8

The typical geometry structures found in image geometry include points, lines, circles, triangles, and polygons as well equations that specify the positions and configurations of the structures. In other words, **image geometry** is an analytic geometry view of images. In digital images, these geometric structures also include image neighbourhoods, image clusters, image segments, image tessellations, collections of image segment centroids, sets of points nearest a particular image point such as a region centroid, image regions gathered together as near sets, adjacent image regions, and the geometry of polygonal image regions.

Another thing to notice that is the topology of digital images. A **topology of digital images** (or **image topology**) is the study of the nearness of pixels to sets of pixels. Such a topological approach to digital images leads to meaningful groupings of the parts of a digital image that includes sets of pixels with border pixels excluded (open sets) or sets of pixels that with border pixels included (closed sets). This basic approach in the study of images is a direct result of A. Rosenfeld's discovery of 4- and 8-neighbourhoods [170] (see, also, [94, 142]).

A **tessellation** of an image is a tiling of the image with polygons. The polygons can have varying numbers of sides.

Example 8.1 **Sample Corner-Based Tessellated Image**.

 (4gon covering most of the fisherman's hat).
An example of a Dirichlet (*aka* Voronoï) tessellated image is the fisherman shown in Fig. 8.1. Matlab script Appendix A.2 is used to produce this image tiling with 233 corners. The red dots ● identify locations of corners in the image. Notice the red dots in the extreme box corners of the image (these are added to provide a more accurate tessellation). Without these box corners, the tiling of the image drifts into infinity. The polygons in this form of tiling have varying numbers of sides. A more accurate tiling of an image image results from what are known as SIFT (scale-invariant feature transform) keypoints. More about this later.   ■

The fact that structured images reveal hidden information in images is the main motivation for tessellating an image. Structured images are then analysed in terms of their component parts such as subsets of image regions, local neighbourhoods, regional topologies, nearness and remoteness of sets, local convex sets and mappings between image structures.

8.1   Image Analysis

Image analysis focuses on various digital image measurements, e.g., pixel size, pixel adjacency, pixel feature values, pixel neighbourhood membership, pixel gradient orientation, pixel gradient magnitude, pixel intensity (either colour channel intensity or greyscale intensity), pixel intensities distribution (histograms), closeness of image neighbourhoods, binning (collecting pixel intensities within a particular range in a bin or bucket) image size, and image resolution (see, e.g., Fig. 8.2 showing HSV pixel intensities). Another important part of image analysis is the detection of pixel features that are invariant to image scaling, translation, rotation and partially invariant to illumination changes and affine or 3D projection. This can be done using D.G. Lowe's SIFT (Scale-Invariant Feature Transform) [115] (see, also, [116]). For more about this, see Sect. 8.8.

Each image pixel has its own geometry with respect to gradient orientation, gradient magnitudes in the x- and y-directions, and what is known as edge strength. The **edge strength** of a pixel equals its gradient magnitude. The implementation of SIFT leads to the detection of what are known as image keypoints, which are those pixels that have different gradient orientations and gradient magnitudes. For more about this, see Sect. 8.6.

Binning provides a basis for the construction of image histograms. An excellent introduction to binning is given by W. Burger and M.J. Burge [21, Sect. 3.4.1] (for more about this, see Sect. 3.1).

Fig. 8.2 HSV pixel intensities

There are a number of important region-based approaches in image analysis are **isodata thresholding** (binarizing images), **Otsu's method** (greyscale image thresholding), **watershed segmentation** (computed using a distance map from foreground pixels to background regions), **maximum Voronoï mesh nuclei** (identifying mesh polygons with the maximum number of sides), and **non-maximum suppression** (finding local maxima by suppressing all pixels that are less likely than their surrounding pixels) [211]. In image analysis, object and background pixels are associated with different adjacencies (neighbourhoods) [3]. There are two basic types of neighbourhoods, namely, adjacency neighbourhoods [102, 170] and topological neighbourhoods [77, 142]. Using different geometries, an adjacency neighbourhood is defined by pixels adjacent to a given pixel. Adjacency neighbourhoods are commonly used in edge detection in digital images.

Fig. 8.3 Fingerprint Pixel Keypoints = ●

8.2   Scene Analysis

Scene analysis focus on the structure of digital images.

An **image scene** is a snapshot of what a camera sees at an instant in time. Image structures include shapes of image objects, dominant image shapes, and image geometry (e.g., subimages bounded by polygons in a tessellated image). Video frames offer a perfect hunting ground for image structures that change or remain fixed over time.

Fig. 8.4 HSV hues

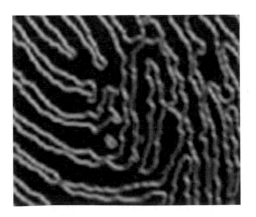

A **visual scene** is a collection of objects in a visual field that captures our attention. In human vision, a **visual field** is the total area in which objects can be seen. A normal visual field is about 60° from the vertical meridian of each eye and about 60° above and 75° below the horizontal meridian. A sample Dirichlet tessellation of a 640×480 digital image containing fisherman scene is shown in Fig. 8.1. Here, the locations of up to 60 image key colour-feature values are the source of sites used to generate the fisherman diagram. The ● indicates the location of a keypoint, i.e., a pixel with a particular gradient orientation (Fig. 8.3).

Example 8.2 **Fingerprint Pixel Keypoints**.
Examples of pixel keypoints (each keypoint pixel has a different gradient orientation and gradient magnitude) are shown as red ● bullets in the fingerprint subimage in Fig. 8.3. ∎

Example 8.3 **Visualizing Edge Pixel Gradients**.
The fingerprint subimage in Fig. 8.4 displays colourized pixel gradients in the HSV colour space, i.e., the three gradients for each pixel are used to select HSV channel values, where

$$\{hue, saturation, value\} = \{orientation, xGradientMag, yGradientMag\}.$$

The edge pixels of a fingerprint are displayed with varying colours and intensities in Fig. 8.2. The HSV channel values for each pixel are determined using the gradient orientation (**H**ue), gradient magnitude in the x-direction (**S**aturation) and gradient magnitude in the y-direction (**V**alue) of each edge pixel are combined to achieve a visualization of the pixel gradient information. To see how this is done, see the Mathematica script 6 in Appendix A.8.5. Try doing the same thing using the RGB and Lab colour spaces. Recall that the CIE Lab color space describes colours visible to the human eye. **Lab** is a 3D color space model, where **L** represents lightness of the colour, the position of the colour between red/magenta and green along an **a** axis, and the position of the colour between yellow and blue along a **b** axis. **Hint**: see Matlab script A.37 and Mathematica script 7 in Appendix A.8.5. ∎

Source of Image Geometry Information

The important thing to notice is that a Voronoï region $V(p)$ is a set of all points nearer to a particular generating point $p \in S$ than to any other generating point in the set of generating points S. Hence, the proximity of each point in the interior of a Voronoï region $V(p)$ is a source of image geometry information about a particular generating point such as a corner, centroid, or keypoint in a digital image scene. ∎

The scale-invariant feature transform (SIFT) transform by David Lowe is usually used for image keypoint detection.

Fig. 8.5 Keypoints $= +$

The foundations for scene analysis are built on the pioneering work by A. Rosenfeld work on digital topology [98, 168–172] (later called digital geometry [94]) and others [39, 99, 102, 104, 105]. The work on digital topology runs parallel with the introduction of computational geometry by M.I. Shamos [175] and F.P. Preparata [158, 159], building on the work on spatial tessellations by G. Voronoi [201, 203]and others [27, 53, 64, 103, 124, 196].

To analyze and understand image scenes, it is necessary to identify the objects in the scenes. Such objects can be viewed geometrically as collections of connected edges (e.g., skeletonizations or edges belonging to shapes or edges in polygons) or image regions viewed as sets of pixels that are in some sense near each other or set of points near a fixed point (e.g., all points near a site (also, seed or generating point) in a Voronoï region [38]). For this reason, it is highly advantageous to associate geometric structures in an image with mesh-generating points (sites) derived from the fabric of an image. Image edges, corners, centroids, critical points, intensities, and keypoints (image pixels viewed as feature vectors) or their combinations provide ideal sources of mesh generators as well as sources of information about image geometry.

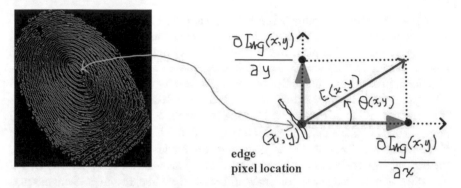

Fig. 8.6 Image geometry: pixel gradient orientation at location (x, y)

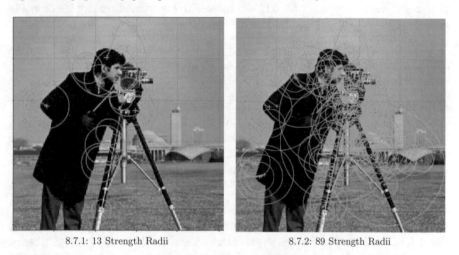

8.7.1: 13 Strength Radii   8.7.2: 89 Strength Radii

Fig. 8.7 Two sets of intensity image edge pixel strengths represented by circle radii magnitudes. The orientation angle of each radius corresponds to the gradient orientation of the circle center keypoints

8.3 Pixel Edge Strength

This section briefly looks at pixel edge strength (also called pixel gradient magnitude).

The edge strength of pixel $Img(x, y)$ (also called the pixel gradient magnitude) is denoted by $E(x, y)$ and defined by

$$E(x, y) = \sqrt{\left(\frac{\partial Img(x, y)}{\partial x}\right)^2 + \left(\frac{\partial Img(x, y)}{\partial y}\right)^2} \quad \textbf{(Pixel edge strength)}$$

$$= \sqrt{G_x(x, y)^2 + G_y(x, y)^2}.$$

Example 8.4 **Intensity Image Pixel Edge Strengths**.
The length of the radius of the large circle surrounding the head in Fig. 8.5 represents the edge strength of the pixel in the center of this circle. The angle of this radius (approximately $75°$) represents the gradient orientation of the central pixel. The circles themselves are called **gradient orientation circles**. Each circle center is called a **keypoint** (called a Speeded-Up Robust Features (**SURF**) point in Matlab, which is a rendition of the original D.G. Lowe Scale Invariant Feature Transform (**SIFT**) point [116]). In Mathematica®, keypoints are called **interest points**. This large head circle comes from the collection of 13 gradient orientation circles in Fig. 8.7.1. Again, for example 89 gradient orientation circles are displayed in Fig. 8.7.2. The SURF points found in an image will each have different edge strength and gradient orientation. Using the edge strength of each pixel, it is possible to use the $\leq$ to induce a partial ordering of the all keypoints found in an image. To experiment finding keypoints in different intensity images, try the Matlab script A.62 in Appendix A.8.3.   ∎

It is possible to control how many SURF points to choose as mesh generators. This choice is important, since typical scenes either in single shot images or, especially, in video frames, there are many different objects.[1]

Example 8.5 **Colour Image Pixel Edge Strengths**.
A collection of 13 gradient orientation circles in a colour image are displayed in Fig. 8.14.2. Again, for example 89 gradient orientation circles are displayed in Fig. 8.14.3. To experiment finding keypoints in different colour images, try the Matlab script A.62 in Appendix A.8.3. In this script, an extra step is needed to convert the colour image to an intensity image.   ∎

Remark 8.6 **Edge Pixel Strength Geometry**.
A sample pixel edge strength is represented by the length of the hypotenuse in see Fig. 8.7.1. This is part of the image geometry shown in Fig. 8.6, illustrated in terms of the edge pixels along the whorls of a fingerprint. Here is a summary of the results for two pixel edge strengths experiments.

cameraman.tif   180 keypoints found (13 and 89 edge strengths displayed in the intensity image in Fig. 8.7).

fisherman.jpg   1051 keypoints found (13 and 89 edge strengths displayed in the intensity image in Fig. 8.14).

The analog of edge pixel strength in 2D images is the length of the radius of a sphere with a keypoint at is center in a 3D image. In either case, keypoints provide a basis for object recognition and solid foundation for the study of image geometric patterns in 2D and 3D images. A common approach in the study of image objects and geometry is to used keypoints as generators of either Voronoï or Delaunay tessellations of images. In either case, the result image mesh reveals clusters of polygons. Recall that every mesh polygon is the nucleus of a mesh nerve. Often, interest (key) points

---

[1] See answer 171744 in http://www.mathworks.com/matlabcentral/answers/ for an approach to selecting the number of SURF points.

tend to cluster around mesh polygons in image regions where the image entropy (and corresponding information levels) is highest. Those high entropy nucleus mesh cluster are good hunting grounds for the recognition of image objects and patterns. Mesh nucleus clusters are examples of Edelsbrunner-Harer nerves [42] (see, also, [148, 150]). ∎

Fig. 8.8 Sample cropped traffic image

8.4 Cropping and Sparse Representations of Digital Images

For complex video frames such as traffic video frames, it is necessary to crop each frame and then select only a portion of the cropped frame to tessellate. By **cropping an image**, we mean *removing the outer parts of an image to isolate and magnify a region of interest*. See, for example, the regions in the sample cropped traffic video frame in Fig. 8.8. For a traffic video, a promising approach is to crop the central part of each frame. See, e.g., P. Knee [95] for common approaches to cropping an image. A **sparse representation** of an image is some form of either a reduction or expansion of the image. Repeated sparse representation results in a sequence of images called a Gaussian pyramid by P.J. Burt and E.H. Adelson [22].

Remark 8.7 **Sparse Representations**.
Sparse representation of digital images is a promising research area (basically, a followup to the approach suggested by P. Knee [95]). See, e.g., P.J. Burt and E.H. Adelson [22] and, more recently, for scalable visual recognition by B. Zhao and E.P. Xing [219]. The article by B. Zhao and E.P. Xing not only presents an

Fig. 8.9 Sample traffic video frame image

Fig. 8.10 Sample cropped traffic video frame image

interesting approach to scalable visual recognition, it also gives an extensive review of the literature for this research area in computer vision. For a Matlab script that illustrates a Gaussian pyramid scheme. A **pyramid scheme** is a construction of a sequence of gradual image changes. The MathWorks approach in a Gaussian pyramid scheme yields two different ways by constructing a sequence of image reductions or a sequence of image expansions, see script A.35 in Appendix A.8.1.   ■

Example 8.8 **Sparse Representations of a Cropped Image**.
A sample traffic video frame is shown in Fig. 8.9. This image is complex and has more information than we want. In the search for interesting shapes, it helps to crop a complex image, selecting that part of image we want to explore. For a sample cropped traffic image, see Fig. 8.10. Next try out reduction and expansion pyramid schemes on the cropped image, using script A.35 in Appendix A.8.1. For the sequence of reduced images, see Fig. 8.11.1 and for the sequence of expanded images, see Fig. 8.11.2. In the sequence of shadow shapes in Fig. 8.12, extracted from the expansion images in Fig. 8.11.1, notice that the second of these shadow shapes in Fig. 8.12.2

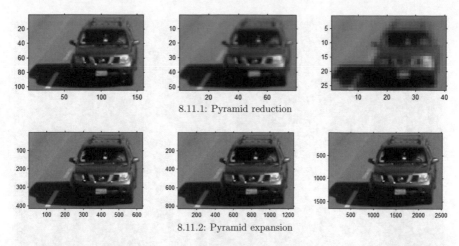

8.11.1: Pyramid reduction

8.11.2: Pyramid expansion

Fig. 8.11   Two sets of sparse representations of a cropped image

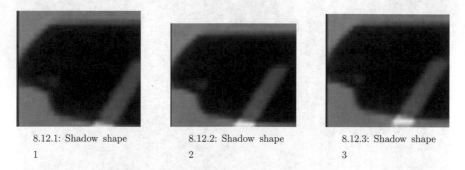

8.12.1: Shadow shape   8.12.2: Shadow shape   8.12.3: Shadow shape
1   2   3

Fig. 8.12   Sequence of auto shadow shapes from Fig. 8.11.2

is clearer than the first shadow shape in Fig. 8.12.1 as well as the third shadow shape in

Fig. 8.12.3. Then ⬛ in Fig. 8.12.2 provides a good laboratory in the study of image object shapes using computational geometry techniques from the previous chapters.   ∎

Example 8.9 **Sparse Representations Using Wavelets**.
Wavelets are a class of functions that localize a given function with respect to space and scaling [208]. For more about wavelets, see Y. Shimizu, Z. Zhang, R. Batres [177]. A sample wavelet based sparse representation pyramid scheme is in Fig. 8.13, using script 4 in Appendix A.8.2.   ∎

Fig. 8.13 Sample sparse representation pyramid scheme using wavelets

8.5   Shape Theory and the Shapes of 2D Image Objects: Towards Image Object Shape Detection

Basically, a **plane shape** like the auto shadow shape in Fig. 8.12.2 is a container for a spatial region in the plane. In the context of shape detection of objects in digital images, the trick is to isolate and compare shapes of interest in a sequence of images such as those found in a video. K. Borsuk was one of the first to suggest studying sequences of plane shapes in his theory of shapes [17]. For an expository introduction to Borsuk's notion of shapes, see K. Borsuk and J. Dydak [18]. Borsuk's initial study of shapes has led to a variety of applications in science (see, e.g., the shape of capillarity droplets in a container by F. Maggi and C. Mihaila [117] and shapes of 2D water waves and hydraulic jumps by M.A. Fontelos, R. Lecaros, J.C. López-Rios and J.H. Ortega [51]). For more about the basics of shape theory, see N.J. Wildberger [210]. For shapes from a physical geometry perspective with direct application in detecting shapes of image objects, see J.F. Peters [145].

Image object shape detection and object class recognition are of great interest in Computer Vision. For example, basic shape features can represented by boundary fragments and shape appearance can be represented by patches such as the

auto shadow shape in the traffic video frame Fig. 8.12.2. This is basic approach to image object shape detection by A. Opelt, A. Pinz and A. Zisserman in [133]. Yet another recent Computer Vision approach to image object shape detection reduces to the problem of finding the contours of an image object, which correspond to object boundaries and symmetry axes. This is the approach suggested by I. Kokkinos and A. Yuille in [97]. A promising approach in image object shape detection in video frames is to track changing image object contours (shapes) and minimizing an energy function that combines region, boundary and shape information. This approach shape detection in videos is given by M.S. Allili and D. Ziou in [5].

Basic Approach in Image Object Shape Detection

In the study of object shapes in a particular image, a good practice is to consider each object shape as a member of a class of shapes. A **class of shapes** is a set of shapes with matching features. Then image object shape detection reduces to checking whether the features of a particular image object shape match the features of a representative of a known class of shapes. The focus here is on shape membership in a particular class of shapes. In other words, a particular shape A is a **member of a known class** $\mathscr{C}$, provided the feature values of shape A match up with feature values of a representative shape in class $\mathscr{C}$. For example, an obvious place to look for a comparable shape feature is shape perimeter length. *Similar shapes have similar perimeter lengths.* If we add shape edge colour to the mix of comparable shape features, then colour image object shapes start dropping into different shape classes, depending on the shape perimeter length and perimeter edge colour of each shape.

8.14.1: Salerno Fisherman   8.14.2: 13 Strength Radii

8.14.3: 89 Strength Radii

Fig. 8.14 Two sets of colour image edge pixel strengths represented by circle radii magnitudes. The orientation angle of each radius corresponds to the gradient orientation of the circle center keypoints

8.6   Image Pixel Gradient Orientation and Magnitude

This section briefly looks at the derivation of pixel gradient orientation and gradient magnitudes.

Let Img be digital image and let $Img(x, y)$ equal the intensity of a pixel at location (x, y). Since $Img(x, y)$ is a function of two variables x, y, we compute the partial derivative of $Img(x, y)$ with respect to x $\left(\text{denoted by } \frac{\partial Img(x,y)}{\partial x}\right)$, which is the **gradient magnitude** of pixel $Img(x, y)$ in the x-direction. The partial derivative $\frac{\partial Img(x,y)}{\partial x}$ is represented, for example, by the ● on the horizontal axis in Fig. 8.6.

Similarly, $\frac{\partial Img(x,y)}{\partial y}$ is the **gradient magnitude** of pixel $Img(x, y)$ in the $y-$ *direction*, which is represented, for example, by the ● on the vertical axis in Fig. 8.6. Let $G_x(x, y)$, $G_y(x, y)$ denote the edge pixel gradient magnitudes in the x- and y- directions, respectively.

Example 8.10 **Two Approaches to Computing an Image Partial Derivative**. Let the Rosenfeld 8-neighbourhood of a 2D pixel intensity $f(x, y)$ (denoted by $Nbhd(f(x, y)))$ be defined by

$$Nbhd(f(x, y)) = \begin{bmatrix} f(x-1, y+1) & f(x, y+1) & f(x+1, y+1) \\ f(x-1, y) & f(x, y) & f(x+1, y) \\ f(x-1, y-1) & f(x, y-1) & f(x+1, y-1) \end{bmatrix} = \begin{bmatrix} 0 & 0 & 0 \\ 0 & 1 & 1 \\ 0 & 1 & 2 \end{bmatrix}.$$

The numbers such as $f(x-1, y) = 0$ $f(x, y) = 1$ $f(x+1, y) = 1$ are pixel intensities with the center of neighbourhood $Nbhd(f(x, y))$ at $f(x, y) = 1$ (very close to black). Next, use Li M. Chen's method of computing the discrete partial derivatives of $f(x, y)$ [29, Sect. 7.2.1, p. 84].

$$\frac{\partial f(x, y)}{\partial x} = f(x+1, y) - f(x, y) = 1 - 1 = 0,$$

$$\frac{\partial f(x, y)}{\partial y} = f(x, y+1) - f(x, y) = 0 - 1 = -1.$$

An alternative to Chen's method is the preferred widely used approach called the Sobel partial derivative given by J.L.R. Herran [78, Sect. 2.4.2, p. 23].

$$\frac{\partial f(x, y)}{\partial x} = \frac{2}{4} f(x+1, y) - f(x-1, y)$$

$$+ \frac{1}{4} f(x+1, y+1) - f(x-1, y+1)$$

$$+ \frac{1}{4} f(x+1, y+1) - f(x-1, y+1) = -\frac{2}{4} - \frac{2}{4} = -1.$$

$$\frac{\partial f(x, y)}{\partial y} = \frac{2}{4} f(x, y+1) - f(x, y-1)$$

$$+ \frac{1}{4} f(x+1, y+1) - f(x+1, y-1)$$

$$+ \frac{1}{4} f(x-1, y+1) - f(x-1, y-1) = \frac{2}{4} + \frac{2}{4} = 1.$$

The Sobel partial derivative is named after I. Sobel [181].   ∎

Fig. 8.15 Sample pixel
gradient orientations

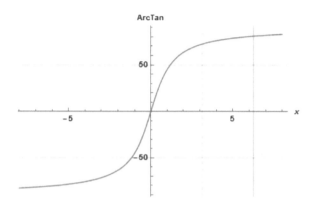

Example 8.11 **Sample arctan values**.
For sample arctan values (imagine pixel gradient orientations), see Fig. 8.15. To
experiment with plotting arctan values, see the Mathematica® notebook 5 in
Appendix A.8.4.   ∎

Let $\vartheta(x, y)$ be the **gradient orientation** angle of the edge pixel $Img(x, y)$ in
image Img. This angle is found by computing the arc tangent of the ratio the edge
pixel gradient magnitudes. Compute $\vartheta(x, y)$ using

$$\vartheta(x, y) = \tan^{-1}\left[\frac{\frac{\partial Img(x,y)}{\partial y}}{\frac{\partial Img(x,y)}{\partial x}}\right] = \tan^{-1}\left[\frac{G_y}{G_x}\right] \text{ (\textbf{Pixel gradient orientation}).}$$

Example 8.12 **Highlighted Gradient Orientations**.
The number of keypoints in an image can be massive. For this reason, visualizing
the locations of image keypoints is difficult. A solution to this problem is to use
colour to highlight the different keypoints in the their locations in an image. Highlighted pixel gradient orientations in a fingerprint are shown in Fig. 8.6. To experiment
with highlighting the gradient orientations of keypoints in different images, see the
Mathematica® notebook 6 in Appendix A.8.5.   ∎

8.16.1: Original image   8.16.2: $k=1.5, \sigma=5.55$

8.16.3: $k=1.5, \sigma=0.98$

Fig. 8.16 DoG images for different values of k, σ

8.7 Difference-of-Gaussians

A Difference-of-Gaussians (DoG) function is defined by convolving a Gaussian with an image at two different scale levels and computing the difference between the pair of convolved images. Let $Img(x, y)$ be an intensity image let $G(x, y, \sigma)$ be a variable scale Gaussian defined by

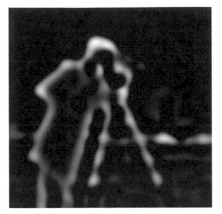

8.17.1: Original image   8.17.2: $k=1.5, \sigma=5.55$

8.17.3: $k=1.5, \sigma=0.98$

Fig. 8.17 Cameraman DoG images for different values of k, σ

$$G(x, y, \sigma) = \frac{1}{2\pi\sigma^2} e^{-\frac{x^2+y^2}{2\sigma^2}}.$$

Let k be a scaling factor and let $*$ be a convolution operation. From D.G. Lowe [116], we obtain a difference-of-Gaussians image (denoted by $D(x, y, \sigma)$ defined by

$$D(x, y, \sigma) = G(x, y, k\sigma) * Img(x, y) - G(x, y, \sigma) * Img(x, y)$$

Then use $D(x, y, \sigma)$ to identify potential interest points that are invariant to scale and orientation.

Example 8.13 **Fisherman DoG Images**.
For the picture of the fisherman in Fig. 8.16.1, the pair of images in Fig. 8.16 display the results of the Difference-of-Gaussian approach for two sets of values for the scaling factor k and standard deviation σ using Matlab script A.38 in Appendix A.8.6.

DoG.1 Use $k = 1.5, \sigma = 5.55$ to produce the DoG image in Fig. 8.16.2.
DoG.2 Use $k = 1.5, \sigma = 0.98$ to produce the DoG image in Fig. 8.16.3.

Smaller standard deviations in a Difference-of-Gaussians (DoGs) lead to better DoG images.   ■

Example 8.14 **Cameraman DoG Images**.
For the picture of the cameraman in Fig. 8.16.1, the pair of images in Fig. 8.17 display the results of the Difference-of-Gaussian approach for two sets of values for the scaling factor k and standard deviation σ using Matlab script A.38 in Appendix A.8.6.

DoG.1 Use $k = 1.5, \sigma = 5.55$ to produce the DoG image in Fig. 8.17.2.
DoG.2 Use $k = 1.5, \sigma = 0.98$ to produce the DoG image in Fig. 8.17.3.

Again, notice that smaller standard deviations in a DoG lead to better DoG images.   ■

8.8   Image Keypoints: D.G. Lowe's SIFT Approach

The Scale-Invariant Feature Transform (SIFT) introduce by D.G. Lowe [115, 116] is a mainstay in solving object recognition as well as object tracking problems. SIFT works in a scale space to capture multiple scale levels and image resolutions. There are four main stages in a SIFT computation on a digital image.

SIFT.1 Use difference-of-Gaussian function to identify potential interest points that are invariant to scale and orientation.
Note: This step is implemented using the approach in Examples 8.13 and 8.14.

SIFT.2 Select keypoints based on measures of their stability. In this case, edge pixel strengths are partially ordered using the relation $\leq$. Choose a number k of keypoints to select and then select the edge pixels that have the highest strengths.
Note: This step is implemented using the approach in Example 8.4 for intensity images, in Example 8.5 for colour images and explained in Remark 8.6.

SIFT.3 One of the features of each keypoint is its gradient orientation (direction). Keypoints distinguished based on their gradient orientation.
Note: Two approaches to computing the partial derivatives needed to find pixel gradient directions are given in Example 8.10.

SIFT.4 Local pixel gradient magnitudes in the x- and y- directions are used to compute pixel edge strengths. **Note**: See Sect. 8.6 for an explanation and examples. ∎

Remark 8.15 **Keypoints, Edge Strength and Mesh Nerves**.
A sample pixel edge strength is represented by the length of the hypotenuse in see Fig. 8.7.1. This is part of the image geometry shown in Fig. 8.6, illustrated in terms of the edge pixels along the whorls of a fingerprint. Here is a summary of the results for two pixel edge strengths experiments.

chipmunk.jpg   860 keypoints found (144 and 233 keypoints displayed in the intensity image in Fig. 8.14).
cycleImage.jpg   2224 keypoints found (144 and 233 keypoints displayed in the intensity image in Fig. 8.7).
carPoste.jpg   902 keypoints found (144 and 233 keypoints displayed in the intensity image in Fig. 8.14).

The analog of edge pixel strength in 2D images is the length of the radius of a sphere with a keypoint at is center in a 3D image. In either case, keypoints provide a basis for object recognition and solid foundation for the study of image geometric patterns in 2D and 3D images. A common approach in the study of image objects and geometry is to used keypoints as generators of either Voronoï or Delaunay tessellations of images (see, e.g., Fig. 8.18 for a Voronoi tessellation of a cycle image using 144 keypoints and Fig. 8.19 for a tessellation using 377 keypoints). In either case, the result image mesh reveals clusters of polygons. Recall

8.18.1: cyle image   8.18.2: 144 keypoints

8.18.3: cyle Voronoï mesh

Fig. 8.18   144 keypoint-generated Voronoï mesh

that every mesh polygon is the nucleus of a mesh nerve. Often, interest (key) points tend to cluster around mesh polygons in image regions where the image entropy (and corresponding information levels) is highest. Those high entropy nucleus mesh cluster are good hunting grounds for the recognition of image objects and patterns. Mesh nucleus clusters are examples of Edelsbrunner-Harer nerves [42] (see, also, [148, 150]) . ∎

8.9 Application: Keypoint Boundaries of Image Mesh Nuclei

This section introduces a practical application of keypoints found in the polygons along the boundary of an image mesh nucleus. This is a continuation of the study of different forms of mesh edgelets introduced in Problem 7.31. For more about edgelets, see Appendix B.5.

Recall that there are at least four different types contour edgelets surrounding the nucleus in a Maximal Nucleus Cluster (MNC). A **nucleus** is the central and most important part of a mesh cluster. For more about contour edgelets, see Appendix B.3. Detection of various mesh nuclei leads to the study of mesh nerves, which are collections of what are known as MNC spokes. Each **spoke** is a combination of a MNC nucleus and an adjacent polgon. A mesh **nerve** is a collection of spoke-like projections centered on a mesh nucleus. Think of a maximal nucleus cluster (MNC) like the one in the Voronoï mesh superimposed on the Poste vehicle outside a train station in Salerno, Italy, shown in Fig. 8.20 and in 8.21.2 as a collection of spokes radiating out from an MNC nucleus. Mesh nerves are useful in the detection and classification of image objects. For more about mesh nerves, see Appendix B.13. The four basic types of MNC contour edgelets are given next.

Types of MNC Contour Perimeters

IP Edgelet   **fine perimeter** (our starting point-call it interior perimeter or briefly IP). This form of nucleus perimeter earns the name **interior**, since this edgelet perimeter is inside the level 1 coarse perimeter S1P.

S1P: Level 1   **coarse perimeter 1** (our starting point-call it *supra* Level 1 perimeter or briefly S1P). This form of nucleus contour edgelet earns the name **supra 1 contour**, since this edgelet consists of line segments between keypoints (or, in general, mesh sites) of the polygons that are adjacent to the IP polygons. A S1P edgelet defines the most primitive of the coarse shapes of an MNC, namely, a S1P shape.

S2P: Level 2   **coarse perimeter 2** (*supra* 2 perimeter or briefly S2P) that contains a supra 1 perimeter. This form of coarse contour edgelet earns the name **supra 2 contour**, since this edgelet consists of line segments between keypoints

8.19.1: 377 keypoints

8.19.2: cyle Voronoï mesh

Fig. 8.19   377 keypoint-generated Voronoï mesh

Fig. 8.20 Fine edgelet in a maximal nucleus cluster (MNC)

of the polygons that are along the outside border of the S1P polygons. A S2P edgelet defines an intermediate coarse shapes of an MNC, namely, a S2P shape.

S3P: Level 3   coarse perimeter 3 (*supra* 3 perimeter or briefly S3P) that contains a S2P and S1P. This form of coarse contour edgelet earns the name **supra 3 contour**, since this edgelet consists of line segments between keypoints of the polygons that are along the outside border of the S2P polygons. A S2P edgelet defines a maximally coarse shapes of an MNC, namely, a S3P shape.

The simplest of the nucleus contours is the edgelet formed by connecting the keypoints inside the Voronoï mesh polygons that are adjacent to an MNC nucleus. This is the **fine nucleus contour** (also called the **fine perimeter**) of an MNC. The sub-image inside a fine contour usually encloses part of an object of interest. The length of a fine contour (nucleus perimeter) traces the shape of small objects and is a source of useful information in fine-grained recognition of an object that has a shape that closely matches the shape of a target object.

Example 8.16 **Fine Edgelet = Interior Perimeter IP**.
An example of an IP (interior perimeter) edgelet is shown in Fig. 8.21.1. This edgelet is formed by the connected blue line segments ●——● using the keypoints surrounding the nucleus. This edgelet reflects the underlying geometry for a maximal nucleus cluster. An *in situ* view of this edgelet is shown in Fig. 8.21.2. The Voronoï mesh

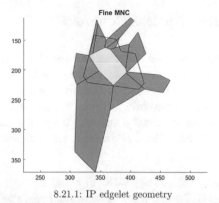

8.21.1: IP edgelet geometry

8.21.2: Image IP

Fig. 8.21   Visualized image geometry via MNC fine contour edgelet

shown in Fig. 8.21.2 is constructed with 89 keypoints. This image edgelet tells us something about the geometry of a subimage containing the maximal nucleus. This geometry appears in the form of a shape described by the IP edgelet. For more about shapes, see B.18. In the best of all possible worlds, this edgelet will enclose an interesting image region that contains some object. ∎

8.10   Supra (Outer) Nucleus Contours

Coarse nucleus contours are useful in detecting the shapes of large image objects. A **coarse nucleus contour** is found by connecting the keypoints inside the Voronoï mesh polygons that are along the border of the fine perimeter polygons of an MNC nucleus. Coarse contours are also called **supra**- or **outer**-contours in an MNC. The length of a coarse contour (nucleus perimeter) traces the shape of medium-sized objects covered by an MNC. The S1P (level 1 supra perimeter) is the innermost MNC coarse contour.

Example 8.17 **Supra Level 1 (S1P) MNC Perimeter**.
The combination of an S1P contour and IP (interior perimeter) contour surrounding an MNC nuclueus is shown in Fig. 8.23.2. In this example, S1P contour is viewed in isolation from the IP contour.

A S1P (coarse perimeter) edgelet is shown in Fig. 8.22. This edgelet is formed by the connected blue line segments ●—● using the keypoints in the polygons along the border of the fine contour polygons surrounding a nucleus polygon. This edgelet gives us the outer shape of an area covered by MNC polygons. An *in situ* view of this S1P edgelet is shown in Fig. 8.23.3. This image edgelet tells us something about the geometry of a subimage covered by a maximal nucleus polygon. This geometry appears in the form of a shape described by the S1P edgelet.

Notice that most of the polygons in the Voronoï mesh covering the image in Fig. 8.23.1 have been suppressed in Fig. 8.23.2. Instead, only the S1P polygons (displayed as red 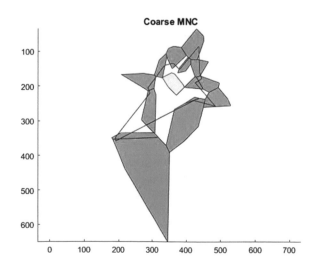 polygons) are shown in in Fig. 8.23.3. These S1P polygons

Fig. 8.22 S1P coarse edgelet geometry in a maximal nucleus cluster (MNC)

8.23.1: Voronoï mesh

8.23.2: S1P edgelet on mesh   8.23.3: S1P edgelet on mesh

8.23.4: S1P edgelet by itself

Fig. 8.23 Visualized image geometry via MNC S1P coarse contour edgelet

Fig. 8.24   Closeup of S1P
nucleus

Fig. 8.24   Closeup of S1P
nucleus

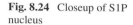

surround a yellow MNC nucleus (for a closeup view of an S1P nucleus,
see Fig. 8.24). Because we are interested in seeing what part of a subimage is covered
by a S1P perimeter, the S1P polygons are suppressed in Fig. 8.23.4. Now only S1P
perimeter is displayed as a sequence of connected green line segments ●——● using
the keypoints as endpoints of each line segment in the S1P. Clearly the S1P contour
shape encloses more of the middle part of the Poste vehicle than a fine IP contour
shape. For more about shapes, see Appendix B.18. In the best of all possible worlds,
this edgelet will enclose an interesting image region that contains some object.   ■

8.11   Quality of a MNC Contour Shape

Quality of an MNC contour shape

The quality of the MNC contour shape will depend on the target shape that we select.
In an object recognition setting, a **target shape** is the shape of an object we wish to
compare with sample shapes in either a single image or in a sequence of video image
frames. The quality of an MNC contour shape is high in cases where the perimeter
of a target shape is close to the length of a sample MNC contour perimeter. In other

words, the **quality of an MNC contour shape** is proportional to the closeness of a target contour shape to a sample contour shape.

8.12 Coarse S2P and S3P (Levels 2 and 3) MNC Contours

This section pushes the envelope for MNC contours by considering Level 2 and level 3 MNC contours, *i.e3.*, S2P and S3P MNC contours. S2P contours are often tightly grouped around S1P contours in an MNC cluster on an image, since the sites (e.g., keypoints or corners) are usually found in the interior of an image rather than along the image borders. This often happens, provided the number of selected sites is high enough.

Example 8.18 **Coarse S1P and S2P Maximal Nucleus Cluster Contours**.
The number of keypoints is 89 in the construction of the Voronoï mesh on the Poste car image in Fig. 8.25. Notice how the keypoints cluster around the driver and monogram on the Poste vehicle as well as around the Poste wheels. So we can expect to find a maximal nucleus cluster in the middle of the Poste car shown in Fig. 8.23.4.

A combination of S1P, S2P and S3P contour edgelets are shown in Fig. 8.26. An S2P contour edgelet is shown in white in Fig. 8.26. Notice how the S2P contour is

Fig. 8.25 Groupings of 89 keypoints on Poste Car Mesh

Fig. 8.26 Sample tightly grouped S2P and S1P contours

tightly grouped around the S1P contour. Here is a summary of the lengths of these contours:

S1P contour length   943.2667 pixels.
S2P contour length   1384.977 pixels.

For object recognition purposes, comparing an S2P contour in a target image with an S2P in a sample image such as a video frame is useful. The caution here is that the tight grouping of the resulting S2P and S1P contours is dependent on the number of keypoints that you choose. A choice of 89 or higher number of keypoints usually produces a good result.   ■

Example 8.19 **Keypoint Mesh with S1P, S2P and S3P Maximal Nucleus Cluster Contours**.
A combination of S1P, S2P and S3P contour edgelets are shown in Fig. 8.27. Now S3P contour is displayed as a sequence of connected red line segments ●──● using

Fig. 8.27   Sample tightly grouped S3P, S2P and S1P contours

the keypoints as endpoints of each line segment in the S3P. Each S3P line segment is drawn between the keypoints in a pair of adjacent polygons along the border of the S2P polygons. Here is a summary of the lengths of these contours:

S1P contour length   943.2667 pixels.
S2P contour length   1384.977 pixels.
S2P contour length   2806.5184 pixels.

Unlike the S2P contour, the line segments in a S3P contour are usually not tightly grouped around the inner contours surrounding the MNC nucleus. This is reflected in the number of pixels in the S3P contour, which is more than double the number of pixels in the S2P contour. The absence of tight grouping reflects the influence of the image edge and corner polygons in the Voronoï mesh.   ∎

8.13   Experimenting with the Number of Keypoints

So far, we have considered a tessellated image containing only one maximal nucleus cluster. By varying the number of generating points (either corners or keypoints some other form of mesh generators), it is possible to vary the number of MNCs

8.28.1: Voronoï mesh

8.28.2: Dual fine contour IP edgelets

Fig. 8.28 Visualized image geometry via dual, overlapping MNCs

8.29.1: Dual coarse S1P and fine IP edgelets

8.29.2: Dual coarse S1P edgelets by themselves

Fig. 8.29   Visualized image geometry via dual MNCs coarse contours

in a tessellated image. The goal is to construct an image mesh that contains either adjacent or overlapping MNCs, which serve as markers of image objects. **Adjacent MNCs** are maximal nucleus clusters in which a polygon in one MNC shares an edge with a polygon in the other MNC. **Overlapping MNCs** occur whenever an entire polygon is common to both MNCs (see, e.g., Figs. 8.28 and 8.29).

After we obtain a Voronoï mesh with multiple MNCs for a selected number of keypoints, the MNCs can either separated (covering different parts of an image) or overlapping. It then is helpful to experiment with either small or very large changes in the number of keypoints in the search for meshes with multiple, overlapping MNCs that are tightly grouped. The ideal situation is to find overlapping MNCs so that the difference in the S1P and S2P contours lengths is small. Let ε be a positive number and let $S1P_c$, $S2P_c$ be the lengths (in pixels). For example, let $\varepsilon = 500$. Then find an MNC so that

$$|S1P_c - S2P_c| < \varepsilon.$$

Notice that neighbouring (in the sense of *close* but neither adjacent nor overlapping) MNCs are possible. **Neighbouring MNCs** are MNCs that are either adjacent, overlapping or separated by at most one polygon.

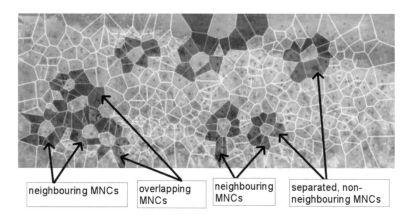

Fig. 8.30 Sample neighouring and non-neighbouring MNCs

Example 8.20 **Neighbouring MNCs**.
Several neighbouring and non-neighbouring MNCs are shown in the keypoint-based Voronoï mesh in Fig. 8.30. In solving an object recognition problem, the shape of an image region of interest is defined by the perimeter of the region, which is compared with the shape of subimage covered by a MNC in a sample image. That is, the length of the contour of a target MNC would be compared with the lengths of each of the contours of MNCs in a sample image or video frame. Similar shapes reveal similar image regions covered by MNCs. ∎

A mesh with neighbouring MNCs can result in contours that cover a region of interest in an image. We illustrate this with a small change in the number of keypoints from Example 8.19, i.e., we select 91 instead of 89 keypoints as generators of a Voronoï diagram superimposed on an image.

Example 8.21 **Edgelet Shapes on Dual, Overlapping MNCs.**
The combination of an S1P contour and IP (interior perimeter) contours surrounding a pair of MNC nuclei is shown in Fig. 8.29.1. Notice that the S1P contour now covers a large portion of the central portion of the Poste vehicle, which is getting closer to what we want for object recognition purposes.

A pair of S2P (coarse perimeter) edgelets (in white) overlapping a pair of S2P edgelets are shown surrounding dual MNC nuclei in Fig. 8.31. These edgelets are formed by the connected green line segments ●——● (S1P coarse contour) and connected white S2P coarse contour using the keypoints in the polygons that include some border polygons. These edgelet gives us the outer shape of an area covered by MNC polygons. These mostly concentric image edgelets tell us something about the geometry of the subimage covered by the dual MNCs.

Surrounding the S2P contours are a pair of S3P edgelets formed by the connected red line segments ●——● (S3P coarse contours). The overlapping S2P and S3P coarse contours covering the Poste vehicle are shown in Fig. 8.32.

Fig. 8.31 Dual coarse S1P and S2P contours on overlapping MNCs

Fig. 8.32   Dual S1P, S2P and S3P contours on overlapping MNCs

Notice that most of the polygons in the Voronoï mesh covering the image in
Fig. 8.23.1 have been suppressed in Fig. 8.23.2. With the selection of 91 keypoints as
mesh generators, we obtain dual yellow nuclei in overlapping MNC nuclei, namely,

 and (for a closeup view of these dual nuclei, see Fig. 8.33). These

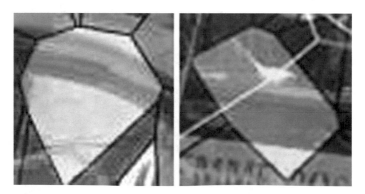

Fig. 8.33   Closeup of dual MNC nuclei

overlapping MNC nuclei are important, since that cover a part of the image where neighbouring keypoints are not only close together but also cover a part of the image where the entropy is highest (in effect, where the information level is highest in this image). ∎

8.14 Coarse Perimeters on Dual MNCs

Example 8.22 **Coarse Contours Surrounding Dual, Overlapping MNCs**.
Because we are interested in seeing what part of a subimage is covered by a S1P perimeter, the S1P polygons are suppressed in Fig. 8.23.4. Now only S1P perimeter is displayed as a sequence of connected green line segments ●——● using the keypoints as endpoints of each line segment in the S1P. Clearly the S1P contour shape encloses more of the middle part of the Poste vehicle than a fine IP contour shape. For more about shapes, see Appendix B.18. In the best of all possible worlds, this edgelet will enclose an interesting image region that contains some object. ∎

Example 8.23 **Keypoint Mesh with S1P, S2P and S3P Maximal Nucleus Cluster Contours**.
A combination of S1P, S2P and S3P contour edgelets are shown in Fig. 8.27. Now S3P contour is displayed as a sequence of connected red line segments ●——● using the keypoints as endpoints of each line segment in the S3P. Each S3P line segment is drawn between the keypoints in a pair of adjacent polygons along the border of the S2P polygons. Here is a summary of the lengths of these coarse contours:

S1P contour length   841.8626 pixels.
S2P contour length   1292.1581 pixels.
S2P contour length   2851.7199 pixels.

Unlike the S2P contour, the line segments in the S3P contour are not tightly grouped around the inner contours surrounding the MNC nucleus. This is reflected in the number of pixels in the S3P contour, which is more than double the number of pixels in the S2P contour. The absence of tight grouping reflects the influence of the image edge and corner polygons in the Voronoï mesh. ∎

8.15 Rényi Entropy of Image MNC Regions

In this section, we call attention to the Rényi entropy of maximal nucleus clusters covering image regions with high information levels.

| Image region with high entropy |.

It is known that the entropy of an image MNC is higher than the entropy of surrounding non-MNC regions [153]. It is also known that Rényi entropy corresponds to the information level of a set of data. For each increase in Rényi entropy there is a corresponding increase in the underlying information level in MNC regions of a Voronoï mesh on a digital image. This result concerning the entropy of the tessellation of digital images stems from a recent study by E. A-iyeh and J.F. Peters [2]. In our case, the Rényi entropy of an MNC corresponds to the information level of that part of an image covered by an MNC.

Let $p(x_1), \dots, p(x_i), \dots, p(x_n)$ be the probabilities of a sequence of events $x_1, \dots, x_i, \dots, x_n$ and let $\beta \geq 1$. Then the Rényi entropy [164] $H_\beta(X)$ of a set of event X is defined by

$$H_\beta(X) = \frac{1}{1-\beta} \ln \sum_{i=1}^{n} p^\beta(x_i) \text{ (Rényi entropy).}$$

Rényi's entropy is based on the work by R.V.L. Hartley [72] and H. Nyquist [129] on the transmission of information. A proof that $H_\beta(X)$ approaches Shannon entropy as $\beta \longrightarrow 1$ is given P.A. Bromiley, N.A. Thacker and E. Bouhova-Thacker in [19], i.e.,

$$\lim_{\beta \longrightarrow 1} \frac{1}{1-\beta} \ln \sum_{i=1}^{n} p^\beta(x_i) = - \sum_{i=1}^{n} p_i \ln p_i.$$

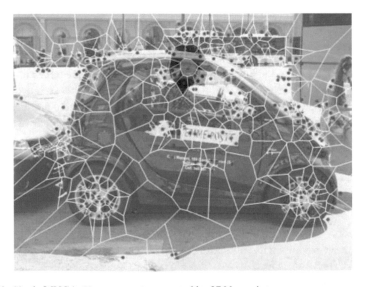

Fig. 8.34 Single MNC in Voronoï mesh generated by 376 keypoints

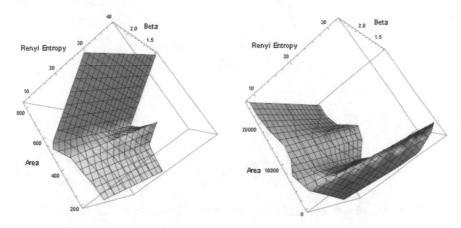

Fig. 8.35 3D plots for MNC and non-MNC entropy

Example 8.24 **MNC versus non-MNC Entropy on the Poste Tessellation with 376 Keypoints**.

A single MNC in a Voronoï mesh with 376 keypoints is shown in Fig. 8.34. 3D plots showing the distribution of Rényi's entropy values with varying β are shown in Fig. 8.35 for the MNC and non-MNC mesh regions. A comparison of the Rényi's entropy values for the MNC and non-MNC region is given in the plot in Fig. 8.36. Observe that the Rényi's entropy values of the MNC region sharply increase and diverge from the non-MNC region. This implies that the information content around the Poste auto driver's head covered by the MNC is higher than the surrounding image regions in this particular Voronoï mesh. ■

Example 8.25 **MNC versus non-MNC Entropy on the Video Frame Tessellation with 145 Keypoints**.

Dual MNCs in a Voronoï mesh with 145 keypoints is shown in Fig. 8.37. 3D plots showing the distribution of Rényi's entropy values with varying β are shown in Fig. 8.38 for the MNC and non-MNC mesh regions. A comparison of the Rényi's entropy values for the MNC and non-MNC region is given in the plot in Fig. 8.39. Observe that the Rényi's entropy values of the MNC regions increase monotonically and are greater than the entropy of the non-MNC regions. This implies that the information content around the front of the train engines covered by the MNCs is higher than the surrounding image regions in this particular Voronoï mesh. ■

The information of order β contained in the observation of the event x_i with respect to the random variable X is defined by $H(X)$. In our case, it is information level of the observation of the quality of a Voronoï mesh cell viewed as random event that is considered in this study.

A main result reported in [2] is the correspondence between image quality and Rényi entropy for different types of tessellated digital images. In other words, the

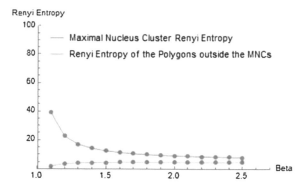

Fig. 8.36 Combined MNC and non-MNC entropy plot for 376-keypoint-based mesh

Fig. 8.37 Dual MNCs in Voronoï mesh generated by 145 keypoints on video frame

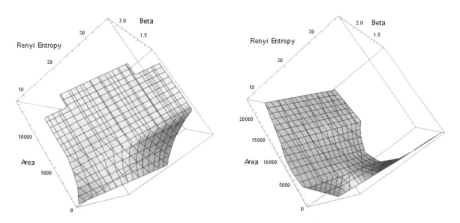

Fig. 8.38 3D Plots for MNC and Non-MNC entropy for a 145 keypoint-based video frame

Fig. 8.39 Combined MNC and Non-MNC entropy plot for 145-keypoint-based mesh

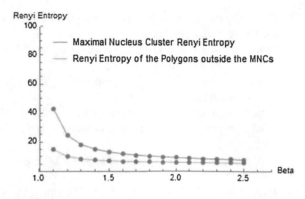

Fig. 8.40 Rényi entropy versus quality of tessellated images

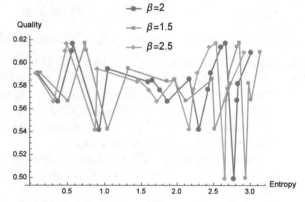

correspondence between the Rényi entropy of mesh cells relative to the quality of the cells varies for different classes of images.

For example, with Voronoï tessellations of images of humans, Rényi entropy tends to be higher for higher quality mesh cells (see, e.g., the plot in Fig. 8.40 for different Rényi entropy levels, ranging from $\beta = 1.5$ to 2.5 in 0.5 increments).

8.16 Problems

Problem 8.26

Let Img be a Voronoï tessellated image using SURF keypoints. Do the following:

1. Select k keypoints, starting with 10 SURF points.
2. Find the maximal nucleus clusters (MNCs) on the Img.
3. Draw the fine IP edgelet geometry (by itself, not on an image). Use blue for the IP line segments. See, for example, 1P edgelet geometry in Fig. 8.21.1.
4. Draw the coarse S1P edgelet geometry (by itself, not on an image). Use blue for the S1P line segments. See, for example, S1P edgelet geometry in Fig. 8.22.

5. Draw the fine IP contour surrounding the MNC nucleus on an image. Use blue for the IP line segments.
6. Draw the coarse S1P contour surrounding the MNC nucleus on an image. Use green for the S1P line segments.
7. Draw the coarse S2P contour surrounding the MNC nucleus on an image. Use white for the S1P line segments.
8. Choose a positive number ε and let $S1Pc, S2Pc$ be the lengths (in pixels) of the level 1 and level 2 MNC contours, respectively. Adjust ε so that

$$|S1P_c - S2P_c| < \varepsilon.$$

9. Repeat Step 1 for $k = 13, 21, 34, 55, 89, 144, 233, 610$ keypoints, until two overlapping or adjacent MNCs are found on the Img.   ∎

Problem 8.27 Select 3 different images. Do the following:

1. Select k keypoints, starting with 10 SURF points.
2. Tessellate a selected image Img, covering it with a Voronoï mesh.
3. Find the maximal nucleus clusters (MNCs) on the Img.
4. Compute Rënyi entropy of each MNC. **Hint**: Let x be the area of a mesh polygon. Compute the probability $P(x) = \frac{1}{x}$, assuming that the occurrence of polygon area in an image tessellation is a random event.
5. Compute Rënyi entropy of the non-MNC region of the Img.
6. Plot the MNC versus the non-MNC image entropies for $\beta = 1.5$ to 2.5 in 0.5 increments.
7. Repeat Step 1 for $k = 13, 21, 34, 55, 89, 144, 233, 610$ keypoints for each of the selected images.   ∎

Chapter 9
Postscript. Where Do Shapes Fit into the Computer Vision Landscape?

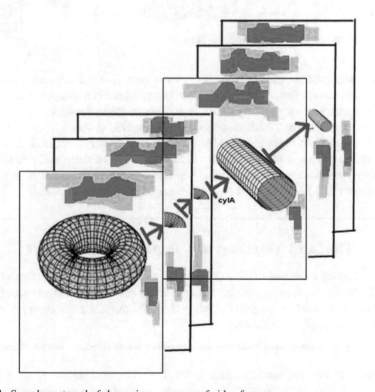

Fig. 9.1 Sample portrayal of shapes in a sequence of video frames

© Springer International Publishing AG 2017

J.F. Peters, *Foundations of Computer Vision*, Intelligent Systems
Reference Library 124, DOI 10.1007/978-3-319-52483-2_9

9.1   Shapes in Natural Scenes

Shapes are elusive creatures that drift in and out of natural scenes that we sometimes perceive, store in memory and record with digital cameras. In a sequence of video frame images, for example, shapes such as the ones shown in Fig. 9.1 sometimes deform into other shapes. In Fig. 9.1, there is a sequence of deformations (represented by $\longmapsto$) like in Fig. 9.2.

Fig. 9.2   Sample sequence of shape deformations

This changing shape phenomenon becomes important in detecting and comparing image object shapes that appear in one form in one frame in a video and reappear in an either a minor or major altered form in another frame in the same video.

In this introduction to the foundations of computer vision, a trio of tools are used to extract from digital images information that is useful in the detection and classification of image objects and their shapes. The three main tools that we use are geometry, topology and algorithms. The end result is an application of what H. Edelsbrunner and J.L. Harer call topics in computational topology [42].

The Inner Workings of Computational Topology

H. Edelsbrunner and J.L. Harer observe that *Geometry gives a concrete face to topological structures, and algorithms offer a means to construct them at a level of complexity that passes the threshold necessary for practical applications.* [42, p. xi].

Image geometry appears in various forms. The archetypal geometric structures in images is in the form of various image regions underlying polygons (Voronoï regions) in Voronoï tessellations and triangles (Delaunay triangular regions) in Delaunay triangulations. A level up in the hierarchy of geometric structures in images is the detection of maximal nucleus clusters (MNCs) containing nuclei and spokes that define mesh nerve structures (for more about this, see Sect. 1.23 and Appendix B.13).

Hidden contours of the nucleus polygon are present in every MNC. Moving further up the hierarchy of geometric structures in images, we find both fine and coarse

Fig. 9.3 Open set $X = \{$ ■, ■, ■, ■, ■, ■ $\}$

edgelets surrounding each MNC nucleus. These are the now the familiar collections of connected straight edges. In a **fine contour**, each straight edge is drawn between generating points along the border of an MNC nucleus polygon.

We have seen numerous examples of fine contours. Taking this image geometry a step further (moving outward along the border of a fine contour), we can identify a course contour surrounding each fine contour. In a **coarse contour**, each straight edge is drawn between generating points along the border of fine contour.

Image topology supplies us with structures useful in the analysis and classification of image regions. The main structure in an image topology is an open set. Basically, an **open set** is a set of elements that does not include the elements on its boundary. In these foundations, open sets first appeared in Sect. 1.2. For more about open sets, see Appendix B.14. Here is another example.

Example 9.1 **Open set of pixels**.
Let a colour image img be represented by the set of picture elements in Fig. 9.3. The picture elements (pixels) are represented by tiny squares. A pixel can be viewed as fat point, i.e., a physical point that has area, which contrasts with a point in the Euclidean plane. An example of an open set is the interior of img:
$X = \{$ ■, ■, ■, ■, ■, ■ $\}$ (Open set).
In this case, a pixel $\{p\}$ (written simply as p) is an open set that belongs to X, provided p has a hue intensity value sufficiently close to one of the hue intensities in X. If the hue of p is blue, then $P = $ ■ is a border pixel of X. The interior $int\, img$ equals X and the blue pixels in img do not belong to X. The set X is an example of an open set in digital geometry. The set X is an example of a **digital open set**. ■

Image topologies are defined on image open sets. An **image topology** is a collection of open sets τ on image open set X with the following properties.

1° The empty set $\varnothing$ is open and $\varnothing$ is in τ.
2° The set X is open and X is in τ.
3° If $\mathcal{A}$ is a sub-collection of open sets in τ, then

$$\bigcup_{B \in \mathcal{A}} B \text{ is a open set in } \tau.$$

In other words, the union of open sets in τ is another open set in τ.
4° If $\mathcal{A}$ is a sub-collection open sets in τ, then

$$\bigcap_{B \in \mathcal{A}} B \text{ is a open set in } \tau.$$

In effect, the intersection of open sets in τ is another open set in τ.   ■

An open set X with a topology τ on it, is called a **topological space**. In other words, the pair (X, τ) is called a **topological space**. A common example of topological space that we have seen repeatedly is the collection of open Voronoï regions. An **open Voronoï region** is a Voronoï region that includes all pixels in its interior and does not include its edges) on a tessellated digital image.

It can be shown that a digital image itself is an open set. In addition, the collection of open Voronoï regions satisfies the properties required for a topology. In that case, we call the topology arising out of a Voronoï tessellated image a digital Voronoï topology. That is, a **digital Voronoï topology** on a tessellated digital image is a collection of open Voronoï regions that satisfy the properties of a topology. For more about topology, see Appendix B.19.

Topology had its beginnings in the 19th century with the work of a number of mathematicians, especially H. Poincaré. The work of K. Borsuk during the 1930s ushered in a paradigm shift in topology in which the focus was the study of shapes. Applied shape theory is a center piece in the foundations of computer vision. For more about shape theory, see Appendix B.18.

In computer vision, shapes are often repeated in a sequence of video frames. And one of the preoccupations of this form of vision is shape-tracking. A shape that occurs in one frame is highly likely to reoccur in a succession of neighbouring frames. The interest lies in the detecting a particular shape (call it the target shape) and then observing the occurrence of a similar shape that is approximately the same as the target shape.

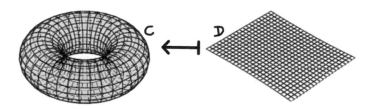

Fig. 9.4 **worldsheet** $\text{wsh}D \mapsto \text{torus}C$

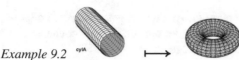

Example 9.2

A sample portrayal of shapes in a sequence of video frames is shown in Fig. 9.1. Over time, the ring torus is the first of the video frames in Fig. 9.1 breaks open and stretches out, eventually assuming a tubular shape. The deformation of one shape into another shape is a common occurrence in the natural world.

Image Object Shape Detection

The trick in image object shape detection is to view the changes in a shape over a sequence of video frames as approximations of an original shape that we might have detected in an initial video frame image.

In an extreme case such as the one in Fig. 9.4, some form of worldsheet rolls up (over time), forming a ring torus. In topological terms, there is a continuous mapping from a planar worldsheet wsh$M$ in $\mathbb{R}^2$ to a ring torus f (wsh$M$) in $\mathbb{R}^3$. A **worldsheet** D (denoted by wsh$D$) is a collection of strings that cover a patch in a natural scene. A **string** is either a wiggly or straight line segment. In string theory, a string is defined by the path followed by a particle moving through space. Another name for such a string is *worldline* [130–132]. The idea of a string works well in explaining the sequences of shapes in video frames in which the paths followed by photons has been recorded by a video camera.

This mapping from a worldsheet to a torus is represented in Fig. 9.4. A *ring torus* is tubular surface in the shape of a doughnut, obtained by rotating a circle of radius r (called the *tube radius*) about an axis in the plane of the circle at distance c from the torus center. Worldsheet wsh$M$ maps to (rolls up into) the tubular surface of a ring torus in 3-space, i.e., there is a continuous mapping from wsh$M$ in 2-space to ring torus surface f (sheet$M$) in 3-space. ∎

9.2 Shape Estimates

This section briefly covers part of the ground for shape estimation. The basic idea is twofold. First, we need some means of measuring the shape of an image object. Second, we to decide when one shape is approximately the same as another shape. For simplicity, we consider only 2D shapes, here.

In the plane, shapes are known by their perimeters and areas. The focus here is on perimeters that are collections of connected straight edges. Recall that edges e, e' are **connected**, provided there is a path between e and e'. A perimeter that is composed of connected straight edges is called an **edgelet**. An edgelet is a **connected edgelet**, provided every pair of straight edges in the edgelet are themselves connected. An image region **shape perimeter** is a connected edgelet.

Algorithm 10: Comparing Image Region Shape Perimeters that are Edgelets

---

Input : Read digital image regions T, R.

Output: shapeSimilarity (Shape perimeters similarity measurement).

1 /* $edgeletT$ equals a shape perimeter in a target image region T */ ;
2 $edgeletT \leftarrow connectedTargetEdges \subset T$;
3 /* $edgeletR$ equals a shape perimeter in a sample image region R */ ;
4 $edgeletR \leftarrow connectedRegionEdges \subset R$;
5 /* ε = upper bound on similarity between shape edgelets */ ;
6 $\varepsilon \leftarrow small + ve\ Real\ Number$;
7 /* Compare shape perimeters: */ ;
8

$$
shapeSimilarity\,(edgeletT, edgeletR) = \begin{cases} 1, & if\ |edgeletT - edgeletR| < \varepsilon, \\ 0, & otherwise. \end{cases}
$$

/* One Shape edgelet approximates another one, provided $shapeSimilarity = 1$ */

9.5.1: Target Drone Video Frame Region   9.5.2: Sample Region in a Drone Video Frame

Fig. 9.5 Target and Sample Drone Video Frame Regions

Since image regions are known by their shape perimeters, it is possible to compare the shape perimeter that encloses an image region containing a target object with the shape perimeter of an image region containing an unknown object. Notice that, after tessellating an image and identifying the maximal nucleus clusters (MNCs) in the tessellated image, each MNC contour surrounding an MNC nucleus polygon is a shape perimeter.

Example 9.3 **Sample Pair of Traffic Drone Video Frame Shape Perimeters**.
A sample pair of drone traffic video frames are shown in Fig. 9.5. To obtain a shape
perimeter from each of these video frames, we do the following:

1° Select video frame images $img1, img2$.
2° Select a set of mesh generating points S.
3° Select a video frame image $img \in \{img1, img2\}$.
4° Superimpose on $img1$ a Voronoö diagram $V(S)$, i.e., **tessellate** img, covering
       img with Voronoö regions $V(s)$, using each generating point (site, seed point)
       $s \in S$.
5° Identify a MNC in the image diagram $V(S)$ (call it $MNC(s)$).
6° Identify coarse edgelet contour $MNCedgelet$ in img (a target MNC shape
       perimeter in a video frame).

9.6.1:Video frame target shape perimeter   9.6.2:Region shape perimeter sample

Fig. 9.6 Pair of Video frame shape perimeters

7° Repeat Step 3, after obtaining a target MNC shape perimeter in img (call it
       $MNCedgeletT$) to obtain a sample video frame image MNC coarse edgelet
       contour $MNCedgeletR$ (a sample MNC shape perimeter in a video frame).
       The result of this step is the production of a pair MNC shape perimeters
       ($MNCedgeletT$ and $MNCedgeletR$) embedded in a pair video frame images.
       An **embedded target shape perimeter** $MNCedgeletT$ is shown in Fig. 9.6.1
       and an **embedded sample region shape perimeter** $MNCedgeletR$ is shown
       in Fig. 9.6.2.
8° Next extract a pair of pure plane shape perimeters from the embedded MNC
       perimeters. **Note**: This is done to call attention to the edgelets whose lengths
       we want to measure and compare.

9° Select shape perimeter $edgelet \in \{MNCedgeletT, MNCedgeletR\}$.

10° Extract a shape perimeter *shape* from *edgelet* (the result of this step is pure plane shape perimeter without the underlying image MNC).

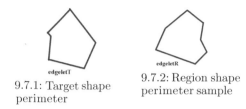

9.7.1: Target shape perimeter

9.7.2: Region shape perimeter sample

Fig. 9.7 Pair of pure planar edgelet-based shape perimeters

11° Repeat Step 9, after obtaining the first MNC shape perimeter (call it *edgeletT*) to obtain a sample MNC shape perimeter *edgeletR* (a sample MNC shape perimeter in a video frame). The result of this step is the production of a pair of pure plane shape perimeters (*edgeletT* and *edgeletR*) embedded in a pair of contours MNCs in Voronoï-tessellated video frame images. An **target shape perimeter** *edgeletT* is shown in Fig. 9.7.1 and a **sample region shape perimeter** *edgeletR* is shown in Fig. 9.7.2.

12° Use *edgeletT* and *edgeletR* as inputs in Algorithm 10 (compute the similarity between the target and sample MNC shape perimeters).

13° Compute the value of *shapeSimilarity* (*edgeletT*, *edgeletR*). ∎

For more about shape boundaries, see Appendix B.18.

Problem 9.4 Shape deformation.
Give three examples of shape deformation in a sequence of video frames. **Hint**: Compare shape perimeters. A change in a shape perimeter happens whenever there is a change in image object shape either due to a change in camera position or a change in a natural scene object. A common example is video that records movements of humans or other animals or birds. ∎

Problem 9.5 Shape perimeters similarity measurement.
Implement Algorithm 10. ∎

Appendix A
Matlab and Mathematica Scripts

This Appendix contains Matlab® and Mathematica® scripts referenced in the chapters. Matlab® R2013b is used to write the Matlab scripts.

A.1 Scripts from Chap. 1

A.1.1 Digital Image Corners

A.1.1: Corners on image   A.1.2: Corners Voronoï mesh on image

Fig. A.1 Sample image corners

```
% script: GeneratingPointsOnImage.m
% image geometry: image corners
% part 1: image corners + Voronoi diagram on image
% part 2: plot image corners + Voronoi diagram by themselves
%
clear all; close all; clc; % housekeeping
%%
img=imread('carRedSalerno.jpg');
g = double(rgb2gray(img)); % convert to greyscale image
```

© Springer International Publishing AG 2017

J.F. Peters, *Foundations of Computer Vision*, Intelligent Systems
Reference Library 124, DOI 10.1007/978-3-319-52483-2

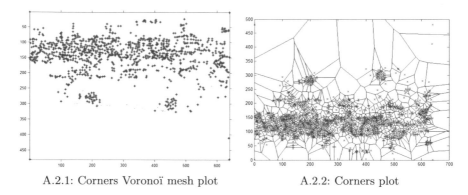

A.2.1: Corners Voronoï mesh plot   A.2.2: Corners plot

Fig. A.2 Sample image corners

```
%
% part 1:
%
cornersMin = corner(g); % min. no. of corners
% identify image boundary corners
box_corners = [1,1;1,size(g,1);size(g,2),1;size(g,2),size(g,1)];
% concatenate image boundary corners & set of interior image corners
cornersMin = cat(1,cornersMin,box_corners);
% set up display of cornersMin on rgb image
figure, imshow(img), ...
       hold on, axis on, axis tight,  % set up corners display on rgb image
plot(cornersMin(:,1), cornersMin(:,2), 'g*');
% set up cornerMin-based Voronoi diagram on rgb image
redCarMesh = figure, imshow(img), ...
       hold on, axis on, axis tight,
voronoi(cornersMin(:,1),cornersMin(:,2),'gx'); % blue edges
% uncomment next line to save Voronoi diagram:
% saveas(redCarMesh,'imageMesh.png'); % save copy of image
%
% part 2:
%
corners = corner(g,1000); % up to 1000 corners
% concatenate image boundary corners & set of interior image corners
corners = cat(1,corners,box_corners);
% plot specified no. of corners:
figure, imshow(g), ...
       hold on, axis on, axis tight,  % set up corners plot
plot(corners(:,1), corners(:,2), 'b*');
% construct corner-based Voronoi diagram
planarMesh = figure
voronoi(corners(:,1),corners(:,2),'bx'); % blue edges
% uncomment next line to save Voronoi diagram:
% saveas(planarMesh,'planarMesh.png'); % save copy of image
```

Listing A.1 Matlab script in `GeneratingPointsOnImage.m` to display corners on a digital image.

Remark A.1 **Views of Corners on a Colour Image**.

Image corners provide a first look at mesh generating points, leading to the study of digital image geometry. Sample image corners plus image boundary corners extracted from a colour image are shown in situ in Fig. A.1.1 with the construction of the corner-based Voronoï mesh superimposed on the image in Fig. A.1.2, produced using Matlab® script A.1. A plot of 1000 image corners plus image boundary corners is given in Fig. A.2.1 and a plot of the corner-based Voronoï mesh is shown in Fig. A.2.2, also using Matlab script A.1. For more about this, see Sect. 1.22.   ∎

A.1.2 Implementation of Voronoï Tessellation Algorithm

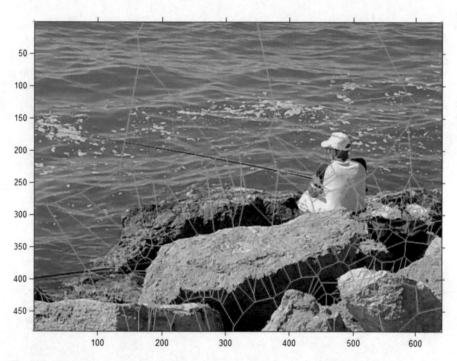

Fig. A.3 Image interior corners plus image boundary corners-based Voronoï mesh

```
% script : VoronoiMeshOnImage .m
% image geometry : overlay Voronoi mesh on image
%
% see http :// homepages . ulb . ac . be/~ dgonze /INFO/ matlab . html
% revised 23 Oct. 2016
clear all; close all; clc; % housekeeping
g=imread('fisherman.jpg');
% im=imread ('cycle.jpg');
% g=imread ('carRedSalerno.jpg');
%%
img = g; % save copy of colour image to make overlay possible
g = double(rgb2gray(g)); % convert to greyscale image
% corners = corner(g); % min. no. of corners
k = 233; % select k corners
corners = corner(g,k); % up to k corners
box_corners = [1,1;1,size(g,1);size(g,2),1;size(g,2),size(g,1)];
```

```
corners = cat(1,corners,box_corners);
vm = figure, imshow(img) ,...
axis on, hold on; % set up image overlay
voronoi(corners(:,1),corners(:,2),'g'); % red edges
% voronoi(corners(:,1),corners(:,2),'g.'); % red edges
% imfinfo('carRedSalerno.jpg')
% figure , mesh(g(300:350,300:350)) ,...
% axis tight ,zlabel('rgb pixel intensity ')
% xlabel('g(300:350)'),ylabel('g(300:350)') % label axes
% saveas(vm,'VoronoiMesh.png'); % save copy of image
```

Listing A.2 Matlab script in `VoronoiMeshOnImage.m` to construct a corner-based Voronoï mesh.

Remark A.2 **Another view of Corner-based Voronoï mesh on a Colour Image.**
Another example of corner-based digital image geometry is given here. A Voronoï mesh derived from image corners plus image boundary corners extracted from a colour image and superimposed on a colour image is shown in Fig. A.3, produced using Matlab® script A.2. For more about this, see Sect. 1.22. ■

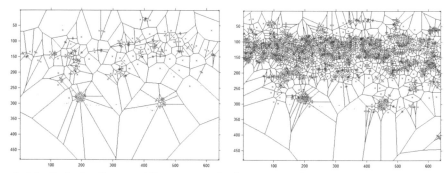

A.4.1: Default Corners Voronoï mesh plot

A.4.2: 2000 Corners Voronoï mesh plot

Fig. A.4 Sample image corners

```
% script: VoronoiMesh1000CarPolygons.m
% image geometry: Voronoi mesh image polygons
% Interior + boundary corners-based Voronoi mesh plot:
% Notice the corner clusters.
%
% Part 1: plot of default interior + boundary corners
% Part 2: plot of up to 2000 interior + boundary corners
%
clear all; close all; clc; % housekeeping
%%
g=imread('carRedSalerno.jpg');
% g=imread('peppers.png');
g = double(rgb2gray(g)); % convert to greyscale image
%
% Part 1
%
```

```
corners = corner(g); % find up min. image corners
size(corners)
% get image box corners
box_corners = [1,1;1,size(g,1);size(g,2),1;size(g,2),size(g,1)];
corners = cat(1,corners,box_corners); % combine corners
figure, imshow(g), hold on; % set up polygon display
voronoi(corners(:,1),corners(:,2),'x'); % display polygons
%
% Part 2
%
corners2000 = corner(g,2000); % find up to 2000 corners
corners2000 = cat(1,corners2000,box_corners); % combine corners
size(corners2000)
figure, imshow(g), hold on; % set up polygon display
voronoi(corners2000(:,1),corners2000(:,2),'x'); % display polygons
% imfinfo('carRedSalerno.jpg')
```

Listing A.3 Matlab script in `VoronoiMesh1000CarPolygons.m` to construct a corner-based Voronoï mesh.

Remark A.3 **Contrasting plots of Corner-based Voronoï meshes**.
The plot of a Voronoï mesh derived from the default number of image corners plus image boundary corners extracted from a colour image is shown in Fig. A.4.1, produced using Matlab script A.3. In addition, the plot of a Voronoï mesh derived from up to 2000 image corners plus image boundary corners extracted from a colour image is shown in Fig. A.4.2, also produced using Matlab script A.3. In this case, the plot of a Voronoï mesh constructed from 200 interior corners contrasts with the plot of a Voronoï mesh constructed from 1338 interior corners. 1338 is the maximum number of corners found in the Salerno auto image in this example. For more about this, see Sect. 1.22. ∎

A.1.3 Implementation of Delaunay Tessellation Algorithm

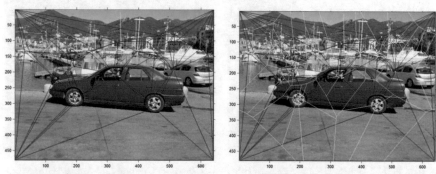

A.5.1: 50 Corner Delaunay triangulation on image

A.5.2: 50 Corner Delaunay triangulation with Voronoï mesh overlay on image

Fig. A.5 Sample 50 corner triangulation with Voronoi mesh overlay on image

```
% script: DelaunayOnImage.m
% image geometry: Delaunay triangles on image
%
% Part 1: default interior + boundary corners—based triangulation
% Part 2: Up to 2000 interior + boundary corners—based triangulation
%
clear all; close all; clc; % housekeeping
%%
g=imread('carRedSalerno.jpg');
% g=imread('8x8grid.jpg');
% g=imread('Fox—2states.jpg');
img = g; % save copy of colour image
g = double(rgb2gray(g)); % convert to greyscale image
%
% Part 1
%
corners = corner(g,50); % default image corners
box_corners = [1,1;1,size(g,1);size(g,2),1;size(g,2),size(g,1)];
corners = cat(1,corners,box_corners); % combined corners
figure, imshow(img), hold on; % set up overlay of mesh on image
% voronoi(corners(:,1),corners(:,2),'x'); % identify polygons
TRI = delaunay(corners(:,1),corners(:,2)); % identify triangles
triplot(TRI,corners(:,1),corners(:,2),'b'); % meshes on image
%
% corner Delaunay triangulation with Voronoi mesh overlay:
%
figure, imshow(img), hold on; % set up overlay of mesh on image
% voronoi(corners(:,1),corners(:,2),'x'); % identify polygons
TRI = delaunay(corners(:,1),corners(:,2)); % identify triangles
triplot(TRI,corners(:,1),corners(:,2),'b'); % meshes on image
voronoi(corners(:,1),corners(:,2),'y'); % identify polygons
%
% Part 2
%
corners1000 = corner(g,2000); % find 1000 image corners
corners1000 = cat(1,corners1000,box_corners); % combined corners
figure, imshow(img), hold on; % set up overlay of mesh on image
% voronoi(corners(:,1),corners(:,2),'x'); % identify polygons
TRI = delaunay(corners1000(:,1),corners1000(:,2)); % identify triangles
triplot(TRI,corners1000(:,1),corners1000(:,2),'b'); % meshes on image
```

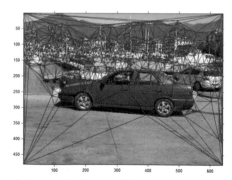

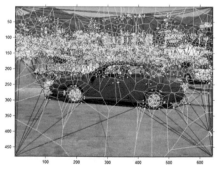

A.6.1: 2000 Corner Delaunay triangula-   A.6.2: 2000 Corner Delaunay triangulation on image   tion with Voronoï mesh overlay on im-
                                           age

Fig. A.6 Sample 2000 corner triangulation with Voronoi mesh overlay on image

```
%
% corner Delaunay triangulation with Voronoi mesh overlay:
%
figure, imshow(img), hold on; % set up overlay of mesh on image
% voronoi(corners(:,1),corners(:,2),'x'); % identify polygons
TRI = delaunay(corners1000(:,1),corners1000(:,2)); % identify triangles
triplot(TRI,corners1000(:,1),corners1000(:,2),'b'); % meshes on image
voronoi(corners1000(:,1),corners1000(:,2),'y'); % identify polygons
% imfinfo('carRedSalerno.jpg')
```

Listing A.4 Matlab script in `DelaunayOnImage.m` to construct a corner-based Delaunay triangulation mesh on image.

Remark A.4 **Contrasting Corner-based Delaunay triangulation on an image**.
The Delaunay triangulation derived from 50 image corners plus image boundary corners extracted from a colour image is shown in Fig. A.5.1, produced using Matlab script A.4. The same Delaunay triangulation with Voronoï mesh overlay on the image is shown in Fig. A.5.1, also produced using Matlab script A.4.

In addition, the Delaunay triangulation derived from up to 2000 image corners plus image boundary corners extracted from a colour image is shown in Fig. A.6.1, produced using Matlab script A.4. In this case, the Delaunay triangulation constructed from 50 interior corners contrasts with the plot of a Delaunay triangulation constructed from 1338 interior corners. Although we have call for 2000 corners, 1338 is the maximum number of corners found in the Salerno auto image in this example. Again the 1338-strong corner-based Delaunay triangulation with Voronoï mesh overlay on the image is shown in Fig. A.6.2, also produced using Matlab script A.4. For more about this, see Sect. 1.22. ■

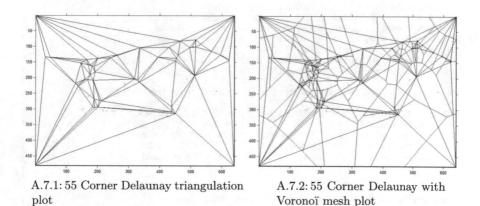

A.7.1: 55 Corner Delaunay triangulation plot

A.7.2: 55 Corner Delaunay with Voronoï mesh plot

Fig. A.7 Sample image mesh plots

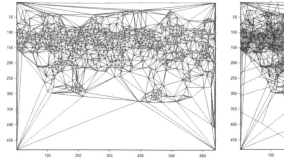 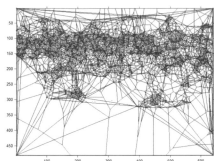

A.8.1: 2000 Corner Delaunay triangula-   A.8.2: 2000 Corner Delaunay with
tion plot   Voronoï mesh plot

Fig. A.8 Sample image triangulation plots

```
% script: DelaunayCornerTriangles.m
% image geometry: Delaunay triangles from image corners
% plus Delaunay triangulation with Voronoi mesh overlay
%
clear all; close all; clc; % housekeeping
%%
g=imread('carRedSalerno.jpg');
% g=imread('Fox-2states.jpg');
img = g; % save copy of colour image
g = double(rgb2gray(g)); % convert to greyscale image
%
% Part 1
%
corners = corner(g,50); % default image corners
box_corners = [1,1;1,size(g,1);size(g,2),1;size(g,2),size(g,1)];
corners = cat(1,corners,box_corners); % combined corners
figure, imshow(g), hold on; % set up overlay of mesh on image
% voronoi(corners(:,1),corners(:,2),'x'); % identify polygons
TRI = delaunay(corners(:,1),corners(:,2)); % identify triangles
triplot(TRI,corners(:,1),corners(:,2),'b'); % meshes on image
%
% 50 corner Delaunay triangulation with Voronoi mesh overlay:
%
figure, imshow(g), hold on; % set up overlay of mesh on image
% voronoi(corners(:,1),corners(:,2),'x'); % identify polygons
TRI = delaunay(corners(:,1),corners(:,2)); % identify triangles
triplot(TRI,corners(:,1),corners(:,2),'b'); % meshes on image
voronoi(corners(:,1),corners(:,2),'r'); % identify polygons
%
% Part 2
%
corners2000 = corner(g,2000); % find 1000 image corners
box_corners = [1,1;1,size(g,1);size(g,2),1;size(g,2),size(g,1)];
corners2000 = cat(1,corners2000,box_corners); % combined corners
figure, imshow(g), hold on; % set up overlay of mesh on image
% voronoi(corners(:,1),corners(:,2),'x'); % identify polygons
TRI2000 = delaunay(corners2000(:,1),corners2000(:,2)); % identify triangles
triplot(TRI2000,corners2000(:,1),corners2000(:,2),'b'); % meshes on image
%
% 2000-corner Delaunay triangulation with Voronoi mesh overlay:
%
figure, imshow(g), hold on; % set up overlay of mesh on image
```

```
% voronoi(corners(:,1),corners(:,2),'x'); % identify polygons
TRI2000 = delaunay(corners2000(:,1),corners2000(:,2)); % identify triangles
triplot(TRI2000,corners2000(:,1),corners2000(:,2),'b'); % meshes on image
voronoi(corners2000(:,1),corners2000(:,2),'r'); % identify polygons
% imfinfo('carRedSalerno.jpg')
```

Listing A.5 Matlab script in `DelaunayCornerTriangles.m` to construct a corner-based Delaunay triangulation mesh by itself.

Remark A.5 **Contrasting Corner-based Delaunay triangulation Voronoï mesh plots**.
The Delaunay triangulation derived from 50 image corners plus image boundary corners extracted from a colour image is shown in Fig. A.7.1, produced using Matlab script A.5. In addition, a Voronoï mesh overlay on the same Delaunay triangulation derived from 50 image corners plus image boundary corners extracted from a colour image is shown in Fig. A.7.2, also produced using Matlab script A.5.

Similarly, the plot of a Delaunay triangulation derived from up to 2000 image corners plus image boundary corners extracted from a colour image is shown in Fig. A.8.1. After that, a Voronoï mesh overlay on the same Delaunay triangulation derived from up to 2000 image corners plus image boundary corners extracted from a colour image is shown in Fig. A.8.2, also produced using Matlab script A.5. For more about this, see Sect. 1.22. ∎

A.1.4   Implementation of Combined Voronoï-Delaunay Tessellation Algorithm

```
% script: DelaunayVoronoiOnImage.m
% image geometry: Delaunay triangles on Voronoi mesh on image
%
clear all; close all; clc; % housekeeping
%%
% Experiment with Delaunay triangulation Voronoi mesh overlays:
g=imread('cycle.jpg');
% g=imread('carRedSalerno.jpg');
img = g; % save copy of colour image
g = double(rgb2gray(g)); % convert to greyscale image
corners = corner(g,50); % find 1000 image corners
box_corners = [1,1;1,size(g,1);size(g,2),1;size(g,2),size(g,1)];
corners = cat(1,corners,box_corners); % combined corners
figure, imshow(img), hold on; % set up overlay of mesh on image
voronoi(corners(:,1),corners(:,2),'y'); % identify polygons
TRI = delaunay(corners(:,1),corners(:,2)); % identify triangles
triplot(TRI,corners(:,1),corners(:,2),'b'); % meshes on image
% imfinfo('cycle.jpg')
% imfinfo('carRedSalerno.jpg')
```

Listing A.6 Matlab script in `DelaunayVoronoiOnImage.m` to construct a Delaunay triangulation on Voronoï mesh overlay on an image.

Remark A.6 **Second Experiment: Corner-based Delaunay triangulation Voronoï mesh overlays**.

The Delaunay triangulation combined with Voronoï mesh each derived from 50
image corners plus image boundary corners extracted from a colour image is shown
in Fig. A.9, produced using Matlab script A.6. For more about this, see Sect. 1.22.
■

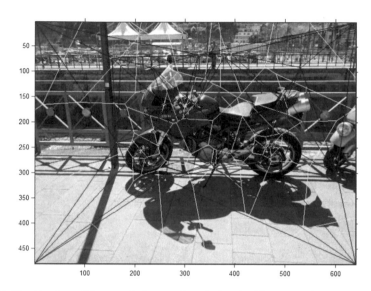

Fig. A.9 Combination of 50 corner Delaunay triangulation plus Voronoï mesh overlay on an image

```
% script: DelaunayOnVoronoi.m
% image geometry: Delaunay triangles on Voronoi mesh polygons
%
clear all; close all; clc; % housekeeping
%%
g=imread('fisherman.jpg'); % input colour image
% g=imread('carRedSalerno.jpg'); % input colour image
g = double(rgb2gray(g)); % convert to greyscale image
corners = corner(g,50); % find up to 50 image corners
box_corners = [1,1;1,size(g,1);size(g,2),1;size(g,2),size(g,1)];
corners = cat(1,corners,box_corners); % box + inner corners
figure, imshow(g), hold on; % set up combined meshes
voronoi(corners(:,1),corners(:,2),'x'); % Voronoi mesh
TRI = delaunay(corners(:,1),corners(:,2)); % Delaunay mesh
triplot(TRI,corners(:,1),corners(:,2),'r'); % combined meshes
% imfinfo('fisherman.jpg')
% imfinfo('carRedSalerno.jpg')
```

Listing A.7 Matlab script in `DelaunayOnVoronoi.m` to construct a Delaunay triangulation
on Voronoi mesh by itself.

Remark A.7 **Third Experiment: Corner-based Delaunay triangulation Voronoï
tessellation overlays**.

The Delaunay triangulation combined with Voronoï tessellation, each derived from
50 image corners plus image boundary corners extracted from a colour image, is

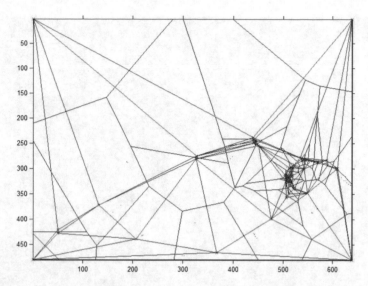

Fig. A.10 Corner-based Delaunay triangulation plus Voronoï mesh overlays on an image

shown in Fig. A.10, produced using Matlab script A.7. For more about this, see Sect. 1.22. ∎

A.1.5   Offline Video Processing Script for Chap. 1

```
% script: offlineVoronoi.m
% OFFLINE VIDEO VORONOI AND DELAUNAY MESH (CORNERS)
% Offline corner−based Voronoi tessellation of video frames
% Example by D. Villar from August 2015 experiment
% Revised version: 15 Dec. 2015, 7 Nov. 2016.
%
close all, clear all, clc % workspace housekeeping
%%
% Initialize input and output videos
videoReader = vision.VideoFileReader('moving_hand.mp4');
videoWriter = vision.VideoFileWriter('offlineVoronoiResult1.avi', ...
    'FileFormat', 'AVI', ...
    'FrameRate',videoReader.info.VideoFrameRate);
videoWriter2 = vision.VideoFileWriter('offlineDelaunayResult1.avi', ...
    'FileFormat', 'AVI', ...
    'FrameRate',videoReader.info.VideoFrameRate);

% Capture one frame to get its size.
videoFrame = step(videoReader);
frameSize = size(videoFrame);

runLoop = true;
frameCount = 0;

disp('Processing video... Please wait.')
```

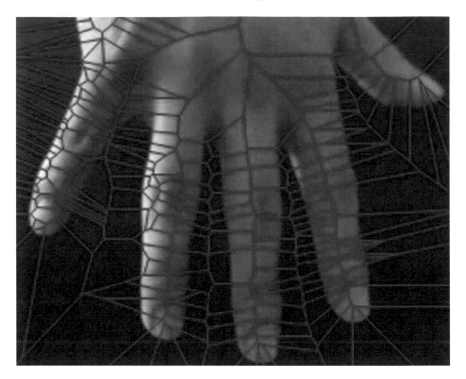

Fig. A.11 Offline corner-based Voronoï tessellation of video frame images

```matlab
% 100 frame video
while runLoop && frameCount < 100
    % Get the next frame and corners
    videoFrame = imresize(step(videoReader), 0.5);
    frameCount = frameCount + 1;
    videoFrameGray = rgb2gray(videoFrame);
    videoFrameGray = medfilt2(videoFrameGray,[5 5]);
    C = corner(videoFrameGray, 300); % get up to 300 frame corners
    [a,b] = size(C);
    % Capture Voronoi tessellation of video frame
    if a > 2
        [VX,VY] = voronoi(C(:,1),C(:,2));
        % Creating matrix of line segments in the form [x_11 y_11 x_12 y_12 ...
        % ... x_n1 y_n1 x_n2 y_n2]
        A = [VX(1,:); VY(1,:); VX(2,:); VY(2,:)];
        A(A>5000) = 5000; A(A<-5000) = -5000;
        A = A';

        % Display Voronoi tessellation of video frame
        videoFrame2 = insertMarker(videoFrame, C, '+', ...
                'Color', 'red');
        videoFrame2 = insertShape(videoFrame, 'Line', A, 'Color', 'red');

        % Display the annotated video frame using the video player object.
        step(videoWriter, videoFrame2);
    else
        step(videoWriter, videoFrame);
    end
```

```
end

disp('Processing complete.')

% Clean up: video housekeeping
release(videoWriter);
disp('offlineVoronoiResult1.mp4 has been produced.')
release(videoWriter2);
disp('offlineDelaunayResult1.mp4 has been produced.')
```

Listing A.8 Matlab script in `offlineVoronoi.m` to construct Voronoï tessellation of video frames offline.

Remark A.8 **Offline Video Manipulation: Corner-based Voronoï tessellation overlays on Video frames**.
Recall that a **Voronoï tessellation** of a digital image is a tiling of the image with Voronoï region polygons. Each Voronoï region polygon is constructed using a generating point (seed, or site). For more about this, see Appendix B.19. An offline Voronoï tessellation of the frames in a video, each derived from 300 image corners extracted from each video frame image, is shown in Fig. A.11, produced using Matlab script A.8. For more about this, see Sects. 1.24 and 1.24.1. ∎

A.1.6 Real-Time Video Processing Script for Chap. 1

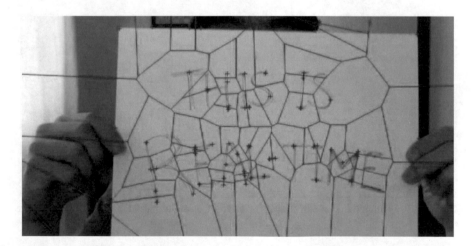

Fig. A.12 Real-time corner-based Voronoï tessellation of video frame images

```
% script: script:realTime1.m
% Real-time Voronoi mesh:
% corner-based tessellation of video frames.
% See lines 32-33.
% Example from D. Villar, July 2015 Compute Vision Experiment.
% Revised 7 Nov. 2016
```

```matlab
%
close all, clear all, clc % housekeeping
%%

% Create the webcam object.
cam = webcam(2);

% Capture one frame to get its size.
videoFrame = snapshot(cam);
frameSize = size(videoFrame);

% Create the video player object.
videoPlayer = vision.VideoPlayer('Position', [100 100 [frameSize(2), frameSize
    (1)]+30]);
videoWriter = vision.VideoFileWriter('realTimeVoronoiResult.mp4', ...
  'FileFormat', 'MPEG4', 'FrameRate', 10);

runLoop = true;
frameCount = 0;

% 100 frame video
while runLoop && frameCount < 100

    % Get the next frame.
    videoFrame = snapshot(cam);
    frameCount = frameCount + 1;
    videoFrameGray = rgb2gray(videoFrame);

    % Voronoi using corners
    C = corner(videoFrameGray, 100);
    [VX,VY] = voronoi(C(:,1),C(:,2));

    % Creating matrix of line segments in the form [x_11 y_11 x_12 y_12 ...
    % ... x_n1 y_n1 x_n2 y_n2]
    A = [VX(1,:); VY(1,:); VX(2,:); VY(2,:)];
    A(A>5000) = 5000; A(A<-5000) = -5000;
    A = A';

    videoFrame = insertMarker(videoFrame, C, '+', ...
            'Color', 'red');
    videoFrame = insertShape(videoFrame, 'Line', A, 'Color', 'red');

    % Display the annotated video frame using the video player object.
    step(videoPlayer, videoFrame);
    step(videoWriter, videoFrame);
end

% Clean up (video camera housekeeping)
clear cam;
release(videoWriter);
release(videoPlayer);
```

Listing A.9 Matlab script in `realTime1.m` to construct Voronoï tessellation of video frames in real-time.

Remark A.9 **Real-time Video Manipulation: Corner-based Voronoï tessellation overlays on Video frames**.

A real-time Voronoï tessellation (tiling) of the frames in a video, each derived from 100 image corners extracted from each video frame image, is shown in Fig. A.12, produced using Matlab script A.9. Remarkably, there is usually little or no noticeable delay between video frames during real-time video frame tiling, since the computation speed of most computers is high. For more about this, see Sect. 1.24.2. ∎

A.2   Scripts from Chap. 2

A.2.1   *Digital Image Pixels*

A.13.1: Sample Colour Image   A.13.2: Sample Greyscale Image

Fig. A.13 Sample images

Fig. A.14 Sample cpselect(g, h) window, g = leaf.jpg, h = leafGrey.jpg

```
% script: inspectPixels.m
% Use cpselect(g,h) to inspect pixels in a raster image
% comment: CTRL<r>, uncomment: CTRL<t>
% Each pixel is represented by a tiny square
clc, close all, clear all  % housekeeping
```

| A.15.1: Zoom in 50% | A.15.2: Zoom in 50% window |

Fig. A.15 Sample images

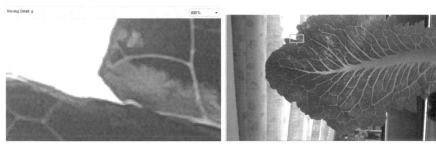

| A.16.1: Zoom in 50% | A.16.2: Zoom in 50% window |

Fig. A.16 Sample images

```
%%
% input a pair of images.
% Choices:
% 1. Input two copies of the same image
% 2. Input two different images.
% Examples:
%
% choice 1:
% g = imread('camera.jpg'); h = imread('camera.jpg');
% g = imread('peppers.png'); h = imread('peppers.png');
% choice 2:
g = imread('naturalTessellation.jpg'); h = imread('imgGrey.jpg');
% use cpselect tool
cpselect(g,h)
```

Listing A.10 Matlab code in `inspectPixels.m`.

Remark A.10 **Inspect regions of a raster image**.
The colour image in Fig. A.13.1 provide input to the sample cpselect() tool GUI shown in Fig. A.14. Using this GUI, the following results are obtained.

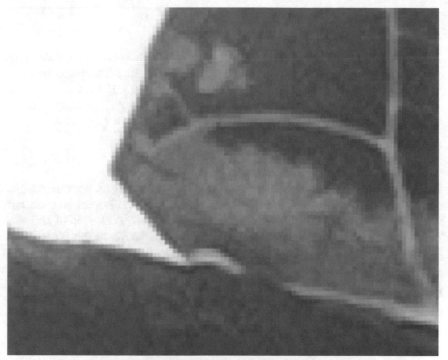

A.17.1: Image region extracted from Fig. A.16.1

A.17.2: Image region extracted from Fig. A.15.1

Fig. A.17 Sample image regions

1° ![greyscale leaf inspection window] is the **pixel inspection window** showing a greyscale image region (zoomed in at 50%) that is displayed in Fig. A.15.2. This inspection window is shown in context in the cpselect display in Fig. A.15.2. A closeup of an image region extracted from Fig. A.16.1 is shown in Fig. A.17.1. In this closeup, the tiny squares representing individual image pixels are clearly seen.

2° ![colour inspection window] is the **pixel inspection window** showing a color image region (zoomed in at 400%) that is displayed in Fig. A.16.2. This inspection window is shown in context in the cpselect display in Fig. A.16.2. A closeup of an image region extracted from Fig. A.15.1 is shown in Fig. A.17.2. In this closeup of a greyscale image region, the tiny squares representing image pixels are not as evident.

For more about this, see Sect. 2.1. ∎

A.2.2 Colour Image Channels

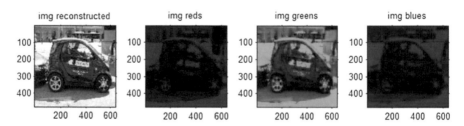

Fig. A.18 Sample colour image *red, green, blue* channels

```
% script: pixelChannels.m
% Display color image channel values
% Script idea from:
% http://www.mathworks.com/matlabcentral/profile/authors/1220757-sixwwwwww
clc, clear all, close all
img = imread('carCycle.jpg'); % Read image
% img = imread('carPoste.jpg'); % Read image
red = img(:,:,1); % Red channel
green = img(:,:,2); % Green channel
blue = img(:,:,3); % Blue channel
rows = size(img,1); columns = size(img,2);
rc = zeros(rows, columns);
justR = cat(3, red, rc, rc);
justG = cat(3, rc, green, rc);
justB = cat(3, rc, rc, blue);
```

```
captureOriginal = cat(3, red, green, blue);
figure, imshow(captureOriginal) ,...
    axis square, axis on;
figure,
subplot(1,4,1),imshow(captureOriginal), ...
    axis square, axis on,title('img reconstructed'),
subplot(1,4,2),imshow(justR), ...
    axis square, axis on,title('img reds'),
subplot(1,4,3),imshow(justG), ...
    axis square, axis on,title('img greens'),
subplot(1,4,4),imshow(justB), ...
    axis square, axis on,title('img blues')
```

Listing A.11 Matlab code in `pixelChannels.m`

Remark A.11 **Sample Red, Green and Blue Colour Channels**.
Pixel colour channels are in Fig. A.18, using script A.11. For more about this, see
Sect. 2.1. ∎

A.2.3 Colour 2 Greyscale Conversion

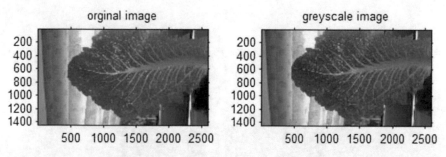

Fig. A.19 Sample colour image ⟼ greyscale image

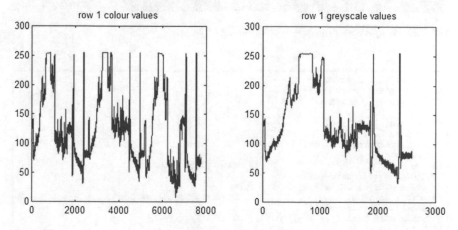

Fig. A.20 Sample colour image ⟼ greyscale pixel intensities

```
% script: rgb2grey.m
% Colour to greyscale conversion.
clc, clear all, close all
%%
img = imread('naturalTessellation.jpg');
% figure, imshow(img),axis on;
imgGrey = rgb2gray(img);
imwrite(imgGrey,'imgGrey.jpg');
figure,
subplot(1,2,1), plot(img(1,:)) ,...  % row 1 colour intensities
    axis square; title('row 1 colour values');
subplot(1,2,2),plot(imgGrey(1,:)) ,...  % row 1 greyscale intensities
    axis square; title('row 1 greyscale values');
figure,
subplot(1,2,1), imshow(img) ,...  % display colour image
    axis on; title('orginal image');
subplot(1,2,2), imshow(imgGrey) ,...  % display greyscale image
    axis on; title('greyscale image');
```

Listing A.12 Matlab code in `rgb2grey.m`.

Remark A.12 **Sample Colour to Greyscale Conversion**.
The result of converting a colour image to a greyscale image is shown in Fig. A.19, using script A.12. Sample plots of colour pixel and greyscale pixel intensities are shown in Fig. A.20. For more about this, see Sect. 2.1. ∎

A.2.4 Algebraic Operations on Pixel Intensities

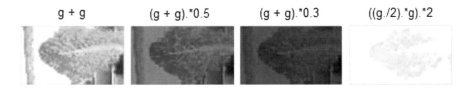

Fig. A.21 Sample colour image ⟼ pixel intensity changes

```
% script: pixelCycle.m
% Sample pixel value changes I.
clc, clear all, close all
%%
g = imread('lcaf.jpg');
% g = imread('carCycle.jpg');
figure, imshow(g),axis on;
figure,
i1 = g + g;                    % add image pixel values
subplot(3,4,1), imshow(i1) ,...
    axis off; title('g + g'); % display sum
i2 = (g + g).*0.5;             % average pixel values
subplot(3,4,2), imshow(i2) ,...
    axis off; title('(g + g).*0.5'); % display average
i3 = (g + g).*0.3,             % 1/3 pixel values
```

```
subplot(3,4,3), imshow(i3) ,...
    axis off; title('(g + g).*0.3');  % display reduced values
i4 = ((g./2).*g).*2;                  % doubled pixel value products
subplot(3,4,4), imshow(i4) ,...
    axis off; title('((g./2).*g).*2');  % display doubled values
```

Listing A.13 Matlab code in `pixelCycle.m`.

Remark A.13 **Sample Algebraic Operations on Pixel Intensities I**.
The result of algebraic operations on pixel intensities is shown in Fig. A.21, using script A.13. For more about this, see Sect. 2.4. ∎

h + 30   h-0.2.*h |h-((h + h).*0.5)| h+((h + h).*0.5)).*2

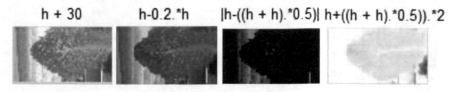

Fig. A.22 Another colour image ⟼ pixel intensity changes

```
% Sample pixel value changes II.
clc, clear all, close all
h = imread('naturalTessellation.jpg');
figure, imshow(h),axis on;
i5 = h + 30;                % pixel values + 30
figure,
subplot(3,4,5), imshow(i5) ,...
    axis off; title('h + 30');  % display augmented image pixels
i6 = imsubtract(h,0.2.*h);  % pixel value differences
subplot(3,4,6), imshow(i6) ,...
    axis off; title('h-0.2.*h');  % display pixel differences
i7 = imabsdiff(h,((h + h).*0.5)); % absolute value of differences
subplot(3,4,7), imshow(i7) ,...
    axis off; title('|h-((h + h).*0.5)|');  % display abs of differences
i8 = imadd(h,((h + h).*0.5)).*2; % summed pixel values doubled
subplot(3,4,8), imshow(i8) ,...
    axis off; title('h+((h + h).*0.5)).*2');  % display doubled sums
```

Listing A.14 Matlab code in `pixelLeaf.m`.

Remark A.14 **Sample Algebraic Operations on Pixel Intensities II**.
The result of algebraic operations on pixel intensities is shown in Fig. A.22, using script A.14. For more about this, see Sect. 2.4. ∎

i9 (0.8).*red   i10 (0.9).*green i11 (0.5).*green i12 (16.5).*blue

Fig. A.23 Yet another colour image ⟼ pixel intensity changes

```
% script: pixelR.m
% Sample pixel value changes III.
clc, clear all, close all
%%
img = imread('leaf.jpg');
% img = imread('CVLab-3.jpg');
figure, imshow(img),axis on;
% set up dummy image
rows = size(img,1); columns = size(img,2);
a = zeros(rows, columns);
% fill dummy image with new red brightness values
figure,
i9 = cat(3,(0.8).*img(:,:,1),a,a);    % changed red intensities
subplot(3,4,9), imshow(i9),...
    axis off; title('i9 (0.8).*red');  % display modified red intensities
% fill dummy image with new green brightness values
i10 = cat(3,a,(0.9).*img(:,:,2),a); % changed green intensities
subplot(3,4,10), imshow(i10),...
    axis off; title('i10 (0.9).*green');  % display newgreen intensities
% fill dummy image with new green brightness values
i11 = cat(3,a,(0.5).*img(:,:,2),a);   % changed green intensities
subplot(3,4,11), imshow(i11),...
    axis off; title('i11 (0.5).*green');  % display new green intensities
i12 = cat(3,a,a,(16.5).*img(:,:,3)); % changed blue intensities
subplot(3,4,12), imshow(i12),...
    axis off; title('i12 (16.5).*blue');  % display new blue intensities
```

Listing A.15 Matlab code in `pixelR.m`.

Remark A.15 **Sample Algebraic Operations on Pixel Intensities III.**
The result of algebraic operations on pixel intensities is shown in Fig. A.23, using
script A.15. For more about this, see Sect. 2.4. ∎

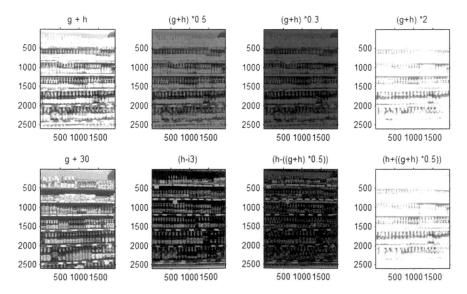

Fig. A.24 Yet another colour image ⟼ pixel intensity changes

```
% script: thaiR.m
% constructing new images from old images
% Sample pixel value changes IV.
clc, clear all, close all
%%
                                          % What's happening?
%g = imread('rainbow.jpg'); h = imread('gems.jpg');
g = imread('P9.jpg'); h = imread('P7.jpg');
i1 = g + h;                                % add image pixel values
subplot(2,4,1), imshow(i1); title('g + h'); % display sum
i2 = (g + h).*0.5;                         % average pixel values
subplot(2,4,2), imshow(i2); title('(g+h).*0.5'); % display average
i3 = (g + h).*0.3;                         % 1/3 pixel values
subplot(2,4,3), imshow(i3); title('(g+h).*0.3'); % display reduced values
i4 = (g + h).*2;                           % doubled pixel value sums
subplot(2,4,4), imshow(i4); title('(g+h).*2'); % display doubled values
i5 = g + 30;                               % pixel value + 30
subplot(2,4,5), imshow(i5); title('g + 30'); % display augmented image pixels
i6 = imsubtract(h,i3);                     % pixel value differences
subplot(2,4,6), imshow(i6); title('(h-i3)'); % display pixel differences
i7 = imabsdiff(h,((g + h).*0.5));          % absolute value of differences
subplot(2,4,7), imshow(i7); title('(h-((g+h).*0.5))'); % display abs of
    differences
i8 = imadd(h,((g + h).*0.5)).*2;           % summed pixel values doubled
subplot(2,4,8), imshow(i8); title('(h+((g+h).*0.5))'); % display doubled sums
```

Listing A.16 Matlab code using thai.m to produce Fig. A.24.

Remark A.16 **Sample Algebraic Operations on Pixel Intensities IV**.
The result of algebraic operations on pixel intensities is shown in Fig. A.24, using
script A.16. For more about this, see Sect. 2.4.   ∎

```
% script: maxImage.m
% Modifying colour channel pixel values using a max intensity
clc, clear all, close all     % housekeeping
%%
g = imread('camera.jpg');     % read colour image
[r,c] = max(g(1,:,1));        % g(r,c) = max red intensity in row 1
h = g(:,:,1) + (0.1).*g(r,c); % add (0.1)max red value to all pixel values
h2 = g(:,:,1) + (0.3).*g(r,c); % add (0.3)max red from all pixel values
h3 = g(:,:,1) + (0.6).*g(r,c); % add (0.6)max red from all pixels
rows = size(g,1); columns = size(g,2);
a = zeros(rows, columns);     % black image
captureR1 = cat(3, h, a, a);  % red channel image
captureR2 = cat(3, h2, a, a); % red channel image
captureR3 = cat(3, h3, a, a); % red channel image
figure, % internal view of a red channel is a greyscale image
subplot(1,3,1), imshow(h),title('g(:,:,1)+(0.1).*g(r,c)');
subplot(1,3,2), imshow(h2),title('g(:,:,1)+(0.3).*g(r,c)');
subplot(1,3,3), imshow(h3),title('g(:,:,1)+(0.6).*g(r,c)');
figure, % external view of a red channel is a colour image
subplot(1,3,1), imshow(captureR1),title('red channel captureR1');
subplot(1,3,2), imshow(captureR2),title('red channel captureR2');
subplot(1,3,3), imshow(captureR3),title('red channel captureR3');
```

Listing A.17 Find max red intensity in row 1 in an image, using maxImage.m

red channel captureR1   red channel captureR2   red channel captureR3

A.25.1: Algebraic operations on camera image colour channel and on greyscale intensities

g(:,:,1)+(0.1).*g(r,c)   g(:,:,1)+(0.3).*g(r,c)   g(:,:,1)+(0.6).*g(r,c)

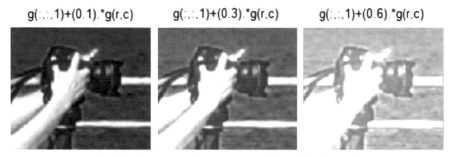

A.25.2: Image region extracted from Fig. A.15.1

Fig. A.25 Sample changes in colour channel and on greyscale intensities

Remark A.17 **Sample Algebraic Operations on Pixel Intensities V.**
The result of algebraic operations on colour channel and on greyscale pixel intensities is shown in Fig. A.25, using script A.17. The result of concatenating the original colour intensities with scaled maximum red channel intensities is shown in Fig. A.25.1. Similarly, The result of concatenating the original greyscale intensities with scaled maximum greyscale intensities is shown in Fig. A.25.2. For more about this, see Sect. 2.4. ∎

A.2.5   Selecting and Displaying Edge Pixel Colour Pixel Intensities

```
% script: imageEdgesOnColorChannel.m
% Edge Colour Channel pixels mapped to new intensities
clc, clear all, close all
%%
img = imread('trains.jpg');
% img = imread('carCycle.jpg');
figure,imshow(img) ,...
    axis square, axis on, title('colour image display');
gR = img(:,:,1); gG = img(:,:,2); gB = img(:,:,3);
imgRGB = edge(rgb2gray(img),'canny'); % greyscale edges in B/W
```

Fig. A.26 Salerno station trains

```
imgR = edge(gR,'canny');      % red channel edges in B/W
imgG = edge(gG,'canny');      % green channel edges in B/W
imgB = edge(gB,'canny');      % blue channel edges in B/W
figure,imshow(imgRGB),...
    axis square, axis on, title('BW edges');
figure,
subplot(1,3,1),imshow(imgR),...
    axis square, axis on, title('R channel edges');
subplot(1,3,2),imshow(imgG),...
    axis square, axis on, title('G channel edges');
subplot(1,3,3),imshow(imgB),...
    axis square, axis on, title('B channel edges');
rows = size(img,1); columns = size(img,2);
a = zeros(rows, columns);     % black image
captureR = cat(3, gR, a, a);  % red channel image
captureG = cat(3, a, gG, a);  % green channel image
captureB = cat(3, a, a, gB);  % red channel image
edgesR = cat(3,imgR,a,a);     % red channel edges image
edgesG = cat(3,a,imgG,a);     % green channel edges image
edgesB = cat(3,a,a,imgB);     % blue channel edges image
edgesBscaled = edgesB+0.2;    % scaled blue edges
edgesRG = cat(3,imgR,imgG,a); % RG technicolor edges
figure,imshow(edgesRG),...
    axis square, axis on, title('technicolor RG edges');
edgesRB = cat(3,imgR,a,imgB); % RB technicolor edges
figure,imshow(edgesRB),...
    axis square, axis on, title('technicolor RB edges');
figure,imshow(captureR),...
    axis square, axis on, title('red channel pixels');
figure,imshow(edgesR),...
```

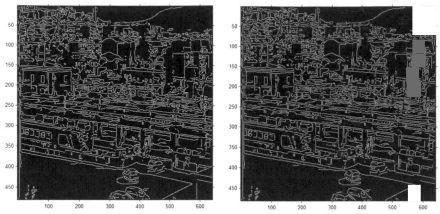

A.27.1: Canny train edges extracted from Fig. A.26

A.27.2: RGB Canny train edges extracted from Fig. A.26

Fig. A.27 Sample Canny train edges in binary and in colour, using Script A.18

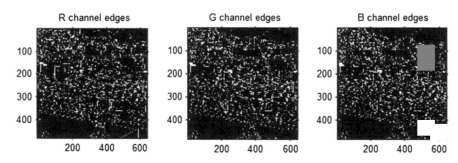

Fig. A.28 Canny edges for each colour image channel, using Script A.18

```
    axis square, axis on, title('red channel edge pixels');
figure,imshow(captureG),...
    axis square, axis on, title('green channel pixels');
figure,imshow(edgesG),...
    axis square, axis on, title('green channel edge pixels');
figure,
subplot(1,2,1),imshow(captureR),...
    axis square, axis on, title('red channel');
subplot(1,2,2),imshow(edgesR),...
    axis square, axis on, title('red edges');
figure,
subplot(1,2,1),imshow(captureG),...
    axis square, axis on, title('green channel');
subplot(1,2,2),imshow(edgesG),...
    axis square, axis on, title('green edges');
figure,
subplot(1,2,1),imshow(captureB),...
    axis square, axis on, title('blue channel');
subplot(1,2,2),imshow(edgesBscaled),...
    axis square, axis on, title('blue edges');
```

Listing A.18 Matlab code in `imageEdgesOnColorChannel.m`.

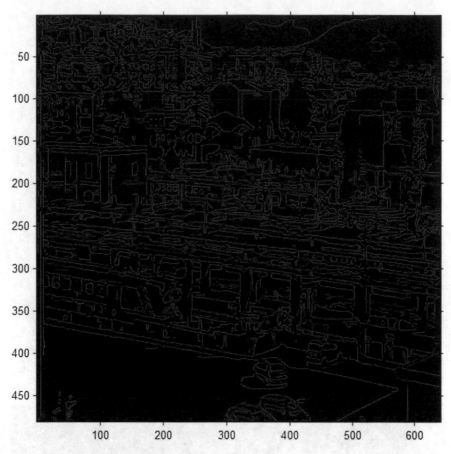

Fig. A.29 Canny edges for combined *red* and *blue colour* channels, using Script A.18

Remark A.18 **Sample Canny Colour Channel Edges**.
Script A.18 produces the following results.

1° Binary Canny edges (see Fig. A.27.1) and RGB Canny edges (see Fig. A.27.2) are extracted from the colour image in Fig. A.26.
2° Figure A.28 displays binary Canny edges for each colour channels extracted from the colour image in Fig. A.26.
3° Binary Canny edges for each colour image channel are shown in Fig. A.28.
4° Canny edges for the combined red and blue colour channels in Fig. A.26 are shown in Fig. A.29.
5° Canny edges for the red colour channels in Fig. A.26 are shown in Fig. A.30.
6° Canny edges for the green colour channels in Fig. A.26 are shown in Fig. A.31.

For more about this, see Sect. 2.5. ∎

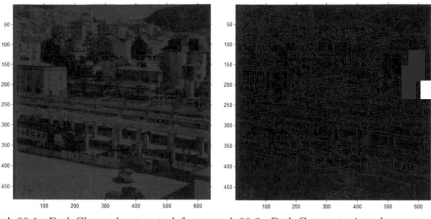

A.30.1: Red Channel extracted from Fig. A.26

A.30.2: Red Canny train edges extracted from Fig. A.26

Fig. A.30 Sample Canny train edges in binary and in colour using Script A.18

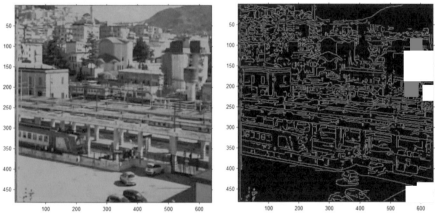

A.31.1: Green Channel extracted from Fig. A.26

A.31.2: Green Canny train edges extracted from Fig. A.26

Fig. A.31 Canny edges for each colour image channel using Script A.18

A.2.6 *Function-Based Pixel Value Changes*

```
% script: cameraPixelsModified.m
% Changing Colour Channel Values.
% Method:   scaled log of channel intensities
%
clc, clear all, close all
%%
img = imread('CNtrain.jpg'); % Read image
% img = imread('carCycle.jpg');
```

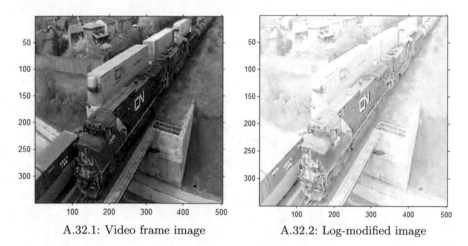

A.32.1: Video frame image   A.32.2: Log-modified image

Fig. A.32 Video frame ⟼ log-modified image

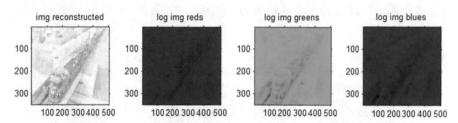

Fig. A.33 Sequence of log-modified video frame colour channel images

```
gR = img(:,:,1); gG = img(:,:,2); gB = img(:,:,3);
% g(:,:) specifies all image pixel intensities
% double(g(:,:)) converts pixel intensities to type double
% let x be a number of type double
% log(x) = natural log of x
% log(double(g(:,:))) computes log all pixel intensities
% 0.2.*log(double(g(:,:))) reduces each pixel channel intensity
% img = stores array to modified pixel channel intensities
imgR = 0.2.*log(double(img(:,:,1)));
imgB = 0.2.*log(double(img(:,:,2)));
imgG = 0.2.*log(double(img(:,:,3)));
rows = size(img,1); columns = size(img,2);
a = zeros(rows, columns);
justR = cat(3, imgR, a, a);
justG = cat(3, a, imgB, a);
justB = cat(3, a, a, imgG);
captureOriginal = cat(3, gR, gG, gB);
figure, imshow(captureOriginal) ,...
    axis square, axis on;
captureModifiedImage = cat(3, imgR, imgB, imgG);
figure, imshow(captureModifiedImage) ,...
    axis square, axis on;
figure,
subplot(1,4,1),imshow(captureModifiedImage), ...
    axis square, axis on,title('img reconstructed'),
subplot(1,4,2),imshow(justR), ...
```

```
    axis square, axis on,title('log img reds'),
subplot(1,4,3),imshow(justG), ...
    axis square, axis on,title('log img greens'),
subplot(1,4,4),imshow(justB), ...
    axis square, axis on,title('log img blues')
```

Listing A.19 Matlab code in `cameraPixelsModified.m`.

Remark A.19 **Sample Log-modified Video Frame Colour Channel images**.
Script A.19 produces the following results.

1° A single Single Video frame image in Fig. A.32.1 mapped to reconstructed log-
    modified image in Fig. A.32.2.
2° A sequence log-modified colour channel images is shown in Fig. A.33.

For more about this, see Sect. 2.6.   ■

A.2.7   *Logical Operations on Images*

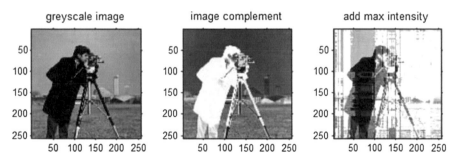

Fig. A.34 Max-intensity modified greyscale image

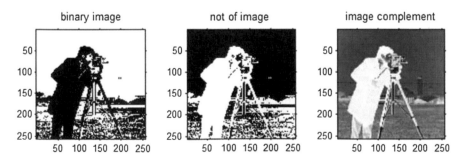

Fig. A.35 Complement versus not of greyscale pixel intensities

```
% script: invert.m
% Greyscale image complement and Logical Not of Binary image
clc, clear all, close all    % housekeeping
%%
g = imread('cameraman.tif');  % read greyscale image
gbinary = im2bw(g);           % convert to binary image
gnot = not(gbinary);          % not of bw intensities
% gbinaryComplement = imcomplement(gbinary);
% gbinaryComplement = imcomplement(gnot);
gbinaryComplement = imcomplement(g);
figure,
subplot(1,3,1), imshow(g) ,...
    axis square, axis on, title('greyscale image');
h = imcomplement(g);          % invert image (complement)
subplot(1,3,2), imshow(h) ,...
    axis square, axis on, title('image complement');
[r,c] = max(g);               % max intensity location
h2 = g + g(r,c);              % max-increased intensities
subplot(1,3,3), imshow(h2) ,...
    axis square, axis on, title('add max intensity');
figure,
subplot(1,3,1), imshow(gbinary) ,...
    axis square, axis on, title('binary image');
subplot(1,3,2), imshow(gnot) ,...
    axis square, axis on, title('not of image');
subplot(1,3,3), imshow(gbinaryComplement) ,...
    axis square, axis on, title(' image complement');
```

Listing A.20 Matlab source invert.m to produce Fig. 2.24.

Remark A.20 **Sample complement, negation, and max-intensity modified greyscale image**.
Script A.20 produces the following results.

1° Pixel maximum intensity provides a basis for modifying greyscale image intensities in Fig. A.34.

2° Logical not versus complement of a greyscale image is shown in Fig. A.35.

For more about this, see Sect. 2.6.   ■

A.3   Scripts from Chap. 3

A.3.1   Pixel Intensity Histograms (Binning)

```
% script: histogramBins.m
% Histogram and stem plot experiment
%
clc, clear all, close all % housekeeping
%%
% This section for colour images
I = imread('trains.jpg'); % sample RGB image
% I = imread('CNtrain.jpg');
% I = imread('fishermanHead.jpg');
% I = imread('fisherman.jpg');
% I = imread('football.jpg');
```

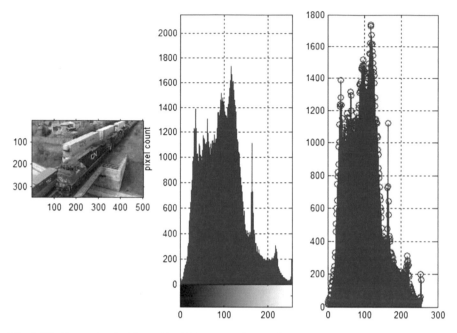

Fig. A.36 Greyscale pixel intensities histogram and stem plot

```
I = rgb2gray(I);
%
% This section for intensity images
%I = imread('pout.tif');
%
% Construct histogram:
%
h = imhist(I);
[counts,x] = imhist(I);
for j=1:size(x)
  [j,counts(j)]
end
% counts
size(counts)
subplot(1,3,1), imshow(I);
subplot(1,3,2), imhist(I),
grid on,
ylabel('pixel count');
subplot(1,3,3), stem(x,counts),
grid on
```

Listing A.21 Matlab source `histogramBins.m`, illustrates binning pixel intensities in an intensity image.

Remark A.21 **Sample greyscale image and stem plot**.
Script A.21 produces the results shown in Fig. A.36. For more about this, see Sect. 3.1.
∎

A.3.2 Pixel Intensity Distributions

Fig. A.37 Sample grid on a colour image

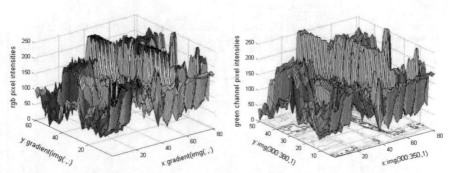

A.38.1: 3D color pixel intensities mesh   A.38.2: 3D color pixel intensities contour mesh

Fig. A.38 Sample pixel 3D views of pixel intensities

```
% script: imageMesh.m
% image geometry: visualizing rgb pixel intensity distribution
%
clear all; close all; clc; % housekeeping
%%
img = imread('trains.jpg'); % sample RGB image
% img=imread('carPolizia.jpg');
figure, imshow(img) ,...
   axis on, grid on, xlabel('x'),ylabel('y');
% img = imcrop(img);
% [r,c] = size(img); % determine cropped image size
% r,c
figure, imshow(img(300:360,300:380)) ,...
   axis on, grid on, xlabel('x'),ylabel('y');
% convert to 64 bit (double precision) format
% surf & surfc need double precision: 64 bit pixel values
img = double(double(img));
% Cr = gradient(img(:,:,1));
% Cg = gradient(img(:,:,2));
% Cb = gradient(img(:,:,2));
% colour channel gradients of manually crop image:
Cr = gradient(img(300:360,300:380,1));
Cg = gradient(img(300:360,300:380,2));
Cb = gradient(img(300:360,300:380,3));
figure;
% vm3D = surf(img(:,:));
vm3D = surf(img(300:360,300:380));
axis tight,zlabel('rgb pixel intensities'),
xlabel('x:gradient(img(:,:)'),ylabel('y:gradient(img(:,:)'); % label axes
saveas(vm3D,'3DcontourMesh.png'); % save copy of image
vm3Dred = figure,
% surfc(img(:,:,1),Cr),
surfc(img(300:360,300:380,1),Cr),
axis tight,zlabel('red channel pixel intensities') ,...
xlabel('x:gradient(img(:,:,1)'),ylabel('y:gradient(img(:,:,1)'); % label axes
vm3Dgreen = figure,
% surfc(img(:,:,2),Cg),
surfc(img(300:360,300:380,2),Cg),
axis tight,zlabel('green channel pixel intensities') ,...
xlabel('x:gradient(img(:,:,2)'),ylabel('y:(img(:,:,2)'); % label axes
vm3Dblue = figure,
% surfc(img(:,:,3),Cb),
surfc(img(300:360,300:380,3),Cb),
axis tight,zlabel('blue channel pixel intensities') ,...
xlabel('x:gradient(img(:,:,3)'),ylabel('y:gradient(img(:,:,3)'); % label axes
saveas(vm3D,'3DcontourMesh.png'); % save copy of image
saveas(vm3Dred,'3DcontourMeshRed.png'); % save copy of red channel contour mesh
saveas(vm3Dgreen,'3DcontourMeshGreen.png'); % save copy of red channel contour
      mesh
saveas(vm3Dblue,'3DcontourMeshRed.png'); % save copy of red channel contour
      mesh
% access and displaying (in the work space) manually cropped image:
% rgb340341 = img(340,341),
% rgb340342 = img(340,342),
% rgb340343 = img(340,343),
% red = img(340:343,1),
% green = img(340:343,2),
% blue = img(340:343,3)
```

Listing A.22 Matlab source `imageMesh.m`.

Remark A.22 **Sample colour image grid, 3D mesh for colour intensities and 3D contour mesh for green channel intensities plots**.

Script A.22 produces the results shown in Fig. A.38.1 and A.38.2 for the colour image with grid overlay shown in Fig. A.37. For more about this, see Sect. 3.1.   ■

A.3.3   Pixel Intensities Isolines

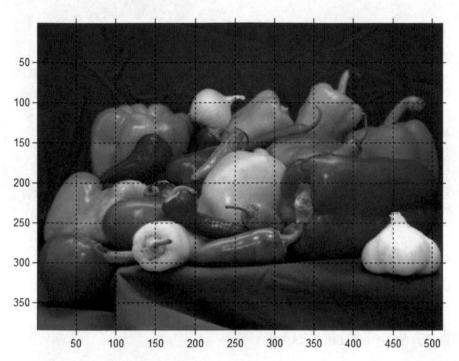

Fig. A.39 Sample grid on a colour image

```
% Source: isolines.m
% Visualisation experiment with isolines
%
clc, close all, clear all  % housekeeping
g = imread('peppers.png');  % read colour image
figure, imshow(g),axis on, grid on;
figure,
contour(g(:,:,1)); % isolines w/o values
figure,
[c,h] = contour(g(:,:,1)), % red channel isolines
clabel(c,h,'labelspacing',80); % isoline label spacing
hold on
set(h,'ShowText','on','TextStep',get(h,'LevelStep'));
colormap jet, title('peppers.png red channel isoline values');
```

Listing A.23 Matlab code in isolines.m to produce the colour channel isolines shown in Fig. A.40.

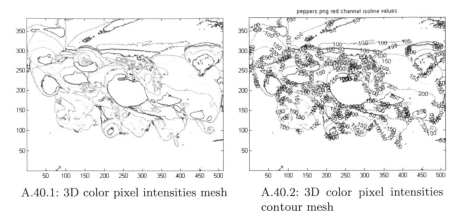

A.40.1: 3D color pixel intensities mesh   A.40.2: 3D color pixel intensities contour mesh

Fig. A.40 Sample colour image isolines with and without labels

Remark A.23 **Sample colour channel isolines with and without labels**.
Script A.23 produces the results shown in Fig. A.40.1 and A.40.2 for the colour image
with grid overlay shown in Fig. A.39. For more about this, see Sect. 3.1.   ■

A.4 Scripts from Chap. 4

The scripts Appendix A.4 are embedded in Chap. 4.

A.5 Scripts from Chap. 5

A.5.1 1D Gaussian Kernel Plots

```
% gaussianSmoothing.m
% Script for 1D Gaussian kernel plots
% Original script by Matthew Brett 6/8/99
% Thanks extended to R. Hettiarachchi for nos correction.
% revised 24 Oct. 2016
clear all, close all, clc

% make vectors of points for the x axis
% minx = 1; maxx = 55; x = minx:maxx; % for discrete plots
% fineness = 1/100;
% finex = minx:fineness:maxx; % for continuous plots

% im = read('peppers.png');
% im = rgb2hsv(im); % use row of im instead of nos variable (below).

%% Let mean u = 0. The formula for 1D Gaussian kernel is defined by
%                   1              ( x^2   )
%      f(x) = ---------------- exp[- --------- ]
%            v*sqrt(2*pi)         ( 2v^2   )
% where v (or sigma) is the standard deviation, and u is the mean.
```

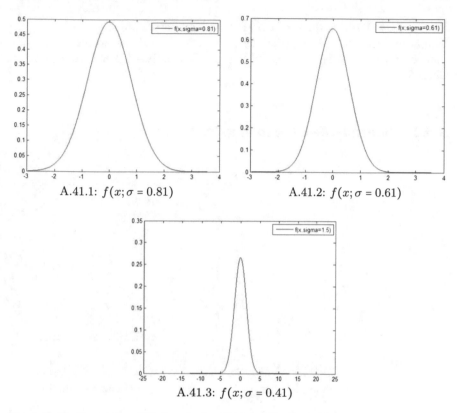

A.41.1: $f(x; \sigma = 0.81)$ A.41.2: $f(x; \sigma = 0.61)$

A.41.3: $f(x; \sigma = 0.41)$

Fig. A.41 Varying widths of 1D Gaussian kernel plots

```
% 1D Gaussian kernel sigma:
%%
sigma1 = 0.41; % 0.51,1.5;
rng('default');
nos = randn(1,100);
fineness = nos/100;
kernx = min(nos):fineness:max(nos);
skerny = 1/(sigma1*sqrt(2*pi)) * exp(-kernx.^2/(2*sigma1^2)); % v = o.51,1,3
figure
plot(kernx, skerny,'r'),...
    legend('f(x;sigma=0.41)','Location','NorthEast');
sigma2 = 0.61; % 1.0;
skerny = 1/(sigma2*sqrt(2*pi)) * exp(-kernx.^2/(2*sigma2^2)); % v = 1,3
figure
plot(kernx, skerny,'r'),...
    legend('f(x;sigma=0.61)','Location','NorthEast');
sigma3 = 0.81; %1.2;
skerny = 1/(sigma3*sqrt(2*pi)) * exp(-kernx.^2/(2*sigma3^2)); % v = 1,3
figure
plot(kernx, skerny,'r'),...
    legend('f(x;sigma=0.81)','Location','NorthEast');
```

Listing A.24 Matlab script in gaussianSmoothing.m to obtain sample 1D Gaussian kernel plots.

Remark A.24 **Varying the width of 1D Gaussian kernel plots**.
Sample plots of the 1D Gaussian kernel function are shown in Fig. A.41 using
Matlab® script A.24. For more about this, see Appendix A.5.2 and Example 5.1 in
Chap. 5.   ■

A.5.2   *Gaussian Kernel Experimenter*

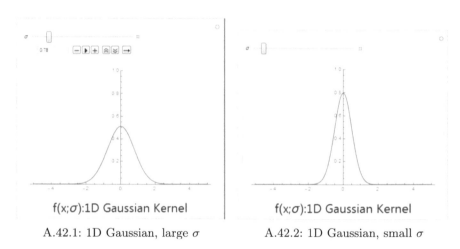

f(x;σ):1D Gaussian Kernel   f(x;σ):1D Gaussian Kernel

A.42.1: 1D Gaussian, large σ   A.42.2: 1D Gaussian, small σ

Fig. A.42 1D Gaussian kernel experiments

Remark A.25 **About the 1D Gaussian kernel**.
Sample plots of 1D Gaussian kernel function are shown in Fig. A.42 using the
Mathematica® Manipulate function. Try doing the same things using Matlab®. Let σ
be the width of the Gaussian kernel plot, centered around 0. The **width** $\sigma > 0$ is called
the **standard deviation** (average distance from the middle of a set of data) and σ^2 is
called the **variance**. The average value or **mean** or **middle** of a set of data is denoted
by μ. In this case, $\mu = 0$. The 1D Gaussian kernel function $f(x; sigma), x \in \mathbb{R}$
(reals) is defined by

$$f(x; \sigma) = \frac{1}{\sigma\sqrt{2\pi}} e^{-\frac{(x-0)^2}{2\sigma^2}} = \frac{1}{\sigma\sqrt{2\pi}} e^{-\frac{x^2}{2\sigma^2}} \text{ (1D Gaussian kernel)}.$$

In the definition of the 1D Gaussian kernel function $f(x; \sigma)$, x is a spatial parameter
and σ is a scale parameter. The semicolon between x and σ separates the two types of
parameters. For x, try letting x range over the pixel intensities in a row or column of
either a colour or grayscale image, letting the scale parameter σ be a small value such
as $\sigma = 0.5$. For more about this, see B.M. ter Haar Romeny in [65]. For other papers

by ter Haar Romeny on computer vision, visualization, and the Gaussian kernel, see
http://bmia.bmt.tue.nl/people/bromeny/index.html.   ■

Mathematica 1   Plotting 1D Gaussian kernel function values.

(*TU Delft 1D Gaussian kernel experimenter. Original script from

2008Biomedical Image − Analysis,

Technische Universiteit Eindhoven, the Netherlands.

Revised 19 Oct. 2016.

*)

Manipulate[Plot[(1/(σSqrt[2π]))Exp[−x^2/(2σ^2)],

{x, −5, 5}, PlotRange->{0, 1}], {{σ, 1}, .2, 4},

FrameLabel → Style["f(x;σ):1D Gaussian Kernel", Large], LabelStyle → Red]

A.5.3   2D Gaussian Kernel Plots

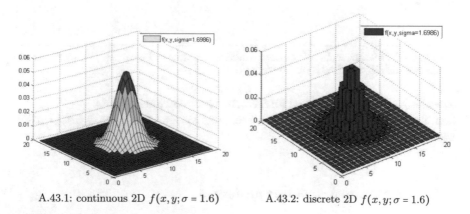

A.43.1: continuous 2D $f(x,y;\sigma = 1.6)$ A.43.2: discrete 2D $f(x,y;\sigma = 1.6)$

Fig. A.43 2D Gaussian kernel experiments

```
% gaussian2DKernelExperiment .m
% Script to produce almost continuous as well as discrete 2D Gaussian kernel
     plots
% Matthew Brett 6/8/99
% revised 24 Oct. 2016

% seed random number generator

clear all, close all, clc

% parameters for Gaussian kernel
```

```
rng('default');
nos = randn(1,100);
fineness = mean(nos);
fineness = fineness*5;
%
FWHM = 4;
sig = FWHM/sqrt(8*log(2))
%
% 2d Gaussian kernel - fairly continuous
Dim = [20 20];
% fineness = 0.55; % .1
[x2d,y2d] = meshgrid(-(Dim(2)-1)/2:fineness:(Dim(2)-1)/2 ,...
    -(Dim(1)-1)/2:fineness:(Dim(1)-1)/2);
gf     = exp(-(x2d.*x2d + y2d.*y2d)/(2*sig*sig));
gf     = gf/sum(sum(gf))/(fineness^2);
figure
colormap hsv
surfc(x2d+Dim(1)/2,y2d+Dim(2)/2,gf) ,...
    legend('f(x,y,sigma=1.6986)','Location','NorthEast');
beta = 1;
brighten(beta)

% 2d Gaussian kernel - discrete
[x2d,y2d] = meshgrid(-(Dim(2)-1)/2:(Dim(2)-1)/2,-(Dim(1)-1)/2:(Dim(1)-1)/2);
gf     = exp(-(x2d.*x2d + y2d.*y2d)/(2*sig*sig));
gf     = gf/sum(sum(gf));
figure
bar3(gf,'r') ,...
    legend('f(x,y,sigma=1.6986)','Location','NorthEast');
axis([0 Dim(1) 0 Dim(2) 0 max(gf(:))*1.2])
axis xy
%
%
%
```

Listing A.25 Matlab script in `gaussian2DKernelExperiment.m` to obtain sample continuous and discrete 2D Gaussian kernel plots.

Remark A.26 **Continuous and discrete 2D Gaussian kernel plots**.
Sample plots of the 2D Gaussian kernel function are shown in Fig. A.43 using Matlab® script A.25. ■

A.5.4   Gaussian Smoothing an Image

```
% script: gaussianFilterSimple.m
% Gaussian filtering (smoothing) a cropped image:
% See, also:
% http://stackoverflow.com/questions/2773606/gaussian-filter-in-matlab
clear all, close all, clc
%%
img = imread('CNtrain.jpg');
% img = imread('tissue.png');
img = imcrop(img); % crop image
% img = img(80 + [1:256],1:256,:);
figure, imshow(img) ,...
    grid on, title('Sample subimage');
%# Create the gaussian filter with hsize = [5 5] and sigma = 2
G = fspecial('gaussian',[5 5],2);
%# Filter (smooth) image
```

Fig. A.44 Cropped image

```
Ig = imfilter(img,G,'same');
%# Display
figure, imshow(Ig) ,...
    grid on, title('Gaussian smoothed image [5 5],2');
G2 = fspecial('gaussian',[3 3],1.2);
%# Filter (smooth) image
Ig2 = imfilter(img,G2,'same');
%# Display
figure, imshow(Ig2) ,...
    grid on, title('Gaussian smoothed image [3 3],1.2');
G3 = fspecial('gaussian',[2 2],0.8);
%# Filter (smooth) image
Ig3 = imfilter(img,G3,'same');
%# Display
figure, imshow(Ig3) ,...
    grid on, title('Gaussian smoothed image [2 2],0.8');
```

Listing A.26 Matlab script in gaussianFilterSimple.m to smooth an image using Gaussian filtering.

Remark A.27 **Continuous and discrete 2D Gaussian kernel plots**.
Sample results of Gaussian kernel are shown in Fig. A.45 of the cropped subimage in Fig. A.44, using script A.26. ∎

A.5.5   *Image Restoration*

```
% script: gaussianFilter.m
% Gaussian blur filter on a cropped image:
% Image courtesy of A.W. Partin.
% Sample application of the deconvreg function.
% Try docsearch deconvreg for more about this.
clear all, close all, clc
```

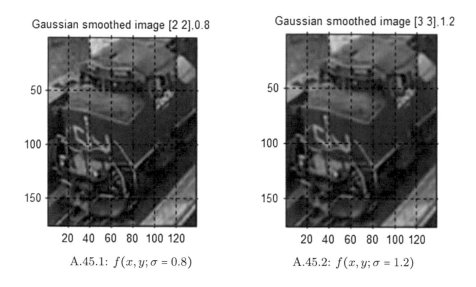

A.45.1: $f(x, y; \sigma = 0.8)$ A.45.2: $f(x, y; \sigma = 1.2)$

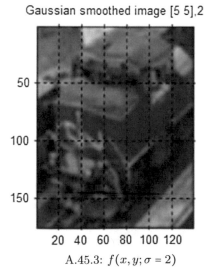

A.45.3: $f(x, y; \sigma = 2)$

Fig. A.45 2D Gaussian smoothing a subimage experiments

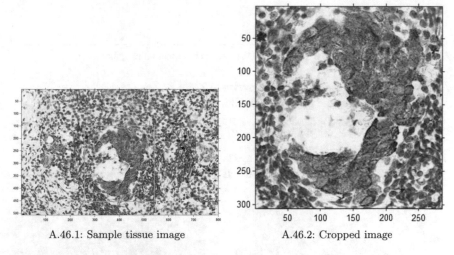

A.46.1: Sample tissue image   A.46.2: Cropped image

Fig. A.46 Tissue image and cropped image for restoration experiments

```
%%
img = imread('tissue.png');
% 1 display tissue sample
figure, imshow(img),...
    grid on, axis tight, title('Tissue image');
% 2 crop image
img = imcrop(img); % tool-based cropping
% img = img(125 + [1:256],1:256,:); % manual cropping
figure, imshow(img),...
    title('Cropped tissue image');
% 3 blur: convolve gaussian smoothed image with cropped image
psf = fspecial('gaussian',11,5); % psf = point spread function
blurred = imfilter(img,psf,'conv'); % convolve img with psf
figure, imshow(blurred),...
    title('Convolved point spread image 1');
% 4 noise: gaussian smooth blurred image
v = 0.02; % suggested v = 0.02,0.002,0.001,0.005
blurredNoisy = imnoise(blurred,'gaussian',0.000,v); % 0 vs. 0.001
figure, imshow(blurredNoisy),...
    title('Blurred noisy image 1');
% 5 restore image (first pass)
np = v*prod(size(img)); % noise power
% output 1: restored image reg1 & output of Lagrange multiplier
[reg1 LAGRA] = deconvreg(blurredNoisy,psf,np);
figure, imshow(reg1),...
    title('Restored image reg1');
% 6 blur: convolve gaussian smoothed image and cropped image
psf2 = fspecial('gaussian',8,5); % psf = point spread function
blurred2 = imfilter(img,psf2,'conv'); % convolve img with psf
figure, imshow(blurred2),...
    title('Convolved point spread image 2');
% 7 noise
v2 = 0.005; % suggested v = 0.02,0.002,0.001,0.005
blurredNoisy2 = imnoise(blurred2,'gaussian',0.001,v2); % 0 vs. 0.001
figure, imshow(blurredNoisy2),...
    title('Blurred noisy image 2');
% 8 restore image (second pass)
np2 = v2*prod(size(img)); % noise power
% output 2: restored image reg2 & output of Lagrange multiplier
```

```
[reg2 LAGRA2] = deconvreg(blurredNoisy2,psf2,np2);
figure, imshow(reg2),...
    title('Restored image reg2');
```

Listing A.27 Matlab script in gaussianFilter.m to deblur an image.

Remark A.28 **Image restoration experiments**.
Script A.27 produces the following results.

1° Figure A.46: Selection of image region in Fig. A.46.1 and cropped image in Fig. A.46.2.

2° Figure A.47: (1) Blur: cropped image convolved with Gaussian smoothed image, (2) Noise: injection of noise in blurred image, and (3) Restoration 1.

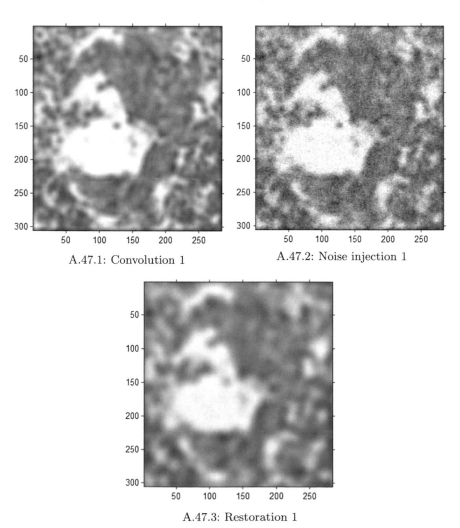

A.47.1: Convolution 1   A.47.2: Noise injection 1

A.47.3: Restoration 1

Fig. A.47 Experiment 1: convolution, noise injection and cropped tissue image restoration

3° Figure A.48: (1) Blur: cropped image convolved with Gaussian smoothed image, (2) Noise: injection of noise in blurred image, and (3) Restoration 2.

For more about this, see Sect. 5.6. ■

A.5.6   Image Corners

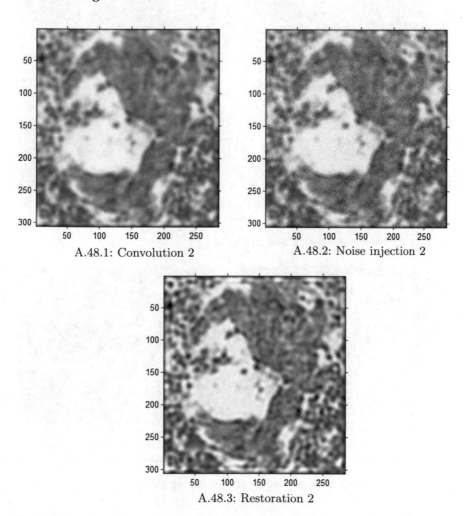

A.48.1: Convolution 2   A.48.2: Noise injection 2

A.48.3: Restoration 2

Fig. A.48 Experiment 2: convolution, noise injection and cropped tissue image restoration

```
% script: imageCorners.m
% imageCorners.m
% Finding image corners, R. Hettiarachchi, 2015
% revised 23 Oct. 2016
```

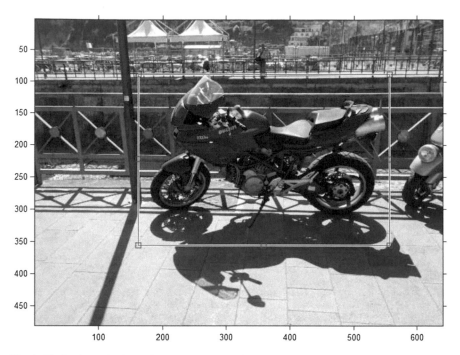

Fig. A.49 Region selection using Matlab script A.28

A.50.1: Cropped Cycle image with up to 50 corners

A.50.2: Cropped cycle image 50 corners-based Voronoi Mesh

Fig. A.50 Image corners & Voronoi mesh on cropped cycle image

```
clear all; close all; clc;
%%
% im=imread('peppers.png');
% im=imread('Carabinieri.jpg');
im=imread('cycle.jpg');
%
figure, imshow(im);
% crop method 1
```

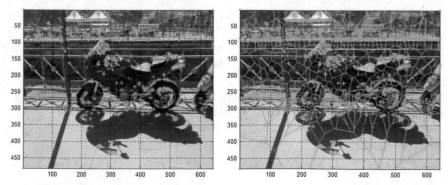

A.51.1: Cycle image with up to 500 corners

A.51.2: Cropped cycle image 500 corners-based Voronoi Mesh

Fig. A.51 Image corners & Voronoi mesh on full cycle image

```matlab
im2 = imcrop(im)
figure, imshow(im2),...
    grid on, title('cropped image');
% crop method 2
% imcrop(im,[xmin ymin width height])
% im2 = imcrop(im,[180 300 300 300]);
% crop method 3
% imcrop(im,[xmin [vertical width] ymin [horizontal width]])
% im2 = im(200 + [1:150],180 + [1:320],:); % crop image
g2=rgb2gray(im2);
[m2,n2]=size(g2);
C2 = corner(g2,50); %find up to 50 corners
%add four corners of the image to C
fc2=[1 1;n2 1; 1 m2; n2 m2];
C2=[C2;fc2];
figure,image(im2), hold on, ...
    grid on, title('corners on cropped image'),
resultOne = plot(C2(:,1), C2(:,2), 'g+');
figure,image(im2), hold on, ...
    grid on, title('Voronoi mesh on cropped image'),
result2 = plot(C2(:,1), C2(:,2), 'g+');
voronoi(C2(:,1),C2(:,2),'g.'); % red edges
% imwrite(result2,'corners2.jpg');

%%
g=rgb2gray(im);
[m,n]=size(g);
C = corner(g,500); %find up to 500 corners
%add four corners of the image to C
fc=[1 1;n 1; 1 m; n m];
C=[C;fc];
figure,image(im), hold on, ...
    grid on, title('corners on whole image'),
resultTwo = plot(C(:,1), C(:,2), 'g+');
figure,image(im), hold on, ...
    grid on, title('Voronoi mesh on whole image'),
result = plot(C(:,1), C(:,2), 'g+');
voronoi(C(:,1),C(:,2),'g.'); % red edges
% imwrite(result,'corners.jpg');
```

Listing A.28 Matlab code in imageCorners.m to produce Fig. A.50.1.

Remark A.29 **Superimposing corners on full as well as on cropped image with corresponding Voronoï tessellations**. A 480 × 640 colour image of Salerno motorcycle image is shown in Fig. A.49. Using the Matlab script A.28, the corners are found in both the full image in Fig. A.51.1 and in a cropped image in Fig. A.50.1. Then, a 500 corner-based Voronoï mesh is found (see Fig. A.51.2) and 50 corner-based Voronoï mesh is found (see Fig. A.50.2). Notice that there are a number different methods that can be used to crop an image (these cropping methods are explained in the comments in script A.28. For more about this, see Sect. 5.13. ∎

A.5.7   *Voronoï Mesh with and Without Image Corners*

A.52.1: Cycle image corners, including image boundary corners

A.52.2: Cycle image corners, excluding image boundary corners

Fig. A.52 Voronoï mesh on cropped cycle image with and without corners

```
% script : VoronoiMeshOnImage .m
% image geometry : overlay Voronoi mesh on image
%
% see http ://homepages.ulb.ac.be/~dgonze/INFO/matlab.html
% revised 23 Oct. 2016
clear all; close all; clc; % housekeeping
g=imread('fisherman.jpg');
% im=imread('cycle.jpg');
% g=imread('carRedSalerno.jpg');
%%
img = g; % save copy of colour image to make overlay possible
g = double(rgb2gray(g)); % convert to greyscale image
% corners = corner(g); % min. no. of corners
k = 233; % select k corners
corners = corner(g,k); % up to k corners
box_corners = [1,1;1,size(g,1);size(g,2),1;size(g,2),size(g,1)];
corners = cat(1,corners,box_corners);
vm = figure, imshow(img) ,...
axis on, hold on; % set up image overlay
```

```
voronoi(corners(:,1),corners(:,2),'g'); % red edges
% voronoi(corners(:,1),corners(:,2),'g.'); % red edges
% imfinfo('carRedSalerno.jpg')
% figure, mesh(g(300:350,300:350)),...
% axis tight,zlabel('rgb pixel intensity')
% xlabel('g(300:350)'),ylabel('g(300:350)') % label axes
% saveas(vm,'VoronoiMesh.png'); % save copy of image
```

Listing A.29 Matlab code in `VoronoiMeshOnImage.m` to produce Fig. A.52.1.

Remark A.30 **Superimposing corners on full as well as on cropped image**.
In this section, the 480×640 colour image of an Italian Carabinieri auto in Fig. A.49 is cropped. Then image corners provide a set of sites that are used to construct a Voronoï mesh on the cropped image. Using the Matlab script A.29, image corners are included in the set of interior image corners are used to construct a Voronoï mesh in Fig. A.52.1, which demonstrates the effectiveness of extreme image corners in producing a fine-grained image mesh. By contrast, see the coarse-grain Voronoï mesh that results from the exclusion of the extreme image corners in the set of interior image corners that are used to construct a Voronoï mesh in Fig. A.52.2. For more about this, see Sect. 5.13. ■

A.6 Scripts from Chap. 6

A.6.1 Finding 2D and 3D Image Centroids

A.53.1: Coin image   A.53.2: Coin centroid

Fig. A.53 Sample 2D image region centroids

Mathematica 2 script: Stanford Bunny.nb: UNISA coin centroid.

(* Digital Image Region Centroid *)

img = ;

c = ComponentMeasurements[img, "Centroid"][[All, 2]]][[1]];

Show[img, Graphics[{Black, PointSize[0.02], Point[c]}]]   ∎

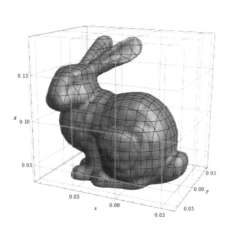

    A.54.1: Stanford bunny   A.54.2: Stanford bunny centroid

Fig. A.54 Sample 3D image region centroids

Mathematica 3 StanfordBunny centroid.

(* 3D Region centroids *)

gr = ExampleData[{"Geometry3D", "StanfordBunny"}];

gm = DiscretizeGraphics[gr];

c2 = RegionCentroid[gm]

Show[Graphics3D[Prepend[First[gr], Opacity[0.6]]],

Graphics3D[{PointSize[0.03], Black, Point@c2}], Axes → True,

LabelStyle → Black, AxesLabel → {x, y, z}, PlotTheme-> "Scientific",

FaceGrids → All,

FaceGridsStyle → Directive[Gray, Dotted]FaceGridsStyle →

Directive[Gray, Dotted]]

   ∎

Remark A.31 **Digital image region centroids**.

A sample 2D image region centroid is shown in Fig. A.53.2 on the image in Fig. A.53.1 using Mathematica® script 2. In Fig. A.53.1, the digital image shown a UNISA coin from a 1982 football tournament in Salerno, Italy. In Fig. A.53.2, the location of the centroid of the UNISA coin is identified with the black dot ●.

A sample 3D image region centroid is shown in Fig. A.54.2 on the 3D Stanford bunny image in Fig. A.54.1 using Mathematica® script 3. The coordinates for the bunny centroid in Fig. A.54.2 are

$$(x, y, z) = (-0.0267934, -0.00829883, 0.0941362).$$

For more about this, see Appendix A.6.2 and Sect. 6.4.   ■

A.6.2 *Another Approach in Finding Image Centroids*

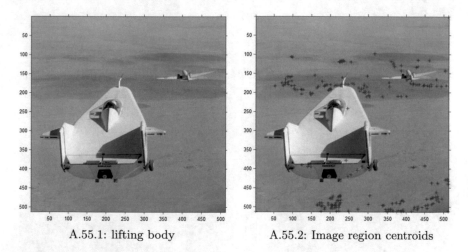

A.55.1: lifting body   A.55.2: Image region centroids

Fig. A.55 Image region centroids

```
% script: findCentroids.m
% centroid-based image Delaunay mesh
clc, clear all, close all
%%
im = imread('fisherman.jpg');
% im = imread('liftingbody.png');
figure,
imshow(im), axis on;
% if size(im,3)==3
%      g=rgb2gray(im);
% end
[m,n]=size(im);
bw = im2bw(im,0.5); % threshold at 50%
bw = bwareaopen(bw,2); % remove objects less 2 than pixels
```

```
stats = regionprops(bw,'Centroid'); % centroid coordinates
centroids = cat(1,stats.Centroid);
fc=[1 1;n 1; 1 m; n m]; % identify image corners
centroids=[centroids;fc];

% superimpose mesh on image
figure,
imshow(im),hold on
plot(centroids(:,1),centroids(:,2),'r+')
hold on;
X=centroids(:,1);
Y=centroids(:,2);

% constuct delaunay triangulation
% TRI = delaunay(X,Y);
% triplot(TRI,X,Y,'y');
```

Listing A.30 Matlab script in `findCentroids.m` to obtain plot of centroids on an image.

Remark A.32 **Region centroids on an image**.

A sample plot of the image region centroids are shown in Fig. A.55.2 on the image in Fig. A.55.1 using Matlab® script A.30. For more about this, see Appendix B.3 and Sect. 6.4. ∎

A.6.3 Finding Image Centroidal Delaunay Mesh

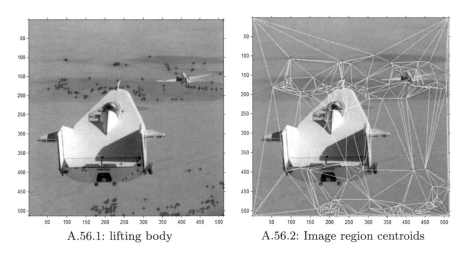

A.56.1: lifting body   A.56.2: Image region centroids

Fig. A.56 Image region centroid-based Delaunay mesh

```
% findCentroidalDelaunayMesh.m
% centroid-based image Delaunay mesh
clc, clear all, close all
%%
im = imread('fisherman.jpg');
% im = imread('liftingbody.png');
```

```
% if size(im,3)==3
%      g=rgb2gray(im);
% end
[m,n]=size(im);
bw = im2bw(im,0.5); % threshold at 50%
bw = bwareaopen(bw,2); % remove objects less 2 than pixels
stats = regionprops(bw,'Centroid'); % centroid coordinates
centroids = cat(1,stats.Centroid);
fc=[1 1;n 1; 1 m; n m]; % identify image corners
centroids=[centroids;fc];

% superimpose mesh on image
figure,
imshow(im),hold on
plot(centroids(:,1),centroids(:,2),'r+')
hold on;
X=centroids(:,1);
Y=centroids(:,2);

% constuct delaunay triangulation
TRI = delaunay(X,Y);
triplot(TRI,X,Y,'y');
```

Listing A.31 Matlab script in `findCentroidalDelaunayMesh.m` to obtain plot of centroid-based Delaunay mesh on an image.

Remark A.33 **Region centroid-based Delaunay triangulation on an image**.
A sample plot of the image region centroid-based Delaunay mesh is shown in Fig. A.56.2 (relative to the region centroids in Fig. A.56.1) using Matlab® script A.31. For more about this, see Sect. 6.4. ∎

A.6.4 Finding Image Centroidal Voronoï Mesh

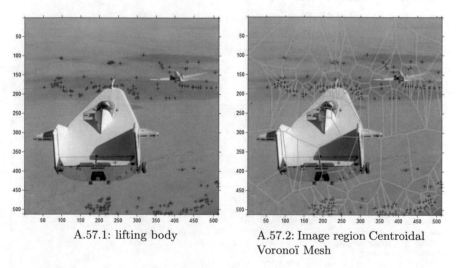

A.57.1: lifting body

A.57.2: Image region Centroidal Voronoï Mesh

Fig. A.57 Image region centroid-based Voronoï mesh

```
% script: findCentroidalVoronoiMesh.m
% centroid−based image Voronoi mesh
clc, clear all, close all
%%
im = imread('fisherman.jpg');
% im = imread('liftingbody.png');
% if size(im,3)==3
%     g=rgb2gray(im);
% end
[m,n]=size(im);
bw = im2bw(im,0.5); % threshold at 50%
bw = bwareaopen(bw,2); % remove objects less 2 than pixels
stats = regionprops(bw,'Centroid'); % centroid coordinates
centroids = cat(1,stats.Centroid);
fc=[1 1;n 1; 1 m; n m]; % identify image corners
centroids=[centroids;fc];
% superimpose mesh on image
figure,
imshow(im),hold on
plot(centroids(:,1),centroids(:,2),'r+')
hold on;
X=centroids(:,1);
Y=centroids(:,2);

% construct Voronoi mesh

[vx,vy] = voronoi(X,Y);
plot(vx,vy,'g-');
```

Listing A.32 Matlab script in `findCentroidalVoronoiMesh.m` to obtain plot of centroid-based Voronoï mesh on an image.

Remark A.34 **Region centroid-based Voronoï mesh on an image**.

A sample plot of the image region centroid-based Voronoï mesh is shown in Fig. A.57.2 (relative to the region centroids in Fig. A.57.1) using Matlab® script A.32. For more about this, see Sect. 6.4. ∎

A.6.5 Finding Image Centroidal Voronoï Superimposed on a Delaunay Mesh

```
% script: findCentroidalVornonoiOnDelaunayMesh.m
% centroid−based image Delaunay and Voronoi mesh
clc, clear all, close all
%%
im = imread('fisherman.jpg');
% im = imread('liftingbody.png');
% if size(im,3)==3
%     g=rgb2gray(im);
% end
[m,n]=size(im);
bw = im2bw(im,0.5); % threshold at 50%
bw = bwareaopen(bw,2); % remove objects less 2 than pixels
stats = regionprops(bw,'Centroid'); % centroid coordinates
centroids = cat(1,stats.Centroid);
fc=[1 1;n 1; 1 m; n m]; % identify image corners
centroids=[centroids;fc];
% superimpose mesh on image
```

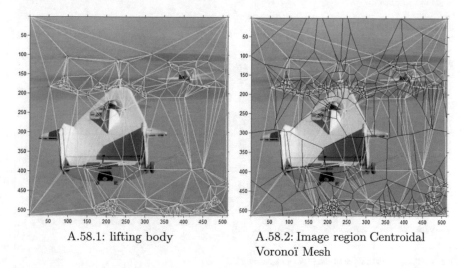

A.58.1: lifting body   A.58.2: Image region Centroidal
                                  Voronoï Mesh

Fig. A.58 Image region centroid-based Voronoï over Delaunay mesh

```
figure,
imshow(im),hold on
plot(centroids(:,1),centroids(:,2),'r+')
hold on;
X=centroids(:,1);
Y=centroids(:,2);

% constuct delaunay triangulation

TRI = delaunay(X,Y);
triplot(TRI,X,Y,'y');

% construct Voronoi mesh

[vx,vy] = voronoi(X,Y);
plot(vx,vy,'k-');
```

Listing A.33 Matlab script in findCentroidalVoronoiOnDelaunayMesn.m to obtain plot of centroid-based Voronoï over Delaunay mesh on an image.

Remark A.35 **Region centroid-based Voronoï over Delaunay mesh on an image.**
A sample plot of the image region centroid-based Voronoï over a Delaunay mesh is shown in Fig. A.58 (relative to the region centroidal Delaunay mesh in Fig. A.58.1) using Matlab® script A.33. For more about this, see Sect. 6.4.   ∎

A.7   Scripts from Chap. 7

A.7.1   *Edgelet Measurements in Voronoï Tessellated Video Frames*

The following is a sample solution to Problem 7.32.

```
% Cropping Voronoi tessellated video frames to explore edgelet metrics
% Solution by Drew Barclay, 2016
%
% Call this script with:
% -videoFile: the name of the video file to extract frames from
% (use '' for webcam)
% -numPoints: the number of corner points to use
%
% example use: Problem1('Train.mp4', 30)
%%%

function Problem1(videoFile, numPoints)
    close all

    %For saving edgelets
    saveFrameNums = [10, 30];

    %Set up directory for stills, target contour, etc.
    [pathstr, name, ext] = fileparts(videoFile);
    savePath = ['./' name '/' int2str(numPoints) 'points'];
    if ~exist(savePath, 'dir')
        mkdir(savePath);
    end

    if strcmp(videoFile, '') %Make anonymous functions for frames/loop cond
        cam = webcam();
        keepGoing = @(frames) frames < 100;
        getFrame = @() snapshot(cam);
    else
        v = VideoReader(videoFile);
        keepGoing = @(frames) hasFrame(v);
        getFrame = @() readFrame(v);
    end

    % Create the video player object, purely for visual feedback.
    videoPlayer = vision.VideoPlayer();
    edgeletPlayer = vision.VideoPlayer();

    frameCount = 0;

    cropRect = [];

    edgelets = {};

    while keepGoing(frameCount)
        % Get the next frame.
        videoFrame = getFrame();

        % Find what to crop if we haven't yet.
        if length(cropRect) == 0
            [videoFrame, cropRect] = imcrop(videoFrame);
            close figure 1; %Close the imcrop figure, which stays around.
        else
            videoFrame = imcrop(videoFrame, cropRect);
        end

        videoFrameG = rgb2gray(videoFrame);
```

```matlab
        frameCount = frameCount + 1;

        [corners, voronoiLines, MNCs] = MakeMeshAndFindMNC(videoFrameG, true,
            numPoints);

        videoFrameT = insertMarker(videoFrame, corners, '+', ...
                'Color', 'red');
        videoFrameT = insertShape(videoFrameT, 'Line', voronoiLines, ...
            'Color', 'red');

        contourFrame = 255 * ones(size(videoFrameT), 'uint8');

        for i = 1:length(MNCs)
            mnc = MNCs{i};
            % draw the fine cluster contour connecting neighbors
            x=mnc(:,1); y=mnc(:,2);
            %arrange points clockwise to get the polygon
            xy = orderPoints(x, y);
            videoFrameT = insertShape(videoFrameT, 'Polygon', ...
                {xy}, 'Color', ...
                {'green'}, 'Opacity', 1); %mark contours

            contourFrame = insertShape(contourFrame, 'Polygon', ...
                {xy}, 'Color', ...
                {'green'}, 'Opacity', 1); %mark contours on their own

            if i == 1
                edgelets{frameCount} = mnc;
            end

            if i == 1 && any(frameCount == saveFrameNums)
                %Save pics, defined above in saveFrameNums
                edgeletPic = insertMarker(videoFrame, mnc, '+', ...
                'Color', 'red');
                edgeletPic = insertShape(edgeletPic, 'Polygon', ...
                    {xy}, 'Color', ...
                    {'green'}, 'Opacity', 1); %mark contours
                edgeletPic = insertShape(edgeletPic, 'FilledPolygon', ...
                    {xy}, 'Color', ...
                    {'green'}, 'Opacity', 0.2); %mark contours
                imwrite(edgeletPic, [savePath 'frame' ...
                    int2str(frameCount) '.png']);
            end
        end

        %Update video player
        pos = get(videoPlayer,'Position');
        pos(3) = size(videoFrame, 2) + 30;
        pos(4) = size(videoFrame, 1) + 30;
        set(videoPlayer,'Position',pos);
        step(videoPlayer, videoFrameT);

        %Cause contour player to 'stick' to the right of the main video
        pos(1) = pos(1) + size(videoFrame, 2) + 30;
        set(edgeletPlayer,'Position',pos);
        step(edgeletPlayer, contourFrame);
end

%We are done now, determine |e_i|
%Note: if code runs slowly, this can be optimized in a few ways
eSize = cellfun(@length,edgelets);
m = {};
for i = 1:length(eSize)
    m{i} = 1;
    %Count how many other edgelets have the same size
    for j = 1:length(eSize)
        if i ~= j && eSize(i) == eSize(j)
```

```matlab
                m{i} = m{i} + 1;
            end
        end
    end
    eProb = cellfun(@(e) 1/length(e), edgelets);

    %Histogram of m_i
    figure, histogram(cell2mat(m)), title('Histogram of m_i');
    xlabel('Frequency'), ylabel('Count at that Frequency');

    %Now, do a compass plot.
    %I have modified this to try and look decent.
    mags = unique(cell2mat(m));
    zs = mags .* exp(sqrt(-1) * (2 * pi * (1:length(mags)) / length(mags)));
    %The above evenly spaces out the magnitudes by making them
    %Complex numbers with a magnitude equal to their frequency values
    %And phases equally spaced
    figure, compass(zs), title('Compass Plot of m_i');

    %TODO: log polar plot

    figure, plot(1:length(eProb), eProb), title('Pr(e_i) vs. e_i');
    xlabel('e_i'), ylabel('Pr(e_i)');

    %3d countour plot
    tri = delaunay(1:length(eSize), cell2mat(m));
    figure, trisurf(tri,1:length(eSize), cell2mat(m), eProb), title('Pr(e_i) vs
        . e_i and m_i');
    xlabel('e_i'), ylabel('m_i'), zlabel('Pr(e_i)');
end

%Take points, order them to the right angle, return [x1,y1,x2,y2...]
function xy = orderPoints(x, y)
    cx = mean(x);
    cy = mean(y);
    a = atan2(y - cy, x - cx);
    [~, order] = sort(a);
    x = x(order);
    y = y(order);
    xy = [x';y'];
    xy = xy(:); %merge the two such that we get [x1,y1,x2..]
    if length(xy) < 6
        %our polygon is a line, dummy value it
        xy = [0 0 0 0 0 0];
    end
end

function [corners, voronoiLines, MNCs]=MakeMeshAndFindMNC(videoFrameG, useSURF,
    numPoints)
    if useSURF
        points = detectSURFFeatures(videoFrameG);
        [features, valid_points] = extractFeatures(videoFrameG, points);
        corners = valid_points.selectStrongest(numPoints).Location;
        corners = double(corners);
    else
        corners = corner(videoFrameG, numPoints);
    end

    voronoiLines = [];
    MNCs = {};

    if (length(corners) < 5)
        return;
    end

    [VX,VY] = voronoi(corners(:, 1), corners(:, 2));
```

```
% Creating matrix of line segments in the form
% [x_11 y_11 x_12 y_12 ... x_n1 y_n1 x_n2 y_n2]
A = [VX(1,:); VY(1,:); VX(2,:); VY(2,:)];
A(A>5000) = 5000; A(A<-5000) = -5000;
A = A';
voronoiLines = A;

%Now find maximal nucleus cluster
[V,C] = voronoin(corners,{'Qbb','Qz'}); %Options added to avoid co-sperical
    error, see matlab documentation
%Limit values, can't draw infinite things
V(V > 5000) = 5000;
V(V < -5000) = -5000;
numSides=cellfun(@length,C);
maxSides=max(numSides);
ind=find(numSides==maxSides);
N=size(corners,1);
for i=1:length(ind)
    xy=[];
    for j=1:N
        if(ind(i)~=j) %Find the corner points which have this edge
            s = size (intersect(C{ind(i)},C{j}));
            if(s(2)>1)%if neighbor voronoi region
                xy=[xy;corners(j,:)]; %keep the xy coords of adjacent
                    polygon
            end
        end
    end
    MNCs{i} = xy;
end
end
```

Listing A.34 Matlab code in `Problem734.m` to obtain edgelet measurements for each Voronoi tessellated video from.

A.8   Scripts from Chap. 8

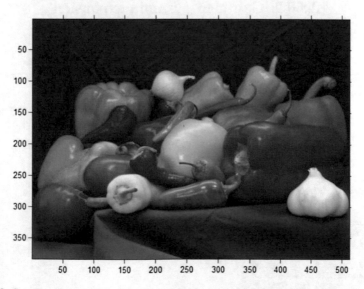

Fig. A.59 Sample colour used in Gaussian pyramid scheme in script A.35

A.8.1   *Gaussian Pyramid Scheme*

Remark A.36 **Pyramid scheme for Gaussian reduction and expansion of a colour.**

A sample colour image in Fig. A.59 is used in a Gaussian pyramid scheme in a sequence of image reductions in Fig. A.60.1 and in a sequence of image expansions in Fig. A.60.2. To experiment with Gaussian reduction and expansion of an image, try Matlab script A.35. ∎

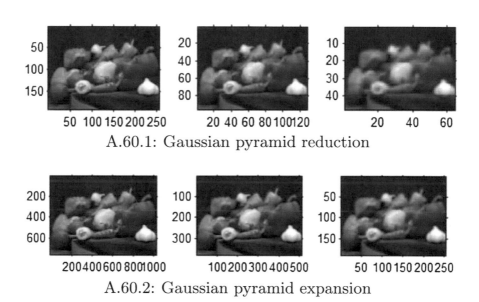

A.60.1: Gaussian pyramid reduction

A.60.2: Gaussian pyramid expansion

Fig. A.60 Sample Gaussian pyramid schemes

```
% pyramidScheme.m
% Gaussian pyramid reduction and expansion of an image
% cf. Section 8.4 on cropping & sparse representations
clear all, close all, clc

im0 = imread('flyoverTraffic.jpg');
% im0 = imread('peppers.png');
% im0 = imread('cameraman.tif');

% Crop (extract) an interestig subimage
im0 = imcrop(im0);
%%

im1 = impyramid(im0,'reduce');
im2 = impyramid(im1,'reduce');
im3 = impyramid(im2,'reduce');

im4 = impyramid(im0,'expand');
im5 = impyramid(im4,'expand');
im6 = impyramid(im5,'expand');
```

```
figure, imshow(im0);
figure,
subplot(1,3,1), imshow(im1);
subplot(1,3,2), imshow(im2);
subplot(1,3,3), imshow(im3);
figure,
subplot(1,3,1), imshow(im4);
subplot(1,3,2), imshow(im5);
subplot(1,3,3), imshow(im6);
```

Listing A.35 Matlab code in `pyramidScheme.m` to produce a Laplacian pyramid scheme as shown in two different ways in Fig. A.62.

A.8.2 *Wavelet Pyramid Scheme*

Fig. A.61 Sample sparse representation using a wavelet

Mathematica 4 Wavelet-based sparseRepresentation.nb pyramid scheme.

(*Sparse Representation Pyramid Scheme Using a Wavelet Transform)

dwd = WaveletBestBasis[DiscreteWaveletPacketTransform[,

Padding → "Extrapolated"]];

imgFunc[img_, {___, 1|2|3}]:=

Composition[Sharpen[#, 0.5]&, ImageAdjust[#, {0, 1}]&, ImageAdjust, Image

Apply[Abs, #1]&][

img]

imgFunc[img_, wind_]:=Composition[ImageAdjust, ImageApply[Abs, #1]&][img];

WaveletImagePlot[dwd, Automatic, imgFunc[#1, #2]&, BaseStyle → Red,

ImageSize → 800] ■

Remark A.37 **Digital image region centroids**.

A sample wavelet-based sparse representation pyramid scheme for a 2D image is shown in Fig. A.61 using Mathematica® script 4. For more about this, see Sect. 8.4. ■

A.62.1: pixel Strength   A.62.2: 21 Strength Radii

Fig. A.62 Sample pixel edge strengths represented by circle radii magnitudes

A.8.3 Pixel Edge Strength

Remark A.38 **Pixel strength**.

Let *Img* be a digital image. Recall that pixel edge strength $E(x, y)$ for a pixel $Img(x, y)$ at location (x, y) is defined by

$$E(x, y) = \sqrt{\left(\frac{\partial Img(x, y)}{\partial x}\right)^2 + \left(\frac{\partial Img(x, y)}{\partial y}\right)^2} \text{ (Pixel edge strength)}$$
$$= \sqrt{G_x(x, y)^2 + G_y(x, y)^2}.$$

The edge strength of the red ● hat pixel in Fig. A.62.1 is represented by the length of the radius of the circle centered on the hat pixel.

A global view of multiple pixel edge strengths is shown in Fig. A.62.2. To experiment with finding the edge strengths of pixels, try Matlab script A.36. ■

```
% pixel edge strength detection
% N.B.: each pixel found is a keypoint
clc; clear all, close all;
% g = imread('cameraman.tif');
% I = g;
g = imread('fisherman.jpg');
I = rgb2gray(g); % necessary step for colour images
points = detectSURFFeatures(I);
% acquire edge pixel strengths
[features,keyPts] = extractFeatures(I,points);
% record number of keypoints found
keyPointsFound = keyPts
% select number pixel edge strengths to display on original image
figure,
imshow(g); hold on;
plot(keyPts.selectStrongest(13),'showOrientation',true),
axis on, grid on;
figure,
imshow(g); hold on;
plot(keyPts.selectStrongest(89),'showOrientation',true),
axis on, grid on;
```

Listing A.36 Matlab code in `pixelEdgeStrength.m` to produce Fig. A.62.

A.8.4 Plotting Arctan Values

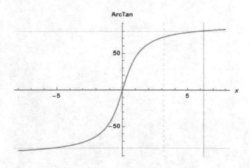

Fig. A.63 Plotting arctan values between 5 and −5

Remark A.39 A sample plot of arctan values is shown in Fig. A.63. Try doing the same things using Matlab®.   ∎

Mathematica 5 Plotting arctan values.

(compute arc tangents *)*

ArcTan[−1]

N[*ArcTan*[−1]/*Degree*]

N[*ArcTan*[+50]/*Degree*]

N[*ArcTan*[−1.5]/*Degree*]

Plot[*ArcTan*[*x*], {*x*, −8, 8}]

Plot[*N*[*ArcTan*[*x*]/*Degree*], {*x*, −8, 8}, *AxesLabel* → {*x*, *ArcTan*},

LabelStyle → *Directive*[*Blue*, *Bold*],

Fig. A.64 Representation of fingerprint pixel intensities

GridLines → {{{*Pi, Dashed*}, {*2Pi, Thick*}},
{{−80, *Orange*}, −.5, .5, {80, *Orange*}}}]

A.8.5 Pixel Geometry: Gradient Orientation and Gradient Magnitude

Remark A.40 Each pixel intensity in Fig. A.64 is a representation of the HSB (Hue Saturation Brightness) colour channel values that correspond to the pixel (gradient orientation (angle), gradient magnitude in the x-direction, gradient magnitude in the x-direction) in a fingerprint. Try Mathematica script 6 to see how the pixel gradient orientations of the pixels varying with each fingerprint. The HSV (Hue Saturation Value) colour space in Matlab is equivalent to the HSB colour space in Mathematica. ■

Mathematica 6 RGB and LAB Views of Fingerprint Gradients.
(* Visualize edge pixel gradient orientations *)

$i =$  ;

orientation = *GradientOrientationFilter*[*i*, 1]*//ImageAdjust*;
magnitude = *GradientFilter*[*i*, 1]*//ImageAdjust*;
ColorCombine[{*orientation, magnitude, magnitude*}, *"HSB"*]

Remark A.41 In Fig. A.65.1, each pixel intensity is a representation of the RGB colour channel values that correspond to the pixel (gradient orientation (angle), gradient magnitude in the x-direction, gradient magnitude in the x-direction) in a fingerprint. In Fig. A.65.2, each pixel intensity is a representation of the LAB colour channel values that correspond to the pixel (gradient orientation (angle), gradient magnitude in the x-direction, gradient magnitude in the x-direction) in a fingerprint. Try Mathematica script 6 for different images to see how the pixel gradients vary with each image. ■

Mathematica 7 Fingerprint Gradients in RGB and LAB.
(* Visualize pixel gradients *)

$i =$;

orientation = *GradientOrientationFilter*[*i*, 1]*//ImageAdjust*;

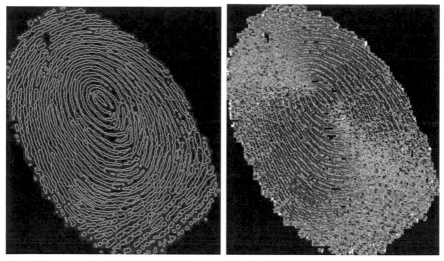

A.65.1: RGB gradients   A.65.2: LAB gradients

Fig. A.65 Colorized image gradients

magnitude = GradientFilter[i, 1]//ImageAdjust;

ColorCombine[{orientation, magnitude, magnitude}, "RGB"]

ColorCombine[{orientation, magnitude, magnitude}, "LAB"]

Remark A.42 In Fig. A.66, a LAB colour space view of an image is given. To experiment with other LAB coloured images, try Matlab script A.37. ■

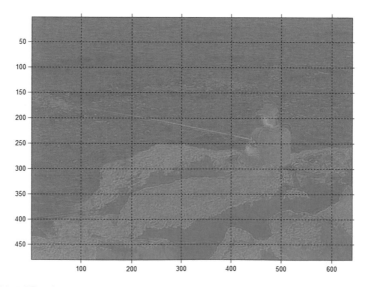

Fig. A.66 LAB colour space view of an image

```
% LAB colour space: Colourize image regions.
% script: LABexperiment.m
clear all; close all; clc; % housekeeping
%%
img = imread('fisherman.jpg');

labTransformation = makecform('srgb2lab');
lab = applycform(img,labTransformation);
figure, imshow(lab), axis, grid on;
```

Listing A.37 Matlab source LABexperiment.m.

A.8.6   Difference-of-Gaussians Image

A.67.1: Fisherman

A.67.2: DoG Image

Fig. A.67 Colour image and DoG image

Remark A.43 The image in Fig. A.67.2 results from a difference of Gaussians convolved with the original image in Fig. A.67.1. Let $Img(x, y)$ be an intensity image let $G(x, y, \sigma)$ be a variable scale Gaussian defined by

$$G(x, y, \sigma) = \frac{1}{2\pi\sigma^2}e^{-\frac{x^2+y^2}{2\sigma^2}}.$$

Let k be a scaling factor and let $*$ be a convolution operation. From D.G. Lowe [116], we obtain a difference-of-Gaussians image (denoted by $D(x, y, \sigma)$ defined by

$$D(x, y, \sigma) = G(x, y, k\sigma) * Img(x, y) - G(x, y, \sigma) * Img(x, y)$$

Then use $D(x, y, \sigma)$ to identify potential interest points that are invariant to scale and orientation. ∎

```
% DoG: Difference of Gaussians.
% script: dogImg.m
clear all; close all; clc; % housekeeping
%%
k = 1.5; % vs. 1.1, 1.5, 33.5
sigma1 =   5.55; % vs. 0.30, 0.98, 0.99, 5.55
sigma2 = sigma1*k;
hsize = [8,8];
% g=imread('cameraman.tif');
g=rgb2gray(imread('fisherman.jpg'));
%%
gauss1 = imgaussfilt(g,sigma1);
gauss2 = imgaussfilt(g,sigma2);
%
dogFilterImage = gauss2 - gauss1; % difference of Gaussians
imshow(dogFilterImage, []), axis, grid on;
% title('DOG Image', 'FontSize', fontSize);
```

Listing A.38 Matlab source `GaussianImageNew.m`.

A.8.7 Image Keypoints and Voronoï Mesh

A.68.1: 21 keypoints   A.68.2: 89 keypoints

Fig. A.68 Image keypoints and Voronoï mesh

Remark A.44 Two views of image geometry are shown in Fig. A.68.

View.1 **Keypoint Gradient Orientation View**. Each of the 21 keypoints in Fig. A.68.1 is the center of a circle. The radius of each circle equals the edge strength of a keypoint.

Example A.45 **Fisherman's hat keypoint**.

: For example, there is a keypoint located on the near side of the fisherman's hat in Fig. A.68.1. The angle of the line segment ——— from the center of the hat keypoint circle to the circumference is identified with the gradient orientation of the keypoint. ∎

Each of the red edges ——— in Fig. A.68.1 is a side of polygon in the Voronoï tessellation derived from one of the keypoints.

View.2 **Display of 89 Keypoints in a Voronoï Tessellation of an Image**. There are 89 keypoints represented by ○s in Fig. A.68.2. In this case, the keypoints are shown without the surrounding circles shown in Fig. A.68.1. Notice that the many of the keypoints are tightly grouped around the fisherman. This suggests a basis for object recognition, which we exploit in Chap. 8. ∎

```
% method: find strongest keypoints in an image and
% use keypoints as generators of Voronoi regions
% script: keypointsExpt5gradients.m
clear all; close all; clc; % housekeeping
% g=imread('peppers.png');
g=imread('fisherman.jpg');
img = g; % save copy of colour image to make overlay possible
g = rgb2gray(g);
% g = double(rgb2gray(g)); % cponvert to greyscale image
pts = detectSURFFeatures(g);
[features,keyPts] = extractFeatures(g,pts);
figure,imshow(img), axis on, hold on;
plot(keyPts.selectStrongest(21),'showOrientation',true);

%% part 1 - voronoi mesh is superimposed on the image using SURF keypoints
%plot voronoi mesh on the image
% XYLoc=keyPts.Location; %for all keypoints - uncomment this and comment lines
      14 and 15 below
strongKey=keyPts.selectStrongest(21);%use only for a selected number of
      strongest keypoints
XYLoc=strongKey.Location;
X=double(XYLoc(:,1));
Y=double(XYLoc(:,2));
voronoi(X,Y,'-r');

%% part 2 - display the keypoints without the surrounding circles and without a
      mesh
%get XY coordinates of key points
% XYLoc=keyPts.Location; %for all keypoints - uncomment this and comment lines
      23 and 24 below
```

```
strongKey=keyPts.selectStrongest(89);%use only for a selected number of
       strongest keypoints
XYLoc=strongKey.Location;
X=double(XYLoc(:,1));
Y=double(XYLoc(:,2));
figure,imshow(img), axis on, hold on;
plot(X,Y,'ro');
% plot(X,Y,'g*');
% voronoi(X,Y,'-r');
%% part 3 overlay mesh on points
voronoi(X,Y,'-g');
%% part 5 keypoints on image
figure,imshow(img), axis on, hold on;
plot(X,Y,'ro');
```

Listing A.39 Matlab script in `keypointsExpt5gradients.m` to display corners on a digital image. Source: **keypointsExpt5gradients.m**

Appendix B
Glossary

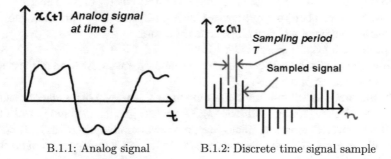

B.1.1: Analog signal   B.1.2: Discrete time signal sample

Fig. B.1 Analog signal and its sampled digital version

B.1   A

A

A/D: Analog-to-digital conversion accomplished by taking samples of an analog signal at appropriate intervals. The A/D process is known as sampling. See Fig. B.1 for an example.

B.2   B

B

bdy A: Set of boundary points of the set A. See, also, **Open set, Closed set**.

© Springer International Publishing AG 2017
J.F. Peters, *Foundations of Computer Vision*, Intelligent Systems
Reference Library 124, DOI 10.1007/978-3-319-52483-2

Bit depth: Bit depth quantifies how many unique colors are available in an image's color palette in terms of the number of 0's and 1's, or "bits," which are used to specify each color.

Example B.1 **Bit Depth**. Digital camera colour image usually has a bit depth of 24 bits with 8-bits per colour.[1]   ■

Blob: A **blob** (binary large object) is a set of path-connected pixels in a binary image. For the details, see Sect. 7.1.3. See, also, **Connected, Path-connected**.

Boundary region of a set: The **boundary region of a set** is the set of points in a boundary region of a set A (denoted by re$A$) that includes the boundary set bdy$A$, i.e., bdy$A \subset$ re$A$. See **Boundary set, Open set, Closed set, Interior of a set**.

Example B.2 **Sample boundary regions of sets**.

Earth atmosphere   Region of space above the earth surface.

Raspberry pi board   The boundary region of a Raspberry pi is its board: https://www.raspberrypi.org/blog/raspberry-pi-3-on-sale/ For example, the boundary region of a binary clock is its enclosure: http://frederickvandenbosch.be/?p=1999

Window glass exterior   Region space surrounding a window pane of glass is the boundary region of the glass.

Electromagnetic spectrum outside visible light   That part of the electromagnetic spectrum below 400 nm (ultraviolet, X-ray and gamma radiation) and above 700 nm (infrared, terahertz, microwave, radio wave radiation) of known photons. Recall that a photon is an elementary particle, which is the quantum of all forms of electromagnetic radiation.

Region outside a closed half space   Set of points on edge of a planar set and constitutes the boundary of interior of the set. See **Half space, polytope**.

2D Image exterior region   Region of the plane outside any 2D image. A Rosenfeld 8-neighbourhood plus the pixels along its borders.

Exterior region of a 2D Image Rosenfeld neighbourhood   All 2D image pixels outside an open Rosenfeld 8-neighbourhood.   ■

Boundary set: The *boundary set* of a nonempty set A (denoted by bdy$A$) is the set of points along the border of and adjacent to a nonempty set. Then bdy$A$ is the set of those points nearest A and not in A. Put another way, let A be any set in the Euclidean plane X. A point $p \in X$ is a boundary point of A, provided the neighbourhood of p(denoted by $N(p)$) intersects both A and the set of all points in X not in A [100, Sect. 1.2, p. 5]. Geometrically, p is on the edge between A and the complement of A in the plane. See **Boundary region, Neighbourhood of a point, Hole**.

Example B.3 **Sample boundary sets**.

---

[1]http://www.cambridgeincolour.com/tutorials/bit-depth.htm.

orange pulp exterior   The skin of an orange is the boundary of the orange interior (the orange pulp).

egg exterior   Egg shell that is the boundary of an egg yoke.

window frame   Window frame surrounding a pane of glass is the boundary of the glass.

empty box   Empty box bounded on each side with a wall. Example: shoe box with no shoes or any else in it.

Plane subset boundary   Set of points on edge of a planar set and constitutes the boundary of interior of the set.

subimage   Any Subimage that includes its boundary pixels. A Rosenfeld 8-neighbourhood plus the pixels along its borders.   ∎

Brightness: Brightness is a relative expression of the intensity of the energy output of a visible light source [67, Sect. 1.4]. It is expressed as a total energy value or as the amplitude at the wavelength of visible light where the intensity is greatest. In the HSV colour space in Matlab®, Value (Lightness in HSL colour space or Intensity in the HSI colour space) is the same as brightness in the HSB colour space in Mathematica®. Brightness in the HSB (or Value in the HSV) colour model is an attribute of a visual sensation in which a visible area appears to emit more or less light [68]. See, also, **Hue, HSV**.

B.3   C

$\mathbb{C}$ Symbol for set of complex numbers. See, also, **Complex numbers, Riemann space**.

Candela: A **Candela** is the SI (Standard International) base unit of luminous intensity, i.e., luminous power per unit solid angle emitted by a point light source in a particular direction. The contribution of each wavelength is weighted by the standard luminosity function [209], defined by

$$E_v = \text{illuminance measurement.}$$
$$r_1 = \text{radius of the limiting aperture.}$$
$$r_2 = \text{radius of the light source.}$$
$$d = \text{physical distance between the light source and aperture.}$$
$$D = \text{distance between an aperture plane and a photometer:}$$
$$= r_1^2 + r_2^2 + d.$$
$$A = \text{area of source aperture.}$$
$$L_v(E_v, D, A) = \frac{E_v D^2}{A} \text{ (Luminosity function).}$$

Example B.4 **Sample luminous intensity distribution for an incandescent light bulb**.

For a sample luminous intensity distribution for an incadescent light bulb, see:
http://www.pozeen.com/support/lighting_basics.html#.WDWH6bIrJXU

The luminosity of each light beam is measured in candela. The luminous intensity distribution provides a fingerprint for each light source.   ■

Centroid: Let X be a set of random events, $x \in X$, $P(x) =$ the probability that event x occurs and let $D(X)$ be the cumulative distribution function (CDF) that the describes the probability that X takes on a value less than or equal to x, em i.e.,

$$D(X) = P(X \leq x) = \sum_{X \leq x} P(x), \text{ and}$$

$$P(x) = D'(x) \text{ (Derivative of } D(X)).$$

Let $\rho : X \to (R)$ be a probability density function on X, $x \in X$. A **centroid** is a center of mass s^* of a region V. It corresponds to a measure of central location for a region, defined by

$$s^* = \frac{\int_V x\rho(x)dx}{\int_V \rho(x)dx}.$$

For more details about region centroids, see Q. Du, V. Faber and M. Gunzburger [38, p. 638]. For the discrete form of centroid, see Sect. 6.4.

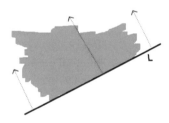

Fig. B.2 Sample closed half space $= L \cup \{$all points above $L\}$

Closed halfspace: Given a line L in an n-dimensional space, a **closed half space** is a closed set of points that includes those points on the line L as well as all points above L. Such a half space is closed because it includes the its boundary points on the line L. This halfspace is also called an **upper half space**. We obtain a **lower half space**, provide the half space includes all points on the line L as well as all points below L. See, also, **Boundary region, Boundary set, Closed set, Polytope, Closed Lower Half Space, Closed Upper Half Space**.

Example B.5 **2D Closed Half Space**.
A sample 2D closed half space is shown in Fig. B.2. This form of a half space lives in the Euclidean plane represented by all vectors in $\mathbb{R}^2$. Try drawing the closed half plane that includes all points $-\infty < x < +\infty$ and $5 \le y \le +\infty$. ∎

Closed lower halfspace: A **closed lower half space** is the set of all points below as well as on a boundary line.

Closed upper halfspace: A **closed upper half space** is the set of all points above as well as on a boundary line.

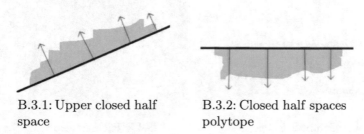

B.3.1: Upper closed half space

B.3.2: Closed half spaces polytope

Fig. B.3 Sample closed *upper* and *lower half spaces*

Example B.6 **Sample 2D Closed Upper and Lower Halfspaces**.
The two types of 2D closed half spaces are shown in Fig. B.3. The half space in Fig. B.3.1 is an example of closed upper half space. This half space is closed, since it includes all planar points above as well as on the line that is the boundary of the half space. The half space in Fig. B.3.2 is an instance of a closed lower half space. In this case, the half space consists of all planar points on and below the indicated line. The line in closed lower half space forms the boundary of the half space. In both cases, each half space is unbounded one side of a line. ∎

Closed set: A set of points that includes its boundary and interior points. Let A be a nonempty set, int$A$ the interior of A, bdy$A$ the boundary of A. Then A is a closed set, provided $A = \text{int}A \cup \text{bdy}A$, i.e., A is closed, provided A is the union of the set of points in its interior and in its boundary.

Example B.7 **Sample closed sets**.

whole orange The skin of an orange plus its pulp.
egg exterior Egg shell plus the egg yoke.
window frame Window frame holding a pane of glass.
Planar subset Set of points in the interior and boundary of a planar set.
subimage The interior of any 2D or 3D subimage plus its boundary pixels. A Rosenfeld 8-neighbourhood plus the pixels along its borders. ∎

C   Complex numbers. $\mathbb{C}$ is the set of complex numbers. Let $a, b \in \mathbb{R}$ be real numbers, $i = \sqrt{-1}$. A complex number $z \in \mathbb{C}$ has the form

$$z = a + bi \ \text{(Complex number)}.$$

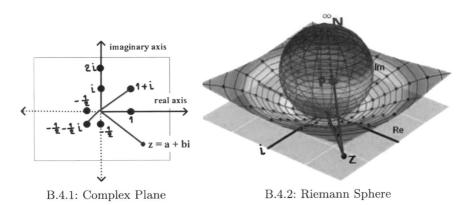

B.4.1: Complex Plane   B.4.2: Riemann Sphere

Fig. B.4 Complex plane with projection to a Riemann sphere

Complex Plane $\mathbb{C}$:   The **complex plane** is the plane of complex numbers (also called the (see z-plane). See, for example, Fig. B.4.1) in Fig. B.4. Points z in the complex plane are of the form $z = a + bi, a, b \in \mathbb{R}, i = \sqrt{-1}$. The origin of the complex plane is at $z = 0 = 0 + 0i$. A **Riemann surface** is the union of the complex planes with infinity (denoted by $\mathbb{C} \cup \{\infty\}$). A projection from the complex plane to a Riemann sphere, which touches the complex z-plane at a point S at $z = 0 = 0 + 0i$ and whose central axis is a line joining the top of the sphere at N diametrically opposite S. A line joining N (at infinity ∞) to a point z pierces the surface of the sphere at the point P (see Fig. B.4.2). For a visual perspective on the complex plane and Riemann sphere, providing insights important for computer vision, see E. Wegert [206]. See, also, **Riemann surface,** $\mathbb{C}, z$

Colour pixel value: Pixel colour intensity or brightness. See **Gamma Correction, RGB, HSV.**

Compact set: Let X be a topological space and let $A \subseteq X$. Briefly, a **cover** of a nonempty set A in a space X is a collection of nonempty subsets in X whose union is A. Put another way, let $\mathscr{C}(A)$ be a cover of A. Then

$$\mathscr{C}(A) = \bigcup_{B \subset X} B \ \text{(Cover of the nonempty set } A\text{)}.$$

The set A is a **compact set**, provided every open covering of A has a finite subcovering [100, S 1.5, p. 17]. See **Cover, Topological space.**

Example B.8 **Compact Picture Sets**.
The interior of every nonempty set of picture points A in the Euclidean plane is compact, since int$A$ has a covering $\mathscr{C}(\mathrm{int}A) = \bigcup_{X \in 2^{\mathrm{int}A}} X$. And $\mathscr{C}(\mathrm{int}A)$ always contains a finite collection that also covers int$A$, *i.e.*, a finite subcover that is an open cover of int$A$. ∎

Complement of a set: Let A be a subset of X. The complement of A (denoted by A^c) is the set of all points in X that are not in A.

Example B.9 Let A be a 2D digital image. Then A^c is the set of all points in plane not in A. ∎

Computational photography (CPh): **CPh** is the use of cameras to capture, record and analyze natural scenes for possible followup actions such as collision-avoidance. Cph records a scene with multiple images with epsilon variation, either by recording a changing scene in a video with some form of video camera such as the web cam. See, for example:
https://www.microsoft.com/accessories/en-au/products/webcams/
Cph also records a scene by varying digital camera parameters (e.g., zoom, focus, landscape, portrait in Nikon® Coolpix AW300 camera) in recording a scene or by recording a changing scene with varying camera parameters such as field of view, depth of field and illumination (see, e.g., [46, Sect. 2.1, p. 245ff]).

Computer Vision: *Computer vision* is the study of the automation of tasks that are performed by the human eye as well as the acquiring, processing, analyzing and understanding of digital images and videos. For downloadable articles in Computer Vision, see the Cornell University e-print service at https://arxiv.org/list/cs.CV/recent.
See, for example, H. Rhodin, C. Richart, D. Casas, E. Insafutdinov, M. Shafiei, H.-P. Seidel, B. Schiele and C. Theobalt on motion capture with two fisheye cameras [165]. See, for example:
http://camerapedia.wikia.com/wiki/Lomographic_Fisheye_Camera
and
https://en.wikipedia.org/wiki/Fisheye_lens
Motion capture is also possible using the **rolling shutter approach** in which a scene is recorded with either a still camera such as a cell phone camera or video camera by scanning across a scene rapidly so that not all parts of a scene are recorded at the same time. See, for example:
http://wikivisually.com/wiki/Rolling_shutter/wiki_ph_id_7. During playback, an entire scene is displayed as a single image, giving the impression that all aspects of a scene have been captured in a single instant. This approach to scene capture contrasts with the common **global shutter** approach in which an entire image frame is captured at the same time.

B.5.1: Subaru EyeSight®️ cameras

B.5.2: EyeSight®️ moving object
tracking

Fig. B.5 Subaru EyeSight® images, courtesy of Winnipeg Subaru

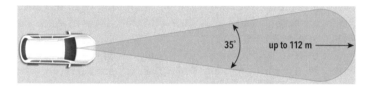

Fig. B.6 Subaru EyeSight® field-of-view, courtesy of Winnipeg Subaru

Example B.10 **Collision Avoidance Vision Systems**.
Vision systems are designed to have a particular field of view. The **field of view** of a
vision system is the extent that an natural scene is observable at any given instant. Both
angle of view and natural scene depth are important field of view measurements. For
example,[2] the dual cameras in the Subaru EyeSight® vision system (see Fig. B.5.1)
is designed for collision avoidance in situations like the one in Fig. B.5.2. The field
of view for the EyeSight vision system is 30° and the depth of its field of view is
112 m (see Fig. B.6). ■

Connectedness A property of sets that share points. A set X is disconnected,
provided that there are disjoint open sets A and B such that $X = A \cup B$, i.e., X
is the union of the disjoint sets A and B. A set X is **connected**, provided X does
not contain disjoint sets whose union is X.

Example B.11 **Disconnected Sets**.
Let a set of line segments be represented by Fig. B.7. Consider the set of points in
line $\overline{pr}$, which is equal the union of the points in the line segments $\overline{pq}$ and $\overline{qr}$ (this
is the set Y in Fig. B.7). And let B be the set of points in the line segment $\overline{s,t}$. From
Fig. B.7, $X = Y \cup B$. Hence, X is disconnected. ■

---

[2]Many thanks to Kyle Fedoruk for supplying the Subaru EyeSight® vision system images.

Fig. B.7 Disconnected set

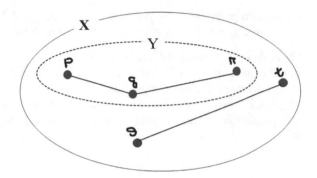

Example B.12 **Disconnected Set of Voronoï Regions**.
Let Voronoï regions A and C be represented by the polygons in Fig. B.8. Notice that
polygons A and B have edges in common. In effect, A and C are not disjoint. Let
$X = A \cup C$. The set X is disconnected, since A and C are disjoint.   ∎

Example B.13 **Connected Sets**.
Let a set Y containing a pair of line segments be represented by Fig. B.7. Let the set
C be the points in line $\overline{p, q}$ and let D be the points in line $\overline{q, r}$. Since $Y = C \cup D$,
the set Y is connected.   ∎

Example B.14 **Connected Set of Voronoï Regions**.
Let Voronoï regions be represented by the polygons in Fig. B.8. Notice that polygons
A and B have a common edge. In effect, A and B are not disjoint. Let $Y = A \cup B$.
The set Y is connected, since A and B are not disjoint.   ∎

Connected line segments   Line segments are connected, provided the line segments have a common endpoint. For example, in Fig. B.7, line segments $\overline{p, q}$ and
$\overline{q, r}$ are connected. See **Disconnected, Path-connected**.

Fig. B.8 Connected Voronoï
regions

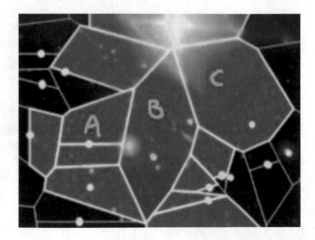

Connected polygons   Polygons are connected, provided the polygons share one or more points. See **Disconnected, Path-connected**.

Example B.15 **Connected Voronoï Polygons.**
For example, the pair of Voronoï regions A and B in Fig. B.8 are connected, since A and B have a common edge. Similarly, the pair of Voronoï regions B and C in Fig. B.8. However, the pair of Voronoï regions A and C in Fig. B.8 are not connected polygons, since A and C have no points in common, i.e., A and C are disjoint sets. ∎

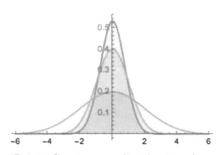
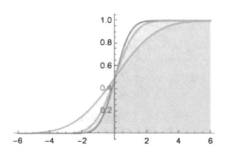

B.9.1: Continuous distribution for the Gaussian kernel $f(x) = \frac{e^{-\frac{x}{2\sigma^2}}}{\sigma\sqrt{2\pi}}$

B.9.2: Continuous Cumulative Density Function (CDF) for the Gaussian kernel

Fig. B.9 Sample continuous distributions

Continuous   A mathematical object X is **continuous**, provided all elements of X are within a neighbourhood of nearby points [173]. Let X be a nonempty set, $x, y \in X$. For example, a function $f : X \longrightarrow \mathbb{R}$ is continuous, provided, whenever x is near y, $f(x)$ is near $f(y)$. Again, for example, let

$$f(x) = \frac{e^{-\frac{x}{2\sigma^2}}}{\sigma\sqrt{2\pi}}, \text{ with standard deviation } \sigma.$$

In the interval $-6 \leq x \leq x$, the distribution of values for $f(x)$ is continuous. This distribution is shown in Fig. B.9.1 and the continuous cumulative density distribution of values for $f(x)$ is shown in Fig. B.9.2. See, also, **Discrete**.

Contour   A **contour** is a collection of connected straight edges surrounding the nucleus of a MNC. The endpoints of contour straight edges are the generating points used to construct the polygons surrounding the MNC nucleus. See **edgelet, contour edgelet, MNC**.

Contour edgelet   A **contour edgelet** is a set of points in surrounding a MNC nucleus. Initially, an edgelet will contain only generating points that are the endpoints of the edges in either a coarse (outer) or fine (inner) contour of a MNC. By joining each pair of generating points nearest other in MNC contour polygons with a line segment, an edgelet gains the points (pixels) on each contour line segment. In tessellated video frames, an edgelet is a set of points in a frame MNC contour.

Cover of a set   A **cover** of a nonempty set A in a space X is a collection of nonempty subsets in X whose union is A. Put another way, let $\mathscr{C}(A)$ be a cover of A. Then

$$\mathscr{C}(A) = \bigcup_{B \subset X} B \text{ (Cover of the nonempty set } A).$$

Example B.16 **Finite cover of a 2D digital image**.
Let img be a 2D digital image, $S \subset Img$ a set of mesh generating points, $V(S)$ a Voronoï mesh on Img. Notice that each point in Img belongs to a Voronoï region $V(s), s \in S$. This means that $Img \subseteq \bigcup_{s \in S} V(s)$. For reason, $V(S)$ covers Img. We can always find a subset $S' \subseteq S$ so that $V(S') \subseteq V(S)$. That is, every Voronoï mesh $V(S')$ is a subcover of $V(S)$. Let s' be a generating point in S'. Then

$$Img \subseteq \bigcup_{s' \in S} V(s') \subseteq \bigcup_{s \in S} V(s)$$

In other words, every mesh cover of Img have a finite sub-mesh that is a cover of Img. ∎

Convex body   A **convex body** is a compact convex set [61, Sect. 3.1, p. 41]. A **proper convex body** has a nonempty interior. Otherwise, a convex body is *improper*. See **Convex set, Compact**.

Convex combination   A unique straight line segment $\overline{p_0 p_1}$, $p_0 \neq p_1$ is defined by two points $p_0 \neq p_1$ so that the line passes through both points [42, Sect. I.4, p. 20]. Each point on $x \in \overline{p_0 p_1}$ can be written as $x = (1-t)p_0 + tp_1$ for some $t \in R$. Notice that $\overline{p_0 p_1}$ is a convex set containing all points on the line (see Fig. B.10.1). **Edelsbrunner–Harer Convex Combination Method**: Then we can construct a convex set containing all points on the triangular face formed by three points after we a third point a_2 to the original set $\{p_0, p_1\}$. That is, we construct a line segment convex set so that is a triangle with a filled triangle face.

> for $t = 0$, we get $x = p_0$,
>
> for $t = 1$, we get $x = p_1$,
>
> for $0 < t < 1$, we get a point in between p_0 and p_1.

A line segment convex set is also a convex hull of two point, since it is the smallest convex set containing the two points. In the case where have more than two points, the above construction is repeated for $\{p_0, p_1, p_2\}$ by adding all points $y = (1-t)p_0 + tp_1$ for some $0 \leq t \leq 1$. The result is a triangle-shaped convex hull with a filled in triangle face (see, e.g., Fig. B.10.2). Using the convex combination method on a set of 4 points, we obtain the convex hull shown in Fig. B.10.3.

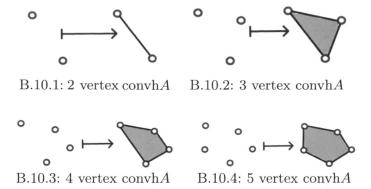

B.10.1: 2 vertex convh$A$ B.10.2: 3 vertex convh$A$

B.10.3: 4 vertex convh$A$ B.10.4: 5 vertex convh$A$

Fig. B.10 Construction of a convex hull of 5 points

Repeating this construction for a set of 5 points, we obtain the convex hull shown in Fig. B.10.4.

In general, starting with a set $\{p_0, p_1, p_2, \ldots, p_k\}$ containing $k + 1$ points, the convex combination construction method can be done in one step, calling $x = \sum_{i=0}^{l} t_i p_i$ a convex combination of the points p_i, provided $\sum_{i=0}^{l} t_i = 1$ and $t_i \geq 0$ for all $0 \leq i \leq k$. In that case, the set of convex combinations is the convex hull of the points p_i. ■

Convex hull   Let A be a nonempty set. The smallest convex set containing the set of points in A is the **convex hull** of A (denoted by convh$A$). G.M. Ziegler [220, p. 3] a method of constructing a convex hull of a set of points K (convh$K$), defined by the intersection of all convex sets that contain K. For practical purposes, let convh$K$ be a 2D convex hull with $K \subset R^2$. Then a 2D convex hull convh$K$ is defined by

$$\text{convh}K = \bigcap \left\{ K' \subseteq \mathbb{R}^2 : K \subseteq K' \text{ with conv}K' \right\} \ (\textbf{Ziegler Method}).$$

A convex hull of a finite nonempty set A is called a **polytope** [11]. An important application of convex hulls is the determination of the shape of a set of points. R.A. Jarvis introduced shape convex hulls [88]. For more about this, see H. Edelsbrunner, D.G. Kirkpatrick and R. Seidel [43]. See, also, **Convex set, Convex combination, Polytope, Shape**.

Example B.17 **Sample 2D and 3D Convex Hull Shapes**.

Sample 2D and 3D convex hulls of points are shown in Fig. B.11. A 7-sided convex hull of 55 randomly selected points in the image plane is shown in Fig. B.11.1. Notice that this convex hull exhibits an important property of all convex hulls, namely, a convex hull is a closed set, e.g., a 2d convex hull is the union of all points in its

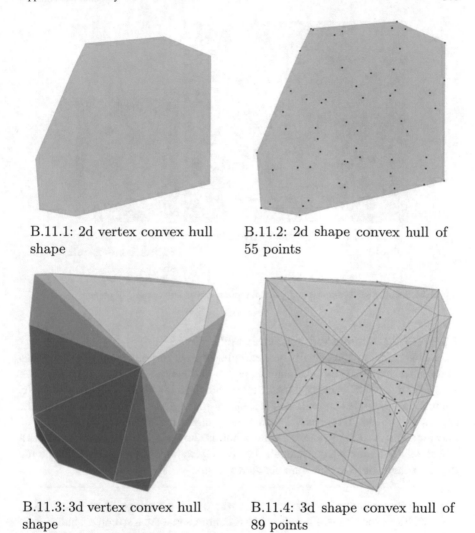

B.11.1: 2d vertex convex hull shape

B.11.2: 2d shape convex hull of 55 points

B.11.3: 3d vertex convex hull shape

B.11.4: 3d shape convex hull of 89 points

Fig. B.11 Sample 2D and 3D convex hulls of sets of points

boundary set and interior set. In the plane, every convex set contains an infinite number of points. The 55 points on the vertices, on the edges and in the interior of the 7-gon shaped convex hull are shown in Fig. B.11.2. A 3D convex hull of 89 points is shown in Fig. B.11.3. The 89 points on the surfaces and in the interior of the sample 3D convex hull are shown in Fig. B.11.4. ∎

B.12.1: MNC Contour Convex hull   B.12.2: Voronoï
shape   region convex hull
                                         shape

Fig. B.12 Sample MNC contour and Voronoï region convex hulls of sets of points

Example B.18 **Sample Voronoï convex hulls**.

Fine MNC contours are usually convex hulls of a set of points, namely, the points in the MNC nucleus plus the remaining points in the interior of the fine contour set. A sample fine MNC contour convex hull is shown in Fig. B.12.1. Notice that MNC contour convex hulls approximate the shape of the region of a digital image that the convex hull covers. Every MNC nucleus is a convex hull of the its vertices plus its interior points. A sample nucleus convex hull is shown in Fig. B.12.2. Sometimes a coarse MNC contour is a convex hull. Try finding examples of coarse MNC contours that are convex hulls and that are not convex hulls. ■

> In a computational geometry setting where a digital image is tessellated with a Voronoï diagram, coarse MNC contours that are also convex hulls provide signatures of the objects covered by the coarse contour. ■

Convex set A nonempty set A in an n-dimensional Euclidean space is a **convex set** (denoted by conv A), provided every straight line segment between any two points in the set is also contained in the set. For example, let A be a nonempty set of points in the Euclidean plane. The set A is a *convex set*, provided

$$(1 - \lambda)x + \lambda y \in A, \text{ for all } x, y \in A, 0 \leq \lambda \leq 1 \text{ (\textbf{Convexity property})}.$$

A nonempty set A is **strictly convex**, provided A is closed and

$$(1 - \lambda)x + \lambda y \in A, \text{ for all } x, y \in A, x \neq y, 0 < \lambda < 1 \text{ (\textbf{Strict Convexity property})}.$$

The earliest appearance of convexity appears in definitions of a convex surface in about 250 B.C. by Archimedes in his Sphere and Cylinder. On Paraboloids, Hyperboloids and Ellipsoids [7]. Archimedes' definitions of a convex surface are given by P.M. Gruber [61, Sect. 3.1]. A good introduction to Archimedes in given by T.L. Heath in [7]. For a complete introduction to convex sets, see P. Mani-Levitska [119]. And for an introduction to convex geometry, see P. M. Gruber and J. M. Wills [62]. For an introduction to convexity in the context of digital images, see J.F. Peters [144]. ∎

Theorem B.19 *A Voronoï region in the Euclidean plane is a proper convex body.*

Proof See J.F. Peters [144, Sect. 11, p. 307].

Example B.20 **Strictly Convex set.**
Let p be a mesh generating point and let V_p (also written $V(p)$ be a Voronoï region. In Fig. B.13, the Voronoï region $A = V_p$ is a strictly convex set, since A is closed and it has the Strict Convexity property. ∎

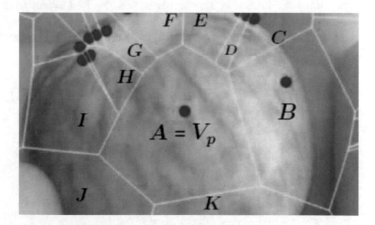

Fig. B.13 Convex regions in an apple Voronoï mesh

Theorem B.21 *Every Voronoï region is a strictly convex set.*

Remark B.22 **Proof sketch** To get the proof started, use pencil and paper to make a sketch of a polygon in a Voronoï mesh. With your sketch in mind, show that a Voronoï region $V(s)$ is a closed set for each mesh generating point s. Take any two points p, q in a Voronoï region $V(s)$. Draw a line segment with endpoints p, q. From the definition of a Voronoï region, argue that all of the points on the line segment $\overline{pq}$ are in $V(s)$. Since $V(s)$ is a closed set and all of the points on each line segment in $V(s)$ are contained $V(s)$, then $V(s)$ is a strictly convex set. ∎

Convexity property   A family of convex sets has the **convexity property**, provided the intersection of any number of sets in the family belongs to the family [182]. See **Convex set**.

Zelins'kyi-Kay-Womble Convexity Structure.
Y.B. Zelins'kyi [216] observes that Solan's view of convexity means that
the set of all subsets of a set is convex. The notion of axiomatic convexity
from Solan and Zelins'kyi has its origins in the 1971 paper by D.C. Kay
and E.W. Womble [91], elaborated by V.V. Tuz [197]. For more about
this in the context of digital images, see J.F. Peters [144, Sect. 1.7, pp.
24–26].   ∎

B.4   D

D11

Data compression: Reduction in the number of bits required to store data (text,
audio, speech, image and video).
Digital video: Time-varying sequence of images captured by digital camera (see,
e.g., S. Akramaullah [4, p. 2]).

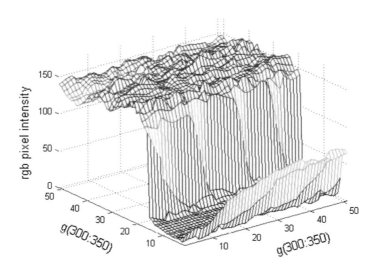

Fig. B.14 3D plot

Dimension: A space X is an n-dimensional space (denoted by $\mathbb{R}^n$), provided each
vector in X has n components. For example, $\mathbb{R}^2$ is a 2D Euclidean space (2D
image plane) and $\mathbb{C}^2$ is a 2D Riemann space.

Example B.23 **2D, 3D, 5D and 5D Euclidean Spaces**. The analog signal in Fig. B.1.1
belongs to a 2D Euclidean space (each vector $v = (x, y)$ in the plane has two
components). The plot in Fig. B.14 belongs to a 3D Euclidean space. Each vector
$v = (x, y, z)$ has 3 components derived from a RGB colour image g (a raster image
in which each pixel is a blend of red, green and blue channel intensities).

In a corner-based tessellation of a static RGB colour image, each geometric
region is described by vectors $(r, g, b, cm, area)$ with at least 5 components: r (red
intensity), b (blue intensity), g (green intensity), cm (centroid) and $area$ (shape area).
In other words, a corner-based mesh on a colour image belongs to a 5D space.

Each geometric region in an RGB video frame image is described by a vector
$(r, g, b, cm, area, t)$ with 6 components: r (red intensity), b (blue intensity), g (green
intensity), cm (centroid), $area$ (shape area), and t (time). In effect, shapes in a video
belong to a space-time 6D space.   ■

Discrete: By **discrete**, we mean that objects such as digital image intensities distinct, separated values. A digital image in a HSV colour space is an example of an
object that is discrete, since its values are both distinct and separated. For another
example of discreteness, see Example 5.5 in Sect. 5.5. Discrete values contrasts
with continuous values such as the values of the function $f(x) = x^2$. See, also,
Continuous in Appendix B.3.

B.5   E

E

$|e_i|$.   Number of edge pixels in edgelet e_i. Initially, $|e_i|$ will equal the number of
mesh generators in a MNC contour. Later, $|e_i|$ will equal the total number of edge
pixels in contour edgelet (denoted by $|\max e_i|$), i.e.,

$$|\mathbf{max}e_i| = \text{no. of all contour edge pixels}, \mathbf{not} \text{ just endpoints.}$$

$\|e_i - e_j\|$.   Compute $D(e_i, e_j) = \max\{\|x - y\| : x \in e_i, y \in e_j\}$, since e_i and e_j
are sets of pixels.

Edgelet.   Set of edge pixels. The term **edgelet** comes from S. Belongie, J. Malik
and J. Puzicha [13, p. 10]. In this work, the notion of an edgelet is extended to
contour edgelet (set of edge pixels in an MNC contour). See **Contour, Contour
edgelet, MNC**. See, especially, Appendix B.3.

Expected value: Let $p(x)$ be the probability of x. Let N be the number of digital
sample values. For the ith digital signal value $x(i)$, $p(x(i)) = \frac{1}{n}$ is its probability.
The expected value of $x(i)$ (denoted by $\langle x(i) \rangle$ or $E[x(i)]$) is

$$\langle x(i) \rangle = \sum_i x(i) p(x(i)) \ (\textbf{Expected value of } x(i))$$

Note: The approximation of a digital signal is its expected value. In the context of video signal analysis, $\hat{x}(i)$ denotes the expected value of the ith digital sample of an analog signal either from an optical sensor in a digital camera or in a video camera. ■

B.6   F

F

Frame: Single digital image either from a conventional non-video camera or a frame is a single image in a video.

B.7   G

G

Gamma γ Correction: Let R be a camera signal in response to incoming light and let γ, a, b be real numbers. Typically, a camera adjusts R using a gamma transform defined by

$$R \longmapsto a R^{\frac{1}{\gamma}} + b \ (\text{Gamma transform}).$$

For the details, see Sect. 2.11. For more about this, see Z.-N. Li, M.S. Drew and J. Liu [111, Sect. 4].

Gamut mapping: Mapping a source signal to a display (e.g., LCD) that meets the requirements of the display. Colour saturation is kept within the boundaries of the destination gamut to preserve relative colour strength and out-of-gamut colours from a source are compressed into the destination gamut. A graphical representation of the destination gamut (nearest colour) and outside gamut (true colour) is shown in https://en.wikipedia.org/wiki/Gamut.
Basically, this is accomplished in practice using gamma correction. See **Gamma γ Correction**.

Generating point: A generating point is a point used to construct a Voronoï region. **Site** and **Generator** are other names for a mesh generating point. Let S be a set of sites, $s \in S$, X a set of points used in constructing a Voronoï mesh, $V(s)$ a Voronoï region defined by

$$V(s) = \{x \in X : \|x - s\| \le \|x - q\| \text{ for all } q \in X\} . \ (\textbf{Voronoï region}).$$

Example B.24 A generating point used to construct a Voronoï region on a 2D digital image is a pixel. Examples are corner and edge pixel. ∎

Geodetic graph: A graph G is a **geodetic graph**, provided, for any two vertices p, q on G, there is at most one shortest path between p and q. A **geodetic line** is a straight line, since the shortest path between the endpoints of a straight line is the line itself. For more about this, see J. Topp [195]. See, also, **Convex hull**.

Example B.25 **Geodetic Graphs**.
A sample geodetic graph with edges drawn between points (vertices) on a map is shown in Fig. B.15.1. Let $p, q_1, q_2, \ldots, q_8$ be vertices in the geodetic graph in Fig. B.15.2. The dotted lines between the pairs of vertices p, q_i are examples of geodetic lines, since the dotted line drawn between the endpoints is the shortest line that can be drawn between the endpoints. Give an example of geodetic graph that is a contour edgelet surrounding the nucleus in a Voronoï MNC. ∎

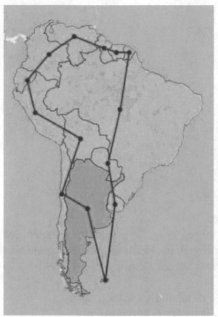 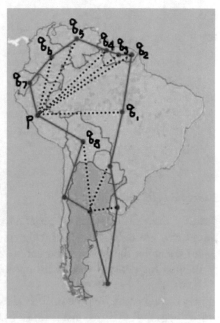

B.15.1: Geodetic graph   B.15.2: Geodetic lines

Fig. B.15 Sample geodetic lines in a geodetic graph

A geodetic graph become interesting when the graph is a convex hull of a set of map points, since a geodetic convex hull approximates the shape of a map region covered by the hull. Try finding a geodetic convex hull by connection the center of mass of the cities in a local region. ∎

B.8   H

H

Halfspace: Given a line L in an n-dimensional vector space $\mathbb{R}^n$ space, a **half space** is a set of points that includes those points on one side of the boundary line L. The points on the line L are included in the half space, provided the half space is closed. Otherwise, the half space is open. See, also, **Boundary region, Boundary set, Open set, Polytope, Closed half space Closed lower half space, Closed upper half space, Open half space, Open lower half space, Open upper half space**.

Hole: A set with an empty interior. In the plane, a portion of the plane with a portion of it missing, i.e., a portion of the plane with a puncture in it (a cavity in the plane). A geometric structure that cannot be continuously shrunk to a point, since there is always part of the structure that is missing. See, also, **Interior, Open set**.

Fig. B.16 Overlapping annuli

Example B.26 **Sample holes.**

donut   Center of a donut, center of a torus.

punctured annulus   Annulus (ring-shaped region) with a hole in its center. See, for example, the overlapping annuli in Fig. B.16. Homology theory provides a method for detecting holes in topological spaces (see, e.g., [101, Sect. 3.2, p. 108ff]).

empty orange skin   Space inside the skin of an orange without its pulp.

empty egg shell   Space inside an Egg shell without the egg yoke.

empty window frame   Space inside a Window frame without a pane of glass.

empty interior set   Set of points in boundary of a planar set with an empty interior.

binary image island   Dark region surround by white pixels in a binary image.

empty subimage   A 2D or 3D subimage that includes its boundary pixels and with no pixels in its interior. The analogue of a subimage that is a hole is a completely black surrounded by a white region. In mathematical morphology,[3] a hole is a foreground region containing dark pixels (white pixels adjacent a hole are the

---

[3] See, e.g., Sect. 6.6.2, p. 146 in http://www.cs.uu.nl/docs/vakken/ibv/reader/chapter6.pdf.

boundary pixels of the hold). A Rosenfeld 8-neighbourhood defines a hole with a dark center surrounded by 7 white pixels. Notice that a hole defines a shape, a contour surrounding a hole. **Question**: Can we say that a circle and a pizza plate have the shape? Or a rectangle and a ruler?

For more about holes in mathematical morphology, see M. Sonka, V. Hlavac and R. Boyle [184, Chap. 13]. For an introduction to holes from a topological perspective, see S.G. Krantz [100, Sect. 1.1] and [101, pp. 1, 95, 108]. ∎

Fig. B.17 Mapping image with holes to image with holes filled

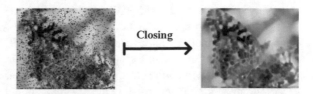

Closing

Example B.27 **MM Closing Operation to remove holes from the foreground of an image**. A mathematical morphology (MM) closing Operation is used to remove holes (pepper noise) from the foreground of an image. Let Img be a 2D image with holes. In effect, **closing** maps Img onto a new image without holes. The following Mathematica script uses the MM closing operation to fill the holes in a noisy image (see, e.g., Fig. B.17).

Mathematica 8 Removing holes (dark specks) from a colour image.
(* Filling holes in an image foreground *)

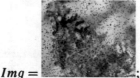

$Img =$;

Closing[Img, 1]

Try writing a Matlab script to remove black holes from a colour (not a binary) image. ∎

Hue: Hue of a colour is the wavelength of the colour within the visible light spectrum at which the energy output is greatest [67, Sect. 1.4]. Hue is a point characteristic of colour, determined at a particular point in the visible light spectrum and measured in nanometers. Let R, G, B be red, green, blue colour. A. Hanbury and J. Serra [68, Sect. 2.2.1, p. 3] define saturation S and hue H' expressions in the following way.

$$S = \begin{cases} \frac{\max\{R,G,B\} - \min\{R,G,B\}}{\max\{R,G,B\}}, & \text{if} \max\{R, G, B\} \neq 0, \\ 0, & \text{otherwise.} \end{cases}$$

and

$$H' = \begin{cases} \text{undefined}, & \text{if } S = 0, \\ \frac{G-B}{\max\{R,G,B\}-\min\{R,G,B\}}, & \text{if } R = \max\{R, G, B\} \neq 0, \\ 2 + \frac{B-R}{\max\{R,G,B\}-\min\{R,G,B\}}, & \text{if } G = \max\{R, G, B\} \neq 0, \\ 4 + \frac{R-G}{\max\{R,G,B\}-\min\{R,G,B\}}, & \text{if } B = \max\{R, G, B\} \neq 0. \end{cases}$$

Then $60^o H'$ equals the hue value in degrees. See **Saturation, Value, HSV**.

Hue Saturation Value (HSV):   The HSV (Hue Saturation Value) theory is commonly used to represent the RGB (Red Green Blue) colour technical model [67]. The HSV colour space was introduced in 1905 by A. Munsell [127] and elaborated in 1978 by G.H. Joblove and D. Greenberg [89] to compensate for technical and hardware limitations for applications in colour display systems. *Hue* is an angular component, *Saturation* is a radial component and *Value* (*Lightness*) is the colour intensity along vertical in the 3D color model that shows the geometry of the HSV colour space in https://en.wikipedia.org/wiki/HSL_and_HSV.
Complete, detailed views of the HSV color model are given by J. Haluška [67] and A. Hanbury and J. Serra [68].

Huffman coding: A lossless data compression algorithm that uses a small number of bits to encode common characters.[4] To see an example of Huffman coding for a digital image, try using the Matlab script A.2 in Appendix A.1 with any digital colour image.

B.9   I

I

i: Imaginary number $i = \sqrt{-1}$. See, also, z, **Complex plane, Riemann surface**.
Int A: Interior of a set nonempty set A, a set without boundary points. See, also, **Open set, Closed set, Boundary set**.

Example B.28 **Sample interior sets**.

orange pulp   Orange without its skin.
Earth subregion   The region of the Earth below its surface.
subimage   Subimage that does not include its boundary pixels. A Rosenfeld 8-neighbourhood that does not include the pixels along its boundaries.   ■

Image quality: See **SSIM, UQI**.
Infimum: The infimum of a set A (denoted by inf A) is the greatest lower bound of the set A.

---

[4]For the details about Huffman coding, see E.W. Weisstein at http://mathworld.wolfram.com/HuffmanCoding.html.

B.10 K

K

Key Frame Selection: Selection of video frames exhibiting the most change.

Example B.29 **Adaptive Key Frame Selection**. Adaptive key frame selection is an approach to efficient video coding suggested by J. Jun et al. in [121]. For example, adaptive video video frame selection is achievable by selecting video frames in which there are significant changes in image tessellation polygons. For example, in a changing scene recorded by a webcam, overlay polygons on video frames temporally close together usually will vary only slightly. By contrast, the areas of the overlay polygons on temporally separated video frames often will vary significantly in recording a changing scene.   ■

Keypoint: A **keypoint** is a feature of a curve (edgelet) in a digital image. The goal of keypoint detection is to identify salient digital image regions such as corners and blobs. Keypoint-based methods extract point features that are stable with view variations, an observation by R. Fabbri and B.B. Kimia [47]. An approach to fast keypoint detection in video sequences is given by L. Baroffio, M. Cesana, A. Redondi and M. Tagliasacchi [9], using the S. Leutenegger, M. Chli and R.Y. Siegwart Binary Robust Invariant Scalable Keypoints (BRISK) algorithm [108]. See **SIFT, SURF**.

B.11 L

L

Luminance.   The light reflected from a surface (denoted by $L(\lambda)$). Let E be the incident illumination, λ wavelength, $R \in [0, 1]$ reflectivity or reflectance of a surface. Then $L(\lambda)$ has a spectrum given by

$$L(\lambda) = E(\lambda)R(\lambda) \text{ cd } m^{-2}.$$

Luminance is measured in SI units of candela per square meter (cd/m^2). **Candela** is the SI (Standard International) base unit of luminous intensity, i.e., luminous power per unit solid angle emitted by a point light source in a particular direction. See, also, **Candela, Photon, Quantum optics**.

B.12   M

M

m_i frequency.   Number of occurrences of edgelets with the same number of edge pixels as edgelet e_i.

$m_1, m_2, \ldots, m_i, \ldots, m_k$.   Frequencies of occurrences of k edgelets.

Maximal Nucleus Cluster (MNC)   A *maximal nucleus cluster* is a cluster of connected polygons with nucleus N is *maximal*, provided N has the highest number of adjacent polygons in a tessellated surface [147]. See **MNC, Nucleus**.

B.18.1: CN train   B.18.2: CN train Mesh

Fig. B.18 Voronoï mesh on CN train video frame image

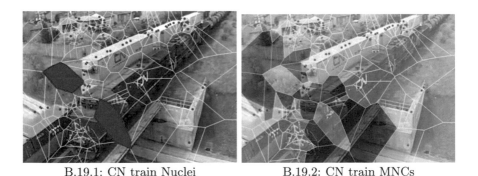

B.19.1: CN train Nuclei   B.19.2: CN train MNCs

Fig. B.19 Max nucleus clusters on CN train video frame image

MNC Nucleus cluster   A **Nucleus cluster** (NC) in a Voronoï mesh on a digital image is a cluster of Voronoï regions that are connected polygons with nucleus N. Every Voronoï region is the *nucleus* of cluster of neighbouring polygons. A nucleus that has the maximal (largest) number of adjacent polygons is the center of a MNC. For the details, see Sect. 7.5. See, also, MNC-based image object shape recognition Methods B.12.

Example B.30 A sample Vornonï mesh on a CN train image is shown in Fig. B.18.2.

A pair of mesh nuclei such as  are shown in Fig. B.19.1. These red nuclei each has the highest number of adjacent polygons. Hence, these red nuclei are the centers of maximal nucleus clusters (MNCs) shown in Fig. B.19.2. ∎

Fig. B.20 Sample MNC circle

MNC-based image object shape recognition Methods   There are three basic methods that can be used to achieve image object shape recognition, namely,

Method 1: Inscribed Circle Perimeter Inside MNC contours. The
circle-perimeter method gets its inspiration from V. Vakil [198]. To begin, choose a keypoint-based Voronoï-tessellated query image Q like the one in Fig. B.20 and a Voronoï-tessellated video frame test image T. Identify an MNC in Q and an MNC in T. Measure the perimeter of circles inscribed in coarse-grained and fine-grained nucleus contours and that are centered on each image MNC nucleus generating point. See, for example, the coarse-grained circle on the MNC in Fig. B.20.

Method 2: Edge Strength.   Again, choose a keypoint-based Voronoï-tessellated query image Q like the one in Fig. B.20 and a Voronoï-tessellated video frame test image T. Identify an MNC in Q with a nucleus $N_Q(q)$ with generating point q and an MNC in T with nucleus $N_T(t)$ with generating point t. Measure and compare the edge strengths of q and t.

Method 3: Circle Perimeter-Edge Strength.   Combine Methods 1 and 2. That is, compare the circle perimeters and edge strengths of a keypoint-based Voronoï-tessellated query image Q like the one in Fig. B.20 and a Voronoï-tessellated video frame test image T.

Table B.1 MNC border circle perimeters compared

image Q	image T	Fine P_Q	Fine P_T	$\lvert P_Q - P_T \rvert$	ε
	...	...	...	...	...
image Q	image T	Coarse P_Q	Coarse P_T	$\lvert P_Q - P_T \rvert$	ε
	...	...	...	...	...

Here are the details.

> Method 1: Inscribed Circle Perimeter Inside MNC contours

1° **Method 1**: Randomly choose one of the keypoints in a fine-grained border about an image MNC.

2° Use the chosen fine-grained border keypoint and the nucleus keypoint as the endpoints of a circle radius (call it $\boldsymbol{r_{fine}}$).

3° On a MNC in the query image Q and test image T, draw a circle with radius $\boldsymbol{r_{fine}}$. Make the circle border colour a bright red.

4° Let $\bigcirc_{Qfine}$ be a fine-grained circle on centered on the nucleus keypoint of an MNC in the query image Q and let $\bigcirc_{Tfine}$ be a fine-grained circle on centered on the nucleus keypoint of an MNC in the test image T. Let P_Q, P_T the lengths of the perimeters of $\bigcirc_{Qfine}$, $\bigcirc_{Tfine}$, respectively. Choose and object recognition threshold $\varepsilon > 0$. Let $\boldsymbol{ObjectRecognized}$ be a Boolean variable. Then

$$ObjectRecognized = \begin{cases} 1, & \text{if } \lvert P_Q - P_T \rvert < \varepsilon, \\ 0, & \text{otherwise} \end{cases}$$

In other words, an object in a query image is recognized, provided the difference $\lvert P_Q - P_T \rvert < \varepsilon$ is small enough.

5° Highlight in a bright green MNC circles in video frames containing recognized a recognized object.

6° In zip file, save the query image and video frame images containing a recognized object.

7° Repeat Step 1 to Step 6, for a coarse-grained border of an MNC. Then, for coarse-grained circles $\bigcirc_{Qcoarse}, \bigcirc_{Tcoarse}$ centered on query and test image MNCs, compute difference the perimeters $\left|P_Q - P_T\right| < \varepsilon$. Save your findings in a Table (see Table B.1).

Table B.2 MNC keypoint edge strengths compared

image Q	image T	P_E	T_E	$|P_E - T_E|$	ε
		...	...	...	...

Method 2: Edge strengths of MNC nucleus generating points

1° Let keypoint(x,y) be the keypoint at location (x, y) in an MNC nucleus. Let $G_x(x, y), G_y(x, y)$ denote the edge pixel gradient magnitudes in the x- and y-directions, respectively.

2° The edge strength of pixel $Img(x, y)$ (also called the pixel gradient magnitude) is denoted by $E(x, y)$ and defined by

$$E(x, y) = \sqrt{\left(\frac{\partial Img(x, y)}{\partial x}\right)^2 + \left(\frac{\partial Img(x, y)}{\partial y}\right)^2} \text{ (\textbf{Pixel edge strength})}$$

$$= \sqrt{G_x(x, y)^2 + G_y(x, y)^2}.$$

Let Q_E, T_E be the edge strengths of the keypoints of the nuclei for the query image Q and test image T, respectively. Then compute

$$Object\, Recognized = \begin{cases} 1, & \text{if } |P_E - T_E| < \varepsilon, \\ 0, & \text{otherwise} \end{cases}$$

In other words, the difference between the edge strengths $|P_E - T_E| < \varepsilon$ is small enough the MNC object in image Q and image T are similar, *recognized*.

3° Summarize your findings for each video in a table (see Table B.2).

MNC spoke.   See **Nerve** in Appendix B.13.

MSSIM   Mean SSIM image quality index. Let X, Y be the rows and columns of an $n \times m$ digital image, row $x_i \in X, 1 \le i \le n$, column $y_j \in Y, 1 \le j \le m$ such that

$$x = (x_1, \ldots, x_n) \,, \, y = (y_1, \ldots, x_m) \,.$$

The Mean SSIM value, measure of overall image quality, is defined by

$$MSSIM(X, Y) = \sum_{\substack{i=1 \\ j=1}}^{n,m} \frac{SSIM\left(x_i, y_j\right)}{nm}.$$

For more about this, see **SSIM**.

B.13   N

N

Fig. B.21 MNC spokes in a mesh nerve

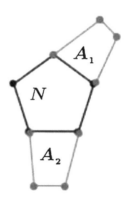

Nanometer: $1\,\text{nm} = 1 \times 10^{-9}\,\text{m}$ or 3.937×10^{-8} in or $10\,\text{Å}$ (Ångströms). RGB colour wavelengths are measured in nanometers

Neighbourhood of a point: Let ε be a positive real number and inf A The neighbourhood of a point p in the plane (denoted by $N(p, \varepsilon)$) is defined by

$$N(p, \varepsilon) = \left\{x \in \mathbb{R}^2 : \|x - p\| < \varepsilon\right\} \text{ (Open Nhbd of } p\text{)}.$$

Notice that $N(p, \varepsilon)$ is an open neighbourhood, since it excludes boundary points of x in the plane. A closed neighbourhood (denoted by $\text{cl}N(p, \varepsilon)$) includes both its interior points and its boundary points, defined by

$$N(p, \varepsilon) = \left\{x \in \mathbb{R}^2 : \|x - p\| \leq \varepsilon\right\} \text{ (Closed Nhbd of } p\text{)}.$$

For more about this, see J.F. Peters [142, Sect. 1.14]. See **Open set, Closed set**.

Nerve: A mesh **nerve** is a collection of spoke-like projections on a mesh nucleus. Think of a maximal nucleus cluster (MNC) as a collection of spokes. Each **spoke** is a combination of a MNC nucleus and an adjacent polygon. An example of two MNC spokes are shown in Fig. B.21, namely, spoke NA_1 and spoke NA_2. Both of the sample spokes share a mesh nucleus N. The study of MNC spokes takes us in the direction of a deeper view of digital image geometry revealed by a keypoint-based nuclei in a Voronoï mesh superimposed on an image. A spoke-based mesh nerve is an example of an Edelsbrunner–Harer nerve [42, 150] .

Noise: Compression of digital video may introduce distortions or noise (also called a *visual artifact*). Noise affects the visual quality perceived by an end-user of either a digital image or video.

Nucleus: A **nucleus** is the central and most important part of a mesh cluster. See, also, the entry on **nerve** in Appendix B.13 and the MNC entry in Appendix B.12.

B.14   O

O

Object tracking: In a digital video sequence, identify moving objects and track them from frame to frame. In practice, each frame is segmented into regions with similar colour and intensity and which are likely to have some motion. This can be accomplished by tessellating each video frame, covering each frame with polygons that are Voronoï regions, then comparing the changes in particular polygons from frame to frame. For more about this, see S.G. Hoggar [83, Sect. 12.8.3, p. 441].

Open lower halfspace: An **open lower half space** is the set of all points below but not on a boundary line.

Open upper halfspace: An **open upper half space** is the set of all points above but not on on a boundary line.

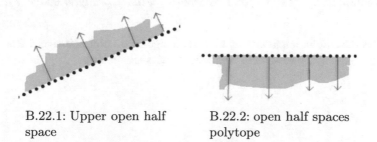

B.22.1: Upper open half space

B.22.2: open half spaces polytope

Fig. B.22 Sample open *upper* and *lower half spaces*

Example B.31 **Sample 2D Open Upper and Lower Halfspaces**.
The two types of 2D open half spaces are shown in Fig. B.22. The half space in Fig. B.22.1 is an example of open upper half space. This half space is open, since it includes all planar points above but not on the dotted line that is the boundary of the half space. The half space in Fig. B.22.2 is an instance of a open lower half space. In this case, the half space consists of all planar below but not on the indicated dotted line. The dotted line in the open lower half space forms the boundary of the half space. In both cases, each half space is unbounded one side of a line. ∎

Open set: A set of points without a boundary.

Example B.32 **Sample open sets**.

orange pulp   Orange without its skin.
egg interior   Egg without its shell.
window glass   Window glass without its frame.
set interior   Interior of any set without a boundary.
subimage   Any Subimage that does not include its boundary pixels. A Rosenfeld 8-neighbourhood is an open set, since it does not include the pixels along its boundaries. ∎

Lemma B.33 *Every 2D digital image is an open set.*

Proof Each point in $\mathbb{R}^2$ with integer coordinates is potentially the location of a digital image pixel. Every 2D digital image is restricted to those pixels within its borders. The borders of a digital image do not include those pixels outside to its borders. The boundary of a digital image A is the set of points outside X and adjacent to A, That is, a digital image does not include its boundary pixels. Hence, a digital image is an open set.

Theorem B.34 *Every subimage in a 2D digital image is an open set.*

Proof Every subimage in a 2D digital image is also a digital image in the Euclidean plane. Hence, from Lemma B.33, every subimage is an open set.

Theorem B.35 *Every Rosenfeld neighbourhood in a 2D digital image is an open set.*

Proof A Rosenfeld neighbourhood is a subimage in 2D digital image. Hence, from Theorem B.34, every Rosenfeld neighbourhood is an open set.

B.15   P

P

Path: sequence $p_1, \ldots, p_i, p_{i+1}, \ldots, p_n$ of n pixels or voxels is a **path**, provided p_i, p_{i+1} are adjacent (no pixels in between p_i and p_{i+1}).

Path-connected: Pixels p and q are **path-connected**, provided there is a path with p and q as endpoints. Image shapes A and B (any polygons) are path-connected, provided there is a sequence $S_1, \ldots, S_i, S_{i+1}, \ldots, S_n$ of n adjacent shapes with $A = S_0$ and $B = S_n$. See **Blob**.

Performance: Speed of video coding process.

Pixel: Smallest component in a raster image. A pixel's dimensions are determined by optical sensor and scene geometry models. Normally, a pixel is the smallest component in the analysis of image scenes. Sub-pixel analysis is possible in very refined views of image scenes. For sub-pixel analysis, see T. Blashke, C. Burnett and A. Pekkarinen [16, Sect. 12.1.3, p. 214].

Pixel intensity: Pixel value or amount of light emitted by a pixel. In a greyscale image, a white pixel emits the maximum amount of light and a black pixel emits zero light. In a colour image, the intensity of a colour channel pixel is its colour brightness.

Photon: A **photon** is an electromagnetic radiation energy packet, formulated by Einstein in 1917 to explain the photoelectric effect. For more about this, see M. Orszag, [134]. See, also, **Quantum optics**.

Planck's constant h The constant $h = 6.6262 \times 10^{-27}$ erg seconds $= 6.6262 \times 10^{-34}$ J s. The energy E of a photon with frequency ν is $E = h\nu$. Let T be the absolute temperature if a source, $c = 2.998 \times 10^{-23}$ m s^{-1} the speed of light, $k = 1.381 \times 10^{-23}$ K^{-1} Boltzmann's constant, λ wavelength. Planck's law for black body radiation (denoted by $B_\nu(T)$) is defined by

$$B(\lambda) = \frac{2h}{c^2} \frac{\lambda^5}{e^{\frac{hc}{k\lambda T}} - 1} \text{ W m}^{-2}\text{m}^{-1}.$$

This is the power emitted by a light source such as the filament of an incandescent light bulb:
https://en.wikipedia.org/wiki/Incandescent_light_bulb.
For the details, see P. Corke [31].

Point   Another name for digital image pixel.

Polytope: Let $\mathbb{R}^n$ be an n-dimensional Euclidean vector space, which is where polytopes live. A **polytope** is a set of points $A \subseteq \mathbb{R}^n$ that is either a convex hull of set of points K (denoted by $convhA(K)$) or the intersection of finitely many closed half spaces in $\mathbb{R}^n$. This view of polytopes is based on [220, p. 5]. Notice that non-convex polytopes are possible, since the intersection of finitely many half spaces may not be the smallest convex set containing a set of points. Polytopes are commonly found in Voronoï tessellations of digital images. See, also, **Convex hull, Convex set, Half space**.

Fig. B.23 Sample 2D
polytopes

B.23.1: Convex   B.23.2: Closed half spaces polytope
hull polytope

Example B.36 **Polytopes.**
The two types of polytopes are shown in Fig. B.23. The polytope in Fig. B.23.1 is a
convex hull of 9 points (represented the black ● dots). The polytope in Fig. B.23.1
is the intersection of 5 closed half spaces.   ∎

B.16   Q

Q

Quality of a digital image: Mean SSIM (MSSIM) index. See **MSSIM, SSIM**.
Quality of an Voronoï region: Recall that a Voronoï region is a polygon with n
  sides. Let $V(s)$ be a Voronoï region and let $Q(V(s))$ be the quality of $V(s)$. The
  quality of $Q(V(s))$ is highest when the polygon sides are equal in length.

Example B.37 Let A be the area of $V(s)$ with 4 sides having lengths l_1, l_2, l_3, l_4.
Then
$$Q(V(s)) = \frac{4A}{l_1^2 + l_2^2 + l_3^2 + l_4^2}.$$

$Q(V(s))$ will vary, depending on the area and the number of sides in a Voronoï region
polygon. Let $Q_i(V(s_i))$ (briefly, Q_i) be the quality of polygon i with $1 \leq i \leq n, n \geq$
1. And let S be set of generating points, $V(S)$ a Voronoï tessellation. Then

$$Q(V(S)) = \frac{1}{n} \sum_{i=1}^{n} Q_i \text{ (Global Mesh Quality Index).}$$

Theorem B.38 [1]
*For any Voronoï-tessellated plane surface, a set of generating points exists for which
the quality of the mesh cells is maximum.*

Quality of an MNC contour shape: The **quality of an MNC contour shape** is
  proportional to the closeness of a target contour shape to a sample contour shape.

In other words, an MNC contour shape is high, provided the difference between a target MNC contour perimeter and a sample MNC contour perimeter is less than some small positive number ε.

Quantum optics: The study of light and interaction between light and matter at the microscopic level. For more about this, see C. Fabre [48]. See, also, **Photon**.

B.17 R

R

$\mathbb{R}$: Set of reals (real numbers).

$\mathbb{R}^2$: Euclidean plane (2-space). 2D digital images live in $\mathbb{R}^2$.

$\mathbb{R}^3$: Euclidean 3-space. 3D digital images live in $\mathbb{R}^3$.

Reality: What we experience as human beings.

RGB: Red Green Blue colour technical model. For the RGB wavelengths based on the CIE (Commission internationale de l'ëclairage: International Commission on Illumination) 1931 color space, see
https://en.wikipedia.org/wiki/CIE_1931_color_space.

Regular polygon: An n-sided polygon in which all sides are the same length and are symmetrically arranged about a common center, which means that a regular polygon is both equiangular and equilateral. For more about this, see E.W. Weisstein [207].

Riemann surface: A **Riemann surface** is a surface that covers the complex plane (z-plane or z-sphere) with sheets. See **Complex plane,** $\mathbb{C}$.

B.18 S

S

Sampling: Extracting samples from an analog signal at appropriate intervals. A continuous analog signal $x_a(t)$ such as the signal from an optical sensor in a digital camera or in a web cam, is captured over a temporal interval t. Let $T > 0$ denote the **sampling period** (duration between samples) and let n be the sample number. The ratio And $x(n)$ is a digital sample of the analog signal $x_a(t)$ at time t, provided

$$x_a(n) = x(nT), \ n^{th} \text{ for some sampling period } T > 0.$$

$$\frac{2\pi}{T} = \text{sampling frequency or sampling rate.}$$

Example B.39 **Sampled Optical Sensor Signal**.
A time-varying analog signal $x(t)$ over time t is shown in Fig. B.1.1. A collection of n digital signal samples $x(n)$ is shown in Fig. B.1.2. Here, the sampling period is T

(duration between sampled signals). Each spike in Fig. B.1.2 represents a sampled signal (either image or video frame).   ■

Fig. B.24 Monotonicity of brightness, saturation and hue from [67]

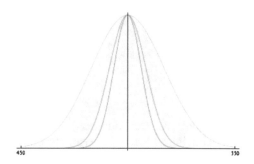

Saturation: *Saturation* is an expression for the relative bandwidth of the visible output from a light source [67, Sect. 1.4]. Notice that saturation is an interval characteristic of colour, determined over an interval, not at a point. Saturation is represented by the steepness of the colour curves in Fig. B.24. Notice that blue has the greatest saturation. The hue of a colour becomes more pure as saturation increases.

Fig. B.25 Polygon edges surrounding a *triangle shape*

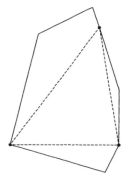

Shape: A shape is the external form or appearance of something. In the context of digital images, a shape is the outline of an area of an image. In terms of subimages, the shape of a subimage is identified with its set of edge pixels called an *shape edgelet* such as the edge pixels along sides of polygon covering an unknown shape (see, e.g., Fig. B.25). A straightforward approach to identifying a subimage shape A is compare the contour edgelet e_A or boundary of A with the contour edgelet e_B of a known shape B. For the details, see Sect. 7.7. For an introduction to shape context descriptors, see M. Eisemann, F. Klose and M. Magnor [44, p. 10]. Another approach to comparing shapes covered by polygons is to compare the total area of the polygons covering one shape with the total area of the polygons covering another shape (see, e.g., the shape measure given by D.R. Lee and G.T. Sallee in [106]).

Example B.40 **MNC shape**.
The fine (or coarse) contour of an MNC in a Voronoï tessellation of a digital image is a set of edge pixels belonging to the line segments between pairs of mesh generators of the polygons along the border of either the nucleus (fine contour case) or along the border polygons of the MNC. Such line segments are called **edgelets**. For example, let edgeletMNC be a fine contour of an MNC in a Voronoï tessellation of a digital image. The shape of an object covered by an MNC is approximated by edgeletMNC.
∎

Shape boundary: In the maximal nucleus cluster (MNC) approach to image object shape recognition, either the shape boundary is approximated by either a coarse or fine contour of an MNC nucleus. This approach is based on an observation by T.M. Apostol and M.A. Mnatsakanian [6], namely, that any planar region can be dissected into smaller pieces that the can rearranged to form any other polygonal region of equal area. For image object shape recognition relative to the shape of a query image object and the shape of test image object, there is a basic requirement, namely, shapes are equivalent provided they have equal areas and equal perimeters. A weakened form of this requirement is that shapes are approximately the same, provided the shapes have either approximately the same perimeter or approximately the same area. The situation becomes more interesting when an MNC contour K is the boundary of a convex hull C of the set of MNC interior points. Let $p(K)$, $p(C)$ be the perimeters of K and C, respectively. From G.D. Chakerian [26], we know that $C = k$, provided $p(K)$, $p(C)$. Extending Chakerian's result, let K_Q, C_Q be the perimeter of the boundary and of the convex hull of an MNC of a query image shape Q with perimeters $p(K_Q)$, $p(C_Q)$ and let K_T, C_T be the perimeter of the boundary and of the convex hull of an MNC of a video frame test image shape T with perimeters $p(K_T)$, $p(C_T)$. Then shape Q is close to shape T, if and only if

$$Q \approx T, \text{ provided } p(K_Q) \approx p(K_T) \text{ and } p(C_Q) \approx p(C_T),$$

i.e., boundary perimeters $p(K_Q)$, $p(K_T)$ are close and convex hull perimeters $p(C_Q)$, $p(C_T)$.
See, also, **Shape, Convex Hull, Convex Set, MNC, Boundary region of a set, Boundary Set** and Jeff Weeks lecture[5] on the shape of space and how the universe could be a Poincaré dodecahedral space: https://www.youtube.com/watch?v=j3BlLo1QfmU.

---

[5]Many thanks to Zubair Ahmad for pointing this out.

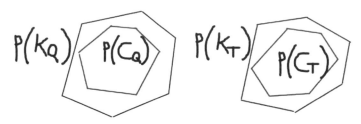

Fig. B.26 Query image Q and test image T boundary and convex hull perimeters

Example B.41 **Comparing query and test image shape convex hulls and boundaries**. In Fig. B.26, query and test image shape boundaries and convex hulls are represented by $p(K_Q)$, $p(K_T)$ and $p(K_T)$, $p(C_T)$, respectively. In each case, the shape boundary contains the shape convex hull. The basic approach is the compare lengths of the boundaries $p(K_Q)$, $p(K_T)$ and convex hull perimeters $p(C_Q)$, $p(C_T)$. Let ε be a positive number. Then the test image shape approximates the query image shape, provided

$$\left| p(K_Q) - p(K_T) \right| \leq \varepsilon \text{ and } \left| p(C_Q) - p(C_T) \right| \leq \varepsilon.$$

The end result is a straightforward approach to image object shape recognition, assuming the boundary length and convex hull perimeter are close enough. ∎

Shape Edgelet: See Example B.40 and Sect. B.18.

SI: Standard International based unit of measurement. There are 7 SI base units, namely,

meter   length. Abbreviation: **m**.
kilogram   mass. Abbreviation: **kg**.
second   time. Abbreviation: **s**.
ampere   electric current. Abbreviation: **i**, e.g., i = 100 amps.
kelvin   temperature. Abbreviation: **K**, $1\,K = 1^{o}C = \frac{9}{5}^{o}F = \frac{9}{5}^{o}R$.
candela   luminous intensity. Abbreviation: **cd**.
mole   amount of substance. Abbreviation: **mol**, e.g., 2 mol of water, 1 mol of dioxygen.

SIFT: Scale-Invariant Feature Transform (SIFT) introduced by D.G. Lowe [115, 116] is a mainstay in solving object recognition as well as object tracking problems.

Similarity Distance: Let A be the set of points on the contour of a sample object and let B be the contour of a known object. The Hausdorff distance [75, Sect. 22, p. 128] between a point x and a set A (denoted by $D(x, A)$) is defined by

$$D(x, A) = \min \{ \|x - a\| : a \in A \} \ \textbf{(Hausdorff point-set distance)}.$$

The similarity distance $D(A, B)$ between the two contours A and B, represented by a set of uniformly sampled points in A and B [60, Sect. 2, p. 29], is defined by

$$D(A, B) = \max \left\{ \max_{a \in A} D(a, B), \max_{b \in B} D(b, A) \right\} \quad \textbf{(Similarity Distance)}.$$

Signal quality: The expected value of a signal compared with an actual signal.

Example B.42 **Signal Quality Measure**.
Mean squared error (MSE). Let $\hat{x}(n)$ denote the approximate n^{th} digital signal value and let N be the number of samples. The expected value of the i_{th} digital signal value x_i (denoted $\hat{x}_i$) is the approximation of x_i. Then $MSE(x)$ is a measure of the quality of a signal $x = (x_1, \ldots, x_n)$ is defined by

$$MSE(x) = \frac{\sum\limits_{i=1}^{N} \left(x_i - \hat{x}_i\right)^2}{N}.$$

In our case, a signal is a vector of pixel intensities such as the row or column intensities in a greyscale digital image.   ∎

Spoke   See **Nerve** in Appendix B.13.
SSIM   Structural similarity image measure. A measure of image structural similarity that compares local patterns of pixel intensities that have been normalized for luminance and contrast, introduced by Z. Wang, A.C. Bovik, H.R. Sheikh and E.P. Simoncelli [205]. SSIM computes structural similarity by comparing image intensities in a row x and column y of the image with n columns and m rows, defined by

$$x = (x_1, \ldots, x_n), y = (y_1, \ldots, y_m).$$

Let μ_x, μ_y be the average pixel intensity in the x and y directions, respectively. Let σ_x, σ_x be image signal contrast in the x and y directions, respectively, defined by

$$\sigma_x = \left(\sum_{i=1}^{n} \frac{(x_i - \mu_x)}{n - 1} \right)^{\frac{1}{2}}, \sigma_y = \left(\sum_{i=1}^{m} \frac{(y_i - \mu_y)}{m - 1} \right)^{\frac{1}{2}}.$$

Let σ_{xy} (used to compute SSIM(x,y)) be defined by

$$\sigma_{xy} = \left(\sum_{i=1}^{n} \frac{(x_i - \mu_x)(y_i - \mu_y)}{n - 1} \right).$$

Let C_1, C_2 be constants used to avoid instability when average intensity values are very close to each other. Then the similarity measure SSIM between signals x and x is defined by

$$SSIM(x, y) = \frac{(2\mu_x\mu_y + C_1)(2\sigma_{xy} + C_2)}{(\mu_x^2 + \mu_y^2 + C_1)(\sigma_x^2 + \sigma_y^2 + C_2)}.$$

See **MSSIM, Quality of a digital image**.

SURF: Speeded up robust features, introduced by H. Bay, A. Ess, T. Tuytelaars and L.V. Gool [10]. SURF is a scale- and rotation-invariant detector and descriptor. SURF integrates the gradient information in an image subpatch. SURF is implement in Matlab.

B.19   T

T

Tessellation: A **tessellation** of an image is a tiling of the image with polygons. The polygons can have varying numbers of sides. See Example 8.1.

Tiling: A **tiling** is a covering of a surface with polygons (traditionally, regular polygons). See **Cover**.

Topological space: A nonempty set X with a topology τ on it, is a **topological space**. See **Topology, Open set**.

Topology: A collection of open sets τ on a nonempty open set X is a *topology* on X [126, Sect. 12, p. 76], [100, Sect. 1.2, p. 1], [128, Sect. 1.6, p. 11], provided

1° The empty set $\emptyset$ is open and $\emptyset$ is in τ.
2° The set X is open and X is in τ.
3° If $\mathcal{A}$ is a sub-collection of open sets in τ, then

$$\bigcup_{B \in \mathcal{A}} B \text{ is a open set in } \tau.$$

In other words, the union of open sets in τ is another open set in τ.

4° If $\mathcal{A}$ is a sub-collection open sets in τ, then

$$\bigcap_{B \in \mathcal{A}} B \text{ is a open set in } \tau.$$

In other words, the intersection of open sets in τ is another open set in τ.   ∎

See **Topological space, Open set**.

> **Brief History of Topology**.
> For a detailed history of topology in four stages (eras), see J. Mil-
> nor [123]. Part 1 of this history traces the origins of topology back to
> 1736 and the first topological statement with L. Euler's solution to the
> seven bridges of Königsberg problem. Part 2 covers the introduction of
> 2-dimensional manifolds in the 19th century, starting with the work by
> S. L'Huilier, 1812–1823, on the surface of a polyhedron in Euclidean
> 3-space that is drilled through with n holes. Part 3 of this history covers
> the study of 3-dimensional manifolds, starting with the 1898 work by
> P. Heegard on the decomposition of closed orientable 3-manifolds as the
> union of two handle-bodies of the same genus and which intersect only
> along their boundaries. Part 4 covers 4-dimensional manifolds, starting
> with the work by A.A. Markov, 1958, and the work by J.H.C. Whitehead
> in 1949.   ∎

B.20   U

U

UQI: Universal Quality Index defined by Z. Wang and A.C. Bovik in [204]. Let
μ_x, μ_y be the average pixel intensity in the x and y directions, respectively. Let
σ_x, σ_x be image signal contrast in the x and y directions, respectively, defined by

$$\sigma_x = \left(\sum_{i=1}^{n} \frac{(x_i - \mu_x)}{n-1} \right)^{\frac{1}{2}}, \sigma_y = \left(\sum_{i=1}^{m} \frac{(y_i - \mu_y)}{m-1} \right)^{\frac{1}{2}}.$$

Let σ_{xy} (used to compute SSIM(x,y)) be defined by

$$\sigma_{xy} = \left(\sum_{i=1}^{n} \frac{(x_i - \mu_x)(y_i - \mu_y)}{n-1} \right).$$

$UQI(x, y)$ is defined by

$$UQI(x, y) = \frac{\left(4\sigma_{xy}\mu_x\mu_y\right)}{\left(\mu_x^2 + \mu_y^2\right)\left(\sigma_x^2 + \sigma_y^2\right)},$$

In [204, Sect. II, p. 81], x is an original image signal and y is a test image signal.
Notice that in a greyscale image, x is a row of pixel intensities and y is a column
of pixel intensities. This is the same as $SSIM(x, x)$, when $C_1 = C_2 = 0$ in the
structural similarity image measure $SSIM(x, x)$. See **SSIM**.

B.21   V

V

Value: The value (brightness) of colour is a relative expression of the intensity of the energy output of visible light source [67, Sect. 1.4].
Viewpoint: A position affording a good view.
Video signal processing: Minimize noise, offline as well as online analysis of video frames using image processing and computer vision methods, and exploit the geometry and temporal nature of the video frames.
Virtual: Almost or nearly as described but not completely based on a strict definition of what is being described.

Virtual reality: Virtual world. Portrayal of a human 3D view of reality. See, for example, L. Valente, E. Clua, A.R. Silva and R. Feijó on live-action virtual reality games [199] based on a mixed reality model. See, for example:
https://en.wikipedia.org/wiki/Category:Mixed_reality
and
https://en.wikipedia.org/wiki/Mixed_reality
and
http://www.pokemongo.com/fr-ca/
Visual quality: Measure of perceived visual deterioration in an output video compared with an original scene. Visual deterioration results from lossy image (e.g., .jpg) or video compression techniques.

B.22   W

W

Webcam: Video camera that streams its images in real-time to a computer network. For applications, see E.A. Vlieg [200].

B.23   X

X

X: Greek letter, pronounced *Kai* as in *Kailua* (*kai lua*) in *Kailua, Hawaii*.
arXiv: For downloadable articles in Physics, Mathematics, Computer Science, Quantitative Biology, Quantitative Finance, Statistics, see the Cornell University e-print service at https://arxiv.org/.

B.24   Z

Z

z: $z = a + bi, a, b \in \mathbb{R}$, a complex number. See **Complex number, Complex plane**.

$\mathbb{Z}$:   Set of integers.

References

1. A-iyeh, E.: Point pattern voronoï tessellation quality and improvement, information and processing: applications in digital image analysis. Ph.D. thesis, University of Manitoba, Department of Electrical and Computer Engineering's (2016). Supervisor: J.F. Peters
2. A-iyeh, E., Peters, J.: Rényi entropy in measuring information levels in Voronoï tessellation cells with application in digital image analysis. Theory Appl. Math. Comput. Sci. **6**(1), 77–95 (2016). MR3484085
3. Aberra, T.: Topology preserving skeletonization of 2d and 3d binary images. Master's thesis, Technische Universität Kaiserslautern, Kaiserslautern, Germany (2004). Supervisors: K. Schladitz, J. Franke
4. Akramaullah, S.: Digital Video Concepts, Methods and Metrics. Quality, Compression, Performance, and Power Trade-Off Analysis, Xxiii+344 pp. Springer, Apress, Berlin (2015)
5. Allili, M., Ziou, D.: Active contours for video object tracking using region, boundary and shape information. Signal Image Video Process. **1**(2), 101–117 (2007). doi:10.1007/s11760-007-0021-8
6. Apostol, T., Mnatsakanian, M.: Complete dissections: converting regions and their boundaries. Am. Math. Mon. **118**(9), 789–798 (2011)
7. Archimedes: sphere and cylinder. On paraboloids, hyperboloids and ellipsoids, trans. and annot. by A. Czwalina-Allenstein. Cambridge University Press, UK (1897). Reprint of 1922, 1923, Geest & Portig, Leipzig 1987, The Works of Archimedes, ed. by T.L. Heath, Cambridge University Press, Cambridge (1897)
8. Baerentzen, J., Gravesen, J., Anton, F., Aanaes, H.: Computational Geometry Processing. Foundations, Algorithms, and Methods. Springer, Berlin (2012). doi:10.1007/978-1-4471-4075-7, Zbl 1252.68001
9. Baroffio, L., Cesana, M., Redondi, A., Tagliasacchi, M.: Fast keypoint detection in video sequences, pp. 1–5 (2015). arXiv:1503.06959v1 [cs.CV]
10. Bay, H., Ess, A., Tuytelaars, T., Gool, L.: Speeded-up robust features (surf). Comput. Vis. Image Underst. **110**(3), 346–359 (2008)
11. Beer, G.: Topologies on Closed and Closed Convex Sets. Kluwer Academic Publishers, The Netherlands (1993)
12. Beer, G., Lucchetti, R.: Weak topologies for the closed subsets of a metrizable space. Trans. Am. Math. Soc. **335**(2), 805–822 (1993)
13. Belongie, S., Malik, J., Puzicha, J.: Matching shapes. In: Proceedings of the IEEE International Conference on Computer Vision (ICCV2001), vol. 1, pp. 454–461. IEEE (2001). doi:10.1109/ICCV.2001.937552

© Springer International Publishing AG 2017   403
J.F. Peters, *Foundations of Computer Vision*, Intelligent Systems
Reference Library 124, DOI 10.1007/978-3-319-52483-2

14. Ben-Artzi, G., Halperin, T., Werman, M., Peleg, S.: Trim: triangulating images for efficient registration, pp. 1–13 (2016). arXiv:1605.06215v1 [cs.GR]
15. Benhamou, F., Goalard, F., Languenou, E., Christie, M.: Interval constraint solving for camera control and motion planning. ACM Trans. Comput. Logic **V**(N), 1–35 (2003). http://tocl.acm.org/accepted/goualard.pdf
16. Blashke, T., Burnett, C., Pekkarinen, A.: Luminaires. In: de Jong, S., van der Meer, F. (eds.) Image Segmentation Methods for Object-Based Analysis and Classification, pp. 211–236. Kluwer, Dordrecht (2004)
17. Borsuk, K.: Theory of Shape. Monografie Matematyczne, Tom 59. [Mathematical Monographs, vol. 59] PWN—Polish Scientific Publishers (1975). MR0418088, Based on K. Borsuk, Theory of Shape, Lecture Notes Series, vol. 28, Matematisk Institut, Aarhus Universitet, Aarhus (1971). MR0293602
18. Borsuk, K., Dydak, J.: What is the theory of shape? Bull. Aust. Math. Soc. **22**(2), 161–198 (1981). MR0598690
19. Bromiley, P., Thacker, N., Bouhova-Thacker, E.: Shannon entropy, Rényi's entropy, and information. Technical report, The University of Manchester, U.K. (2010). http://www.tina-vision.net/docs/memos/2004-004.pdf
20. Broomhead, D., Huke, J., Muldoon, M.: Linear filters and non-linear systems. J. R Stat. Soc. Ser. B (Methodol.) **54**(2), 373–382 (1992)
21. Burger, W., Burge, M.: Digital Image Processing. An Algorithmic Introduction Using Java, 2nd edn, 811 pp. Springer, Berlin (2016). doi:10.1007/978-1-4471-6684-9
22. Burt, P., Adelson, E.: The Laplacian pyramid as a compact image code. IEEE Trans. Commun. **COM–31**(4), 532–540 (1983)
23. Camastra, F., Vinciarelli, A.: Machine Learning for Audio, Image and Video Analysis, Xvi + 561 pp. Springer, Berlin (2015)
24. Canny, J.: Finding edges and lines in images. Master's thesis, MIT, MIT Artificial Intelligence Laboratory (1983). ftp://publications.ai.mit.edu/ai-publications/pdf/AITR-720.pdf
25. Canny, J.: A computational approach to edge detection. IEEE Trans. Pattern Anal. Mach. Intell. **8**, 679–698 (1986)
26. Chakerian, G.: A characterization of curves of constant width. Am. Math. Mon. **81**(2), 153–155 (1974)
27. Chan, M.: Topical curves and metric graphs. Ph.D. thesis, University of California, Berkeley, CA, USA (2012). Supervisor: B. Sturmfels
28. Chaudhury, K., Munoz-Barrutia, A., Unser, M.: Fast space-variant elliptical filtering using box splines, pp. 1–42 (2011). arXiv:1003.2022v2
29. Chen, L.M.: Digital Functions and Data Reconstruction. Digital-Discrete Methods, Xix+207 pp. Springer, Berlin (2013). doi:10.1007/978-1-4614-5638-4
30. Christie, M., Olivier, P., Normand, J.M.: Camera control in computer graphics. Comput. Graph. Forum **27**(8), 2197–2218 (2008). https://www.irisa.fr/mimetic/GENS/mchristi/Publications/2008/CON08/870.pdf
31. Corke, P.: Robitics, Vision and Control. Springer, Berlin (2013). doi:10.1007/978-3-642-20144-8
32. Danelljan, M., Häger, G., Khan, F., Felsberg, M.: Coloring channel representations for visual tracking. In: Paulsen, R., Pedersen, K. (eds.) SCIA 2015. LNCS, vol. 9127, pp. 117–129. Springer, Berlin (2015)
33. Delaunay, B.D.: Sur la sphère vide. Izvestia Akad. Nauk SSSR, Otdelenie Matematicheskii i Estestvennyka Nauk **7**, 793–800 (1934)
34. Deza, E., Deza, M.M.: Encyclopedia of Distances. Springer, Berlin (2009)
35. Dirichlet, G.: Über die reduktion der positiven quadratischen formen mit drei unbestimmten ganzen zahlen. Journal für die reine und angewandte **40**, 221–239 (1850). MR
36. Drucker, S.: Intelligent camera control for graphical environments. Ph.D. thesis, Massachusetts Institute of Technology, Media Arts and Sciences (1994). http://research.microsoft.com/pubs/68555/thesiswbmakrs.pdf. Supervisor: D. Zeltzer

37. Drucker, S.: Automatic conversion of natural language to 3d animation. Ph.D. thesis, University of Ulster, Faculty of Engineering (2006). http://www.paulmckevitt.com/phd/mathesis.pdf. Supervisor: P. McKevitt

38. Du, Q., Faber, V., Gunzburger, M.: Centroidal voronoi tessellations: applications and algorithms. SIAM Rev. **41**(4), 637–676 (1999). MR1722997

39. Eckhardt, U., Latecki, L.J.: Topologies for the digital spaces $\mathbb{Z}^2$ and $\mathbb{Z}^3$. Comput. Vis. Image Underst. **90**(3), 295–312 (2003)

40. Edelsbrunner, H.: Geometry and Topology of Mesh Generation, 209 pp. Cambridge University Press, Cambridge (2001)

41. Edelsbrunner, H.: A Short Course in Computational Geometry and Topology, 110 pp. Springer, Berlin (2014)

42. Edelsbrunner, H., Harer, J.: Computational Topology. An Introduction, Xii+110 pp. American Mathematical Society, Providence (2010). MR2572029

43. Edelsbrunner, H., Kirkpatrick, D., Seidel, R.: On the shape of a set of points in the plane. IEEE Trans. Inf. Theory **IT-29**(4), 551–559 (1983)

44. Eisemann, M., Klose, F., Magnor, M.: Towards plenoptic raumzeit reconstruction. In: Cremers, D., Magnor, M., Oswald, M., Zelnik-Manor, L. (eds.) Video Processing and Computational Video, pp. 1–24. Springer, Berlin (2011). doi:10.1007/978-3-642-24870-2

45. Escolano, F., Suau, P., Bonev, B.: Information Theory in Computer Vision and Pattern Recognition. Springer, Berlin (2009)

46. Nielson, F. (ed.): Emerging Trends in Visual Computing, Xii+388 pp. Springer, Berlin (2008)

47. Fabbri, R., Kimia, B.: Multiview differential geometry of curves, pp. 1–34 (2016). arXiv:1604.08256v1 [cs.CV]

48. Fabre, C.: Basics of quantum optics and cavity quantum electrodynamics. Lect. Notes Phys. **531**, 1–37 (2007). doi:10.1007/BFb0104379

49. Favorskaya, M., Jain, L., Buryachenko, V.: Digital video stabilization in static and dynamic situations. In: Favorskaya, M., Jain, L. (eds.) Intelligent Systems Reference, vol. 73, pp. 261–310. Springer, Berlin (2015)

50. Fechner, G.: Elemente der Psychophysik, 2 vols. E.J. Bonset, Amsterdam (1860)

51. Fontelos, M., Lecaros, R., López-Rios, J., Ortega, J.: Stationary shapes for 2-d water-waves and hydraulic jumps. J. Math. Phys. **57**(8), 081,520, 22 pp. (2016). MR3541857

52. Frank, N., Hart, S.: A dynamical system using the Voronoi tessellation. Am. Math. Mon. **117**(2), 92–112 (2010)

53. Gardner, M.: On tessellating the plane with convex polygon tiles. Sci. Am. 116–119 (1975)

54. Gaur, S., Vajpai, J.: Comparison of edge detection techniques for segmenting car license plates. Int. J. Comput. Appl. Electr. Inf. Commun. Eng. **5**, 8–12 (2011)

55. Gersho, A., Gray, R.: Vector Quantization and Signal Compression. Kluwer Academic Publishers, Norwell (1992). ISBN: 0-7923-9181-0

56. Gersho, A., Gray, R.: Vector Quantization and Signal Compression, Xii + 732 pp. Kluwer Academic Publishers, Boston (1992)

57. Gonzalez, R., Woods, R.: Digital Image Processing. Prentice-Hall, Upper Saddle River, NJ 07458 (2002). ISBN: 0-20-118075-8

58. Gonzalez, R., Woods, R.: Digital Image Processing, 3rd edn, Xxii + 954 pp. Pearson Prentice Hall, Upper Saddle River (2008)

59. Gonzalez, R., Woods, R., Eddins, S.: Digital Image Processing Using Matlab®, Xiv + 609 pp. Pearson Prentice Hall, Upper Saddle River (2004)

60. Grauman, K., Shakhnarovich, G., Darrell, T.: Coloring channel representations for visual tracking. In: Comaniciu, R.M.D.S.D., Kanatani, K. (eds.) Statistical Methods in Video Processing (SMVP) 2004. LNCS, vol. 3247, pp. 26–37. Springer, Berlin (2004)

61. Gruber, P.: Convex and discrete geometry, Grundlehren der Mathematischen Wissenschaften, vol. 336, Xiv+578 pp. Springer, Berlin (2007). MCS2000 52XX, 11HXX, ISBN: 978-3-540-71132-2, MR2335496

62. Gruber, P.M., Wills, J.M. (eds.): Handbook of Convex Geometry. North-Holland, Amsterdam (1993) vol. A: lxvi+735 pp.; vol. B: ilxvi and 7371438 pp. ISBN: 0-444-89598-1, MR1242973

63. Grünbaum, B., Shephard, G.: Tilings and Patterns, Xii+700 pp. W.H. Freeman and Co., New York (1987). MR0857454
64. Grünbaum, B., Shepherd, G.: Tilings with congruent tiles. Bull. (New Ser.) Am. Math. Soc. **3**(3), 951–973 (1980)
65. ter Haar Romeny, B.: Computer vision and mathematica. Comput. Vis. Sci. **5**(1), 53–65 (2002). MR1947476
66. Hall, E.: The Silent Language. Doubleday, Garden City (1959)
67. Haluška, J.: On fields inspired with the polar HSV – RGB theory of colour, pp. 1–16 (2015). arXiv:1512.01440v1 [math.HO]
68. Hanbury, A., Serra, J.: A 3d-polar coordinate colour representation suitable for image analysis. Technical report, Vienna University of Technology (2003). http://cmm.ensmp.fr/~serra/notes_internes_pdf/NI-230.pdf
69. Haralick, R.: Digital step edges from zero crossing of second directional derivatives. IEEE Trans. Pattern Anal. Mach. Intell. **PAMI-6**(1), 58–68 (1984)
70. Haralick, R., Shapiro, L.: Computer and Robot Vision. Addison-Wesley, Reading (1993)
71. Harris, C., Stephens, M.: A combined corner and edge detector. In: Proceedings of the 8th Alvey Vision Conference, pp. 147–151 (1988)
72. Hartley, R.: Transmission of information. Bell Syst. Tech. J. **7**, 535 (1928)
73. Hassanien, A., Abraham, A., Peters, J., Schaefer, G., Henry, C.: Rough sets and near sets in medical imaging: a review. IEEE Trans. Info. Technol. Biomed. **13**(6), 955–968 (2009). doi:10.1109/TITB.2009.2017017
74. Hausdorff, F.: Grundzüge der Mengenlehre, Viii + 476 pp. Veit and Company, Leipzig (1914)
75. Hausdorff, F.: Set Theory, trans. by J.R. Aumann, 352 pp. AMS Chelsea Publishing, Providence (1957)
76. Henry, C.: Near sets: theory and applications. Ph.D. thesis, University of Manitoba, Department of Electrical and Computer Engineering (2010). http://130.179.231.200/cilab/. Supervisor: J.F. Peters
77. Henry, C.: Arthritic hand-finger movement similarity measurements: tolerance near set approach. Comput. Math. Methods Med. **2011**, 1–14 (2011). doi:10.1155/2011/569898
78. Herran, J.: Omnivis: 3d space and camera path reconstruction for omnidirectional vision. Master's thesis, Harvard University, Mathematics Department (2010). Supervisor: Oliver Knill
79. Hettiarachchi, R., Peters, J.: Voronoï region-based adaptive unsupervised color image segmentation, pp. 1–2 (2016). arXiv:1604.00533v1 [cs.CV]
80. Hidding, J., van de Weygaert, R., G. Vegter, B.J., Teillaud, M.: The sticky geometry of the cosmic web, pp. 1–2 (2012). arXiv:1205.1669v1 [astro-ph.CO]
81. Hlavac, V.: Fundamentals of image processing. In: Cristóbal, H.T.G., Schelkens, P. (eds.) Optical and Digital Image Processing. Fundamentals and Applications, pp. 25–48. Wiley-VCH, Weinheim (2011)
82. Hoggar, S.: Mathematics of Digital Images. Cambridge University Press, Cambridge (2006). ISBN: 978-0-521-78029-2
83. Hoggar, S.: Mathematics of Digital Images, Xxxii + 854 pp. Cambridge University Press, Cambridge (2006)
84. Holmes, R.: Mathematical foundations of signal processing. SIAM Rev. **21**(3), 361–388 (1979)
85. Houit, T., Nielsen, F.: Video stippling, pp. 1–13 (2010). arXiv:1011.6049v1 [cs.GR]
86. Jacques, J., Braun, A., Soldera, J., Musse, S., Jung, C.: Understanding people in motion in video sequences using Voronoi diagrams. Pattern Anal. Appl. **10**, 321–332 (2007). doi:10.1007/s10044-007-0070-1
87. Jähne, B.: Digital Image Processing, 6th revised, extended edn. Springer, Berlin (2005). ISBN: 978-3-540-24035-8 (Print) 978-3-540-27563-3 (Online)
88. Jarvis, R.: Computing the shape hull of points in the plane. In: Proceedings of the Computer Science Conference on Pattern Recognition and Image Processing, pp. 231–241. IEEE (1977)
89. Joblove, G., Greenberg, D.: Color spaces for computer graphics. In: Proceedings of the 5th Annual Conference on Computer Graphics and Interactive Techniques, pp. 20–25. Association for Computing Machinery (1978)

90. Karimaa, A.: A survey of hardware accelerated methods for intelligent object recognition on camera. In: Świątek, J., Grzech, A., Świątek, P., Tomczak, J. (eds.) Advances in Systems Science, vol. 240, pp. 523–530. Springer, Berlin (2013)

91. Kay, D., Womble, E.: Automatic convexity theory and relationships between the carathèodory, helly and radon numbers. Pac. J. Math. **38**(2), 471–485 (1971)

92. Kim, I., Choi, H., Yi, K., Choi, J., Kong, S.: Intelligent visual surveillance-A survey. Int. J. Control Autom. Syst. **8**(5), 926–939 (2010)

93. Kiy, K.: A new real-time method of contextual image description and its application in robot navigation and intelligent control. In: Favorskaya, M., Jain, L. (eds.) Intelligent Systems Reference, vol. 75, pp. 109–134. Springer, Berlin (2015)

94. Klette, R., Rosenfeld, A.: Digital Geometry. Geometric Methods for Digital Picture Analysis. Morgan Kaufmann Publishers, Amsterdam (2004)

95. Knee, P.: Sparse representations for radar with Matlab examples. Morgan & Claypool Publishers (2012). doi:10.2200/S0044ED1V01Y201208ASE010

96. Kohli, P., Torr, P.: Dynamic graph cuts and their applications in computer vision. In: Cipolla, G.F.R., Battiato, S. (eds.) Computer Vision, pp. 51–108. Springer, Berlin (2010)

97. Kokkinos, I., Yuille, A.: Learning an alphabet of shape and appearance for multi-class object detection. Int. J. Comput. Vis. **93**(2), 201–225 (2011). doi:10.1007/s11263-010-0398-7

98. Kong, T., Roscoe, A., Rosenfeld, A.: Concepts of digital topology. Special issue on digital topology. Topol. Appl. **46**(3), 219–262 (1992). Am. Math. Soc. MR1198732

99. Kong, T., Rosenfeld, A.: Topological Algorithms for Digital Image Processing. North-Holland, Amsterdam (1996)

100. Krantz, S.: A Guide to Topology, Ix + 107 pp. The Mathematical Association of America, Washington (2009)

101. Krantz, S.: Essentials of topology with applications, Xvi+404 pp. CRC Press, Boca Raton (2010). ISBN: 978-1-4200-8974-5. MR2554895

102. Kronheimer, E.: The topology of digital images. Special issue on digital topology. Topol. Appl. **46**(3), 279–303 (1992). MR1198735

103. Lai, R.: Computational differential geometry and intinsic surface processing. Ph.D. thesis, University of California, Los Angeles, CA, USA (2010). Supervisors: T.F. Chan, P. Thompson, M. Green, L. Vese

104. Latecki, L.: Topological connectedness and 8-connectedness in digital pictures. Comput. Vis. Graph. Image Process. **57**, 261–262 (1993)

105. Latecki, L., Conrad, C., Gross, A.: Preserving topology by a digitization process. J. Math. Imaging Vis. **8**, 131–159 (1998)

106. Lee, D., Sallee, G.: A method of measuring shape. Geogr. Rev. **60**(4), 555–563 (1970)

107. Leone, F., Nelson, L., Nottingham, R.: The folded normal distribution. TECHNOMETRICS **3**(4), 543–550 (1961). MR0130737

108. Leutenegger, S., Chli, M., Siegwart, R.: Brisk: binary robust invariant scalable keypoints. In: Proceedings of the 2011 IEEE International Conference on Computer Vision, pp. 2548–2555. IEEE (2011)

109. Li, L., Wang, F.Y.: Advanced Motion Control and Sensing for Intelligent Vehicles. Springer, Berlin (2007)

110. Li, N.: Retrieving camera parameters from real video images. Master's thesis, The University of British Columbia, Computer Science (1998). http://www.iro.umontreal.ca/~poulin/fournier/theses/Li.msc.pdf

111. Li, Z.N., Drew, M., Liu, J.: Color in Image and Video. Springer, Berlin (2014). doi:10.1007/978-3-319-05290-8_4

112. Lin, Y.J., Xu, C.X., Fan, D., He, Y.: Constructing intrinsic Delaunay triangulations from the dual of geodesic Voronoi diagrams, pp. 1–32 (2015). arXiv:1605.05590v2 [cs.CG]

113. Lindeberg, T.: Edge detection and ridge detection with automatic scale selection. Int. J. Comput. Vis. **30**(2), 117–154 (1998)

114. Louban, R.: Image Processing of Edge and Surface Defects. Materials Science, vol. 123. Springer, Apress (2009). See pp. 9–29 on edge detection

115. Lowe, D.: Object recognition from local scale-invariant features. In: Proceedings of the 7th IEEE International Conference on Computer Vision, vol. 2, pp. 1150–1157 (1999). doi:10. 1109/ICCV.1999.790410

116. Lowe, D.: Distinctive image features from scale-invariant keypoints. Int. J. Comput. Vis. **60**(2), 91–110 (2004). doi:10.1023/B:VISI.0000029664.99615.94

117. Maggi, F., Mihaila, C.: On the shape of capillarity droplets in a container. Calc. Var. Partial Differ. Equ. **55**(5), 122 (2016). MR3551302

118. Mahmoodi, S.: Scale-invariant filtering design and analysis for edge detection. R. Soc. Proc.: Math. Phys. Eng. Sci. **467**(2130), 1719–1738 (2011)

119. Mani-Levitska, P.: Characterizations of convex sets. Handbook of Convex Geometry, vol. A, B, pp. 19–41. North-Holland, Amsterdam (1993). MR1242975

120. Marr, D., Hildreth, E.: Theory of edge detection. Proc. R. Soc. Lond. Ser. B **207**(1167), 187–217 (1980)

121. Mery, D., Rueda, L. (eds.): Advances in Image and Video Technology, Xviii+959 pp. Springer, Berlin (2007)

122. Michelson, A.: Studies in Optics. Dover, New York (1995)

123. Milnor, J.: Topology through the centuries: low dimensional manifolds. Bull. (New Ser.) Am. Math. Soc. **52**(4), 545–584 (2015)

124. Gavrilova, M.L. (ed.): Generalized Voronoi Diagrams: A Geometry-Based Approach to Computational Intelligence, Xv + 304 pp. Springer, Berlin (2008)

125. Moselund, T.: Introduction to Video and Image Processing. Building Real Systems and Applications, Xi + 227 pp. Springer, Heidelberg (2012)

126. Munkres, J.: Topology, 2nd edn., Xvi + 537 pp. Prentice-Hall, Englewood Cliffs (2000), 1st edn. in 1975. MR0464128

127. Munsell, A.: A Color Notation. G. H. Ellis Company, Boston (1905)

128. Naimpally, S., Peters, J.: Topology with Applications. Topological Spaces via Near and Far, Xv + 277 pp. World Scientific, Singapore (2013). American Mathematical Society. MR3075111

129. Nyquist, H.: Certain factors affecting telegraph speed. Bell Syst. Tech. J. **3**, 324 (1924)

130. Olive, D.: Algebras, lattices and strings 1986. Unification of fundamental interactions. Proc. R. Swed. Acad. Sci. Stockh. **1987**, 19–25 (1987). MR0931580

131. Olive, D.: Loop algebras, QFT and strings. Proc. Strings Superstrings, Madr. **1987**, 217–2858 (1988). World Scientific Publishing, Teaneck, NJ. MR1022259

132. Olive, D., Landsberg, P.: Introduction to string theory: its structure and its uses. Physics and mathematics of strings. Philos. Trans. R. Soc. Lond. 329, pp. 319–328 (1989). MR1043892

133. Opelt, A., Pinz, A., Zisserman, A.: Learning an alphabet of shape and appearance for multi-class object detection. Int. J. Comput. Vis. **80**(1), 16–44 (2008). doi:10.1007/s11263-008-0139-3

134. Orszag, M.: Quantum Optics. Including Noise Reduction, Trapped Ions, Quantum Trajectories, and Decoherence. Springer, Berlin (2016). doi:10.1007/978-3-319-29037-9

135. Ortiz, A., Oliver, G.: Detection of colour channels uncoupling for curvature-insensitive segmentation. In: F.P. et al. (ed.) IbPRIA 2003. LNCS, vol. 2652, pp. 664–672. Springer, Berlin (2003)

136. Over, E., Hooge, I., Erkelens, C.: A quantitative method for the uniformity of fixation density: the Voronoi method. Behav. Res. Methods **38**(2), 251–261 (2006)

137. Pal, S., Peters, J.: Rough Fuzzy Image Analysis. Foundations and Methodologies. CRC Press, Taylor & Francis Group, London: ISBN: 13: 9781439803295. ISBN: **10**, 1439803293 (2010)

138. Paragios, N., Chen, Y., Faugeras, O.: Handbook of Mathematical Models in Computer Vision. Springer, Berlin (2006)

139. Perona, P., Malik, J.: Scale-space and edge detection using anisotropic diffusion. IEEE Trans. Pattern Anal. Mach. Intell. **12**(7), 629–639 (1990)

140. Peters, J.: Proximal Delaunay triangulation regions, pp. 1–4 (2014). arXiv:1411.6260 [math-MG]

141. Peters, J.: Proximal Voronoï regions, pp. 1–4 (2014). arXiv:1411.3570 [math-MG]

142. Peters, J.: Topology of Digital Images - Visual Pattern Discovery in Proximity Spaces. Intelligent Systems Reference Library, vol. 63, Xv + 411 pp. Springer, Berlin (2014). Zentralblatt MATH Zbl 1295 68010

143. Peters, J.: Proximal Voronoï regions, convex polygons, & Leader uniform topology. Adv. Math. **4**(1), 1–5 (2015)

144. Peters, J.: Computational Proximity. Excursions in the Topology of Digital Images. Intelligent Systems Reference Library, vol. 102, Viii + 445 pp. Springer, Berlin (2016). doi:10.1007/978-3-319-30262-1

145. Peters, J.: Two forms of proximal physical geometry. axioms, sewing regions together, classes of regions, duality, and parallel fibre bundles, pp. 1–26 (2016). To appear in Adv. Math.: Sci. J., vol. 5 (2016). arXiv:1608.06208

146. Peters, J., Guadagni, C.: Strong proximities on smooth manifolds and Voronoi diagrams. Adv. Math. Sci. J. **4**(2), 91–107 (2015). Zbl 1339.54020

147. Peters, J., İnan, E.: Rényi entropy in measuring information levels in Voronoï tessellation cells with application in digital image analysis. Theory Appl. Math. Comput. Sci. **6**(1), 77–95 (2016). MR3484085

148. Peters, J., İnan, E.: Strongly proximal Edelsbrunner-Harer nerves. Proc. Jangjeon Math. Soc. **19**(2), 563–582 (2016)

149. Peters, J., İnan, E.: Strongly proximal Edelsbrunner-Harer nerves in Voronoï tessellations. Proc. Jangjeon Math. Soc. **19**(3), 563–582 (2016). arXiv:1604.05249v1

150. Peters, J., İnan, E.: Strongly proximal Edelsbrunner-Harer nerves in Voronoï tessellations, pp. 1–10 (2016). arXiv:1605.02987v3

151. Peters, J., Naimpally, S.: Applications of near sets. Notices Am. Math. Soc. **59**(4), 536–542 (2012). doi:10.1090/noti817.MR2951956

152. Peters, J., Puzio, L.: Image analysis with anisotropic wavelet-based nearness measures. Int. J. Comput. Intell. Syst. **2**(3), 168–183 (2009). doi:10.1016/j.ins.2009.04.018

153. Peters, J., Tozzi, A., İnan, E., Ramanna, S.: Entropy in primary sensory areas lower than in associative ones: the brain lies in higher dimensions than the environment. bioRxiv **071977**, 1–12 (2016). doi:10.1101/071977

154. Poincaré, H.: La Science et l'Hypothèse. Ernerst Flammarion, Paris (1902). Later ed.; Champs sciences, Flammarion, 1968 & Science and Hypothesis, trans. by J. Larmor, Walter Scott Publishing, London, 1905; cf. Mead Project at Brock University. http://www.brocku.ca/MeadProject/Poincare/Larmor_1905_01.html

155. Poincaré, J.: L'espace et la géomètrie. Revue de m'etaphysique et de morale **3**, 631–646 (1895)

156. Poincaré, J.: Sur la nature du raisonnement mathématique. Revue de méaphysique et de morale **2**, 371–384 (1894)

157. Pottmann, H., Wallner, J.: Computational Line Geometry. Springer, Berlin (2010). doi:10.1007/978-3-642-04018-4. MR2590236

158. Preparata, F.: Convex hulls of finite sets of points in two and three dimensions. Commun. Assoc. Comput. Mach. **2**(20), 87–93 (1977)

159. Preparata, F.: Steps into computational geometry. Technical report, Coordinated Science Laboratory, University of Illinois (1977)

160. Prewitt, J.: Object Enhancement and Extraction. Picture Processing and Psychopictorics. Academic Press, New York (1970)

161. Prince, S.: Computer Vision. Models, Learning, and Inference, Xvii + 580 pp. Cambridge University Press, Cambridge (2012)

162. Pták, P., Kropatsch, W.: Nearness in digital images and proximity spaces. In: Proceedings of the 9th International Conference on Discrete Geometry, LNCS **1953**, 69–77 (2000)

163. Ramakrishnan, S., Rose, K., Gersho, A.: Constrained-storage vector quantization with a universal codebook. IEEE Trans. Image Process. **7**(6), 785–793 (1998). MR1667391

164. Rényi, A.: On measures of entropy and information. In: Proceedings of the 4th Berkeley Symposium on Mathematical Statistics and Probability, vol. 1, pp. 547–547. University of California Press, Berkeley, California (2011). Math. Sci. Net. Review. MR0132570

165. Rhodin, H., Richart, C., Casas, D., Insafutdinov, E., Shafiei, M., Seidel, H.P., Schiele, B., Theobalt, C.: Egocap: egocentric marker-less motion capture with two fisheye cameras, pp. 1–11 (2016). arXiv:1609.07306v1 [cs.CV]
166. Roberts, L.: Machine perception of three-dimensional solids. In: Tippett, J. (ed.) Optical and Electro-Optical Information Processing. MIT Press, Cambridge (1965)
167. Robinson, M.: Topological Signal Processing, Xvi+208 pp. Springer, Heidelberg (2014). ISBN: 978-3-642-36103-6. doi:10.1007/978-3-642-36104-3. MR3157249
168. Rosenfeld, A.: Distance functions on digital pictures. Pattern Recognit. 1(1), 33–61 (1968)
169. Rosenfeld, A.: Digital Picture Analysis, Xi + 351 pp. Springer, Berlin (1976)
170. Rosenfeld, A.: Digital topology. Am. Math. Mon. 86(8), 621–630 (1979). Am. Math. Soc. MR0546174
171. Rosenfeld, A., Kak, A.: Digital Picture Processing, vol. 1, Xii + 457 pp. Academic Press, New York (1976)
172. Rosenfeld, A., Kak, A.: Digital Picture Processing, vol. 2, Xii + 349 pp. Academic Press, New York (1982)
173. Rowland, T., Weisstein, E.: Continuous. Wolfram Mathworld (2016). http://mathworld.wolfram.com/Continuous.html
174. Ruhrberg, K.: Seurat and the neo-impressionists. In: Art in the 20th Century, pp. 25–48. Benedict Taschen Verlag, Koln (1998)
175. Shamos, M.: Computational geometry. Ph.D. thesis, Yale University, New Haven, Connecticut, USA (1978). Supervisors: D. Dobkin, S. Eisenstat, M. Schultz
176. Sharma, O.: A methodology for raster to vector conversion of colour scanned maps. Master's thesis, University of New Brunswick, Department of Geomatics Engineering (2006). http://www2.unb.ca/gge/Pubs/TR240.pdf
177. Shimizu, Y., Zhang, Z., Batres, R.: Frontiers in Computing Technologies for Manufacturing Applications. Springer, London (2007). ISBN: 978-1-84628-954-5
178. Slotboom, B.: Characterization of gap-discontinuities in microstrip structures, used for opto-electronic microwave switching, supervisor: G. Brussaard. Master's thesis, Technische Universiteit Eindhoven (1992). http://alexandria.tue.nl/extra1/afstversl/E/394119.pdf
179. Smith, A.: A pixel is not a little square (and a voxel is not a little cube), vol. 6. Technical report, Microsoft (1995). http://alvyray.com/Memos/CG/Microsoft/6_pixel.pdf
180. Sobel, I.: Camera models and perception. Ph.D. thesis, Stanford University, Stanford (1970)
181. Sobel, I.: An Isotropic 3x3 Gradient Operator, Machine Vision for Three-Dimensional Scenes, pp. 376–379. Freeman, H., Academic Press, New York (1990)
182. Solan, V.: Introduction to the axiomatic theory of convexity [Russian with English and French Summaries], 224 pp. Shtiintsa, Kishinev (1984). MR0779643
183. Solomon, C., Breckon, T.: Fundamentals of Digital Image Processing. A Practical Approach with Examples in Matlab, X + 328 pp. Wiley-Blackwell, Oxford (2011)
184. Sonka, M., Hlavac, V., Boyle, R.: Image Processing, Analysis and Machine Vision. Springer, Berlin (1993). doi:10.1007/978-1-4899-3216-7
185. Sonka, M., Hlavac, V., Boyle, R.: Image Processing, Analysis, and Machine Vision, 829 pp. Cengage Learning, Stamford (2008). ISBN: -13 978-0-495-24438-7
186. Stahl, S.: The evolution of the normal distribution. Math. Mag. 79(2), 96–113 (2006). MR2213297
187. Stijns, E., Thienpont, H.: Fundamentals of photonics. In: Cristóbal, G., Schelkens, P., Thienpong, H. (eds.) Optical and Digital Image Processing, pp. 25–48. Wiley, Weinheim (2011). ISBN: 978-3-527-40956-3
188. Stijns, E., Thienpont, H.: Fundamentals of photonics. In: Cristóbal, H.T.G., Schelkens, P. (eds.) Optical and Digital Image Processing. Fundamentals and Applications, pp. 25–48. Wiley-VCH, Weinheim (2011)
189. Sya, S., Prihatmanto, A.: Design and implementation of image processing system for lumen social robot-humanoid as an exhibition guide for electrical engineering days, pp. 1–10 (2015). arXiv:1607.04760

190. Szeliski, R.: Computer Vision. Algorithms and Applications, Xx + 812 pp. Springer, Berlin (2011)
191. Takita, K., Muquit, M., Aoki, T., Higuchi, T.: A sub-pixel correspondence search technique for computer vision applications. IEICE Trans. Fundam. **E87-A**(8), 1913–1923 (2004). http://www.aoki.ecei.tohoku.ac.jp/research/docs/e87-a_8_1913.pdf
192. Tekdas, O., Karnad, N.: Recognizing characters in natural scenes. A feature study. CSci 5521 Pattern Recognition, University of Minnesota, Twin Cities (2009). http://rsn.cs.umn.edu/images/5/54/Csci5521report.pdf
193. Thivakaran, T., Chandrasekaran, R.: Nonlinear filter based image denoising using AMF approach. Int. J. Comput. Sci. Inf. Secur. **7**(2), 224–227 (2010)
194. Tomasi, C.: Cs 223b: introduction to computer vision. Matlab and images. Technical report, Stanford University (2014). http://www.umiacs.umd.edu/~ramani/cmsc828d/matlab.pdf
195. Topp, J.: Geodetic line, middle and total graphs. Mathematica Slovaca **40**(1), 3–9 (1990). https://www.researchgate.net/publication/265573026_Geodetic_line_middle_and_total_graphs
196. Toussaint, G.: Computational geometry and morphology. In: Proceedings of the First International Symposium for Science on Form, pp. 395–403. Reidel, Dordrecht (1987). MR0957140
197. Tuz, V.: Axiomatic convexity theory [Russian]. Rossiĭskaya Akademiya Nauk. Matematicheskie Zametki [Math. Notes and Math. Notes] **20**(5), 761–770 (1976)
198. Vakil, V.: The mathematics of doodling. Am. Math. Mon. **118**(2), 116–129 (2011)
199. Valente, L., Clua, E., Silva, A., Feijó, R.: Live-action virtual reality games, pp. 1–10 (2016). arXiv:1601.01645v1 [cs.HC]
200. Vlieg, E.: Scratch by Example. Apress, Berlin (2016). doi:10.1007/978-1-4842-1946-1_10. ISBN: 978-1-4842-1945-4
201. Voronoi, G.: Sur une fonction transcendante et ses applications à la sommation de quelque séries. Ann. Sci. Ecole Norm. Sup. **21**(3) (1904)
202. Voronoï, G.: Nouvelles applications des paramètres continus à la théorie des formes quadratiques. J. für die reine und angewandte Math. **133**, 97–178 (1907). JFM 38.0261.01
203. Voronoï, G.: Nouvelles applications des paramètres continus à la théorie des formes quadratiques. J. für die reine und angewandte Math. **134**, 198–287 (1908). JFM 39.0274.01
204. Wang, Z., Bovik, A.: A universal image quality index. IEEE Signal Process. Lett. **9**(3), 81–84 (2002). doi:10.1109/97.995823
205. Wang, Z., Bovik, A., Sheikh, H., Simoncelli, E.: Image quality assessment: from error visibility to structural similarity. IEEE Trans. Image Process. **13**(4), 600–612 (2004)
206. Wegert, E.: Visual Complex Functions. An Introduction to Phase Portraits, Xiv + 359 pp. Birkhäuser, Freiburg (2012). doi:10.1007/978-3-0348-0180-5
207. Weisstein, E.: Regular polygon. Wolfram Mathworld (2016). http://mathworld.wolfram.com/RegularPolygon.html
208. Weisstein, E.: Wavelet. Wolfram Mathworld (2016). http://mathworld.wolfram.com/Wavelet.html
209. Wen, B.J.: Luminance meter. In: Luo, M. (ed.) Encyclopedia of Color Science and Technology, pp. 824–886. Springer, New York (2016). doi:10.1007/978-1-4419-8071-7
210. Wildberger, N.: Algebraic topology: a beginner's course. University of South Wales (2010). https://www.youtube.com/watch?v=Ap2c1dPyIVo&index=40&list=PL6763F57A61FE6FE8
211. Wirjadi, O.: Models and algorithms for image-based analysis of microstructures. Ph.D. thesis, Technische Universität Kaiserslautern, Kaiserslautern, Germany (2009). Supervisor: K. Berns
212. Witkin, A.: Scale-space filtering. In: Proceedings of the 8th International Joint Conference on Artificial Intelligence, pp. 1019–1022. Karlsruhe, Germany (1983)
213. Xu, L., Zhang, X.C., Auston, D.: Terahertz beam generation by femtosecond optical pulses in electo-optic materials. Appl. Phys. Lett. **61**(15), 1784–1786 (1992)
214. Yung, C., Choi, G.T., Chen, K., Lui, L.: Trim: triangulating images for efficient registration, pp. 1–13 (2016). arXiv:1605.06215v1 [cs.GR]
215. Zadeh, L.: Theory of filtering. J. Soc. Ind. Appl. Math. **1**(1), 35–51 (1953)

216. Zelins'kyi, Y.: Generalized convex envelopes of sets and the problem of shadow. J. Math. Sci. **211**(5), 710–717 (2015)
217. Zhang, X., Brainard, D.: Estimation of saturated pixel values in digital color imaging. J. Opt. Soc. Am. A **21**(12), 2301–2310 (2004). http://color.psych.upenn.edu/brainard/papers/Zhang_Brainard_04.pdf
218. Zhang, Z.: Affine cameral. In: Ikeuchi, K. (ed.) Computer Vision. A Reference Guide, pp. 19–20. Springer, Berlin (2014)
219. Zhao, B., Xing, E.: Sparse output coding for scalable visual recognition. Int. J. Comput. Vis. **119**, 60–75 (2016). doi:10.1007/s11263-015-0839-4
220. Ziegler, G.: Lectures on Polytopes. Springer, Berlin (2007). doi:10.1007/978-1-4613-8431-1

Author Index

© Springer International Publishing AG 2017
J.F. Peters, *Foundations of Computer Vision*, Intelligent Systems
Reference Library 124, DOI 10.1007/978-3-319-52483-2

Subject Index

© Springer International Publishing AG 2017
J.F. Peters, *Foundations of Computer Vision*, Intelligent Systems
Reference Library 124, DOI 10.1007/978-3-319-52483-2

Printed in the United States
By Bookmasters